Molecular Plant Path

Molecular Plant Pathology

M. Dickinson
School of Biosciences, University of Nottingham, Nottingham, UK

First published 2003

A CIP catalogue record for this book is available from the British Library.

ISBN 1 85996 044 8

BIOS Scientific Publishers
Taylor & Francis Group
11 New Fetter Lane, London EC4P 4EE
and
29 West 35th Street,
New York, NY 10001-2299, USA
Tel: (+1) 212 216 7800, Fax: (+1) 212 564 7854

BIOS Scientific Publishers is a member of the Taylor & Francis Group.

Production Editor: Andrew Watts
Typeset by Charon Tec Pvt. Ltd, Chennai, India

Contents

Abbreviations

A	adenine
ABA	abscisic acid
ABC	ATP-binding cassette
ACC	1-aminocyclopropane-1-carboxylate
ACMV	*African cassava mosaic virus*
ADC	arginine decarboxylase
AFLP	amplified fragment length polymorphism
AOS	active oxygen species
avr	*avirulence genes*
BBTV	*Banana bunchy top virus*
BCTV	*Beet curly top virus*
BIBAC	binary bacterial artificial chromosome
BMV	*Brome mosaic virus*
BSMG	*Barley stripe mosaic virus*
BSV	*Banana streak virus*
BTH	benzathiodioazole
CaMV	*Cauliflower mosaic virus*
CAPS	cleaved amplified polymorphic sequence
CC	coiled coil
CHS	chalcone synthases
CCMV	*Cowpea chlorotic mottle virus*
CCR	central conserved region
CDPK	calcium-dependent protein kinase
CPMV	*Cowpea mosaic virus*
CRP	catabolite activator protein
CWA	cell-wall apposition
CWDE	cell-wall-degrading enzyme
DAG	diacylglycerol
DAS	double-antibody sandwich
DHN	dihydroxynaphthalene
DHPLC	denaturing high-performance liquid chromatography
DI	defective interfering
DMI	demethylase inhibitor
ds	double-stranded
ELISA	enzyme-linked immunosorbent assay
EMS	ethylmethane sulphonate
EPS	exopolysaccharide/ extracellular polysaccharide
EREBP	ethylene response element-binding protein
EST	expressed sequence tag
FITC	fluorescein isothiocyanate
FRET	fluorescence resonance energy transfer
G	guanine
G+C	guanine plus cytosine
GEAR	genetically engineered acquired resistance
GIP	glutamine amidotransferase/ indoleglycerolphosphate synthase/phosphoribosyl-anthranilate
GPCR	G-protein coupled receptor
GUS	glucuronidase
HABS	high-affinity binding site
HC	helper component
HR	hypersensitive response
hrc	*h*ypersensitivity *r*esponse, pathogenicity and *c*onserved genes
HRGP	hydroxyproline-rich glycoprotein
hrp	*h*ypersensitivity *r*esponse and *p*athogenicity genes
IAA	indole-3-acetic acid
IAM	indoleacetamide
IGS	intragenic spacers
IpyA	indolepyruvic acid
ISR	induced systemic resistance
ITR	inverted terminal repeat
ITS	internal transcribed spacers
LINE	long interspersed nuclear element
LPS	lipopolysaccharide

LRR	leucine-rich repeat
LTR	long terminal repeat
LZ	leucine zippers
MALDI TOF	matrix-assisted laser desorption/ionisation–time of flight
MAPK	mitogen-activated protein kinase
MBC	methyl-benzimidazole-carbamate
MeJA	methyl jasmonate
MFS	major facilitator superfamily
MHC	major histocompatibility complex
MP	movement protein
NBS	nucleotide-binding site
NO	nitric oxide
NOS	nitric oxide synthase
OCT	ornithine carbamoyltrans-ferase
ODC	ornithine decarboxylase
ORF	open reading frame
PA	phosphatidic acid
PAI	pathogenicity island
PAL	phenylalanine ammonia lyase
PAMP	pathogen-associated molecular pattern
PAP	pokeweed antiviral protein
PCD	programmed cell death
PG	polygalacturonase
PGIP	polygalacturonase inhibitor protein
PK	protein kinase
PKC	protein kinase C
PL	pectate lyase
PME	pectin methylesterase
PPV	*Plum pox virus*
PR	pathogenesis-related
PRP	proline-rich protein
PSbMV	*Pea seed-borne mosaic virus*
PTA	plate-trapped antigen
PTGS	post-transcriptional gene silencing
PVX	*Potato virus X*
PWL	pathogenicity on weeping lovegrass
QTL	quantitative trait loci
RAPD	randomly amplified polymorphic DNA
RBR	retinoblastoma-related
RdRp	RNA-dependent RNA polymerase
REMI	restriction enzyme-mediated insertion
RFLP	restriction fragment length polymorphism
RGA	resistance gene analogue
RIP	ribosome-inactivating protein
RITC	rhodamine isothiocyanate
RNAi	RNA interference
ROI	reactive oxygen intermediates
ROS	reactive oxygen species
SAGE	serial analysis of gene expression
SAM	sphinganine-analogue mycotoxin
SAR	systemic acquired resistance
SAS	systemic acquired silencing
SCSV	*Subterranean clover stunt virus*
SINE	short interspersed nuclear element
siRNP	small interfering ribonucleo-protein
SNP	single nucleotide poly-morphism
SOD	superoxide dismutase
ss	single-stranded
SSH	suppression subtractive hybridisation
SSLP	simple sequence length polymorphism
SSR	simple sequence di, tri and tetranucleotide repeats
TAC	transformation-competent artificial chromosome
TAS	triple-antibody sandwich
Tβl	tabtoxine-β-lactan
T-cms	Texas cytoplasmic male sterility factor
Ti	tumour inducing
TMV	*Tobacco mosaic virus*
TSWV	*Tomato spotted wilt virus*
TVCV	*Tobacco vein-clearing virus*
TYMV	*Turnip yellow mosaic virus*
UTR	untranslated
VIGS	virus-induced gene sequencing
YAC	yeast artificial chromosome
ZYMV	*Zucchini yellow mosaic virus*

The fundamentals of plant pathology

1

1.1 The concept of plant disease

So much of our existence and our society depends on the ability of plants to harness light and produce oxygen and organic matter. Domestication of plants for agriculture resulted in many of the great civilisations of the past, Asian civilisations based on rice, Middle Eastern on wheat and barley, and American on maize. Over the past few thousand years, more than half of the suitable land on Earth and virtually all of the most fertile land, has been converted for agricultural use. Agriculture today is a global business, and a necessity for the production of food, drinks and other vital commodities such as building materials, fibres, clothing, drugs and medicines. New products from, and uses for, plants are constantly being sought and developed, and plants are crucial for maintaining the environment, both globally in maintaining our atmosphere, and locally in the form of recreational facilities. Today, plants dominate our lives and economy, just as they have in all civilisations.

Mankind is not alone in the need to live off plants. Indeed, since plants first colonised land around 460 million years ago, they have probably been the main nutrient source for microbes such as fungi. Most of these microbes are saprophytic, living off nutrients released from dead and decaying plants, but many have also found ways to tap into living plants for their own growth and development. Some of these are considered beneficial for agriculture, such as the nitrogen-fixing *Rhizobia*, or the mutualistic mycorrhizal fungi that often enhance nutrient uptake. However, it is when the interactions between plants and microbes infringe on our food supply and environment that we consider the organisms to be pathogens and the result to be disease. So just as the field of medicine has developed to understand and combat diseases on humans, plant pathology has filled this role in agriculture, horticulture and forestry.

1.2 The causal agents

1.2.1 Fungi

Of the more than 74 000 known species of fungi that have been described, the majority are saprophytes, living off dead and decaying organic matter. A few cause human, animal, fish and insect diseases. However, there are more than 10 000 that can parasitise living plants to cause varying degrees of damage. Some have developed a biotrophic lifestyle, in which they obtain nutrients from living host tissue, and reduce plant vigour and yield through the diversion of nutrients for their own growth and development (see Section 3.1). Other fungi exhibit a necrotrophic lifestyle in which they utilise toxins or cell-wall-degrading enzymes to kill plant cells and then metabolise the nutrients that are released. Physical damage to plants is a prerequisite for these fungi. There are also many fungi that use a combination of strategies, the hemi-biotrophs. These will initially adopt a biotrophic infection and subsequently cause more significant damage and cell death to plants as the infection progresses and sporulation commences.

All plant species are susceptible to fungal infections, and there are many fungi for which their only hosts are living plants (obligate pathogens). Other fungi can colonise plants but are also able to survive as saprophytes on dead tissue. In some cases these fungi must colonise plants for part of their life cycle, for example apple scab *Venturia inaequalis* and maize smut *Ustilago maydis*, but others may be able to survive and reproduce exclusively as saprophytes and merely use living plants as an alternative source of nutrients.

The spread of fungi from plant to plant is generally through transmission of spores. These may be carried long distances by wind and air currents, such as occurs for the rusts and powdery mildews, or they may be deposited in the soil and remain viable but inactive until triggered to germinate, often through detection of the presence of potential host plants in the vicinity. The life cycles of fungi and the different spore-producing stages that they go through are often complex and vary greatly between species. Such information is beyond the scope of this book, and readers are encouraged to refer to other texts, in particular 'The Biology of Fungi' by Ingold and Hudson (1993) and 'Plant Pathology' by Agrios (1997), for more detailed information on the life cycles of these and other plant pathogens. *Table 1.1* lists some of the major species of fungi in their phyla. Whilst there are a few examples of plant pathogens in the Chytridiomycota and Zygomycota, the majority belong to the filamentous class of the Ascomycota (the filamentous ascomycetes) or to the Basidiomycota. The nomenclature of fungi can be further complicated by the presence of both imperfect (asexual) and perfect (sexual) stages to the life cycles, and fungi for which no perfect stage has yet been identified can be classified in the Deuteromycotina. Some ascomycete fungi are commonly referred to by the name of their anamorph (imperfect stage), whilst in others the teleomorph (perfect stage) is more commonly used, particularly where no anamorphic stage has been identified.

1.2.2 The Oomycota

True fungi are organisms that produce mycelia containing glucans and chitin, and lack chloroplasts. Recently, molecular evidence has indicated that a number of organisms that were once considered as fungi because of their similar growth style and infection strategies (despite the fact that they lacked chitin in their cell walls), are in fact in a different kingdom, the Chromista, related to the brown algae and diatoms. These organisms, the Oomycota, contain the water moulds, white rusts and downy mildews and include such devastating and notorious plant pathogens as potato late blight *(Phytophthora infestans)*. The oomycetes have a sexual resting stage (the oospores) and produce flagellated zoospores in zoosporangia and the flagella are used by the spores for motility through moisture in the soil. *Table 1.2* lists some of the more common plant-pathogenic Oomycota.

1.2.3 Protozoa

A small number of plant diseases are caused by protozoa. In particular, *Plasmodiophora brassicae* causes a serious clubroot disease on many crucifers

Table 1.1 Significant plant pathogens of the kingdom Fungi

	Genus (Anamorphs in brackets)	Diseases/Symptoms
Phylum:	ASCOMYCOTA	
Subphyla:	Taphrinomycotina	
	Taphrina	Peach leaf curl
Subphyla:	Saccharomycotina	
	No significant plant pathogens	
Subphyla:	Pezizomycotina	
Class:	Sordariomycetes	
	Cryphonectria	Chestnut blight
	Gaeumannomyces	Take-all on cereal roots/stem bases
	Magnaporthe (Pyricularia)	Rice blast
	Ophiostoma	Dutch elm
	Gibberella (Fusarium)	Vascular wilts, root rots, ear blights
	Nectria (Fusarium)	stem rots, seed infections
	Claviceps	Ergot on grasses
	Epichloë	Choke disease on grasses
	Ceratocystis	Oak wilt, cankers in stone fruits and trees
	Glomerella (Colletotrichum)	Anthracnose, mainly on fruits and beans
	(Verticillium)	Wilts
Class:	Dothidiomycetes	
	(Alternaria)	Leaf spots
	Cochliobolus (Bipolaris)	Blights
	Leptosphaeria (Phoma)	Black/foot rot/stem canker of crucifers
	(Cladosporium)	Tomato leaf mould
	Venturia	Apple scab
	Mycosphaerella	
	(Septoria)	Leaf spots and blotches
	(Cercospora)	Sigatoka disease of bananas/leaf spots
	(Rhynchosporium)	Barley scald
Class:	Leotiomycetes	
	Blumeria	Powdery mildews
	Erysiphe	
	Uncinula	
	Leveillula	
	Botryotinia (Botrytis)	Grey moulds and rots
	Monilinia	Post harvest diseases of fruits
	Sclerotinia	White mould of vegetables
	Diplocarpon	Black spot of rose
	Tapesia	Eyespot of grasses
Phylum:	BASIDIOMYCOTA	
	Ustilago	Smuts
	Tilletia	Bunts/smuts
	Hemileia	Coffee rust
	Melampsora	Flax/poplar rusts
	Puccinia	Cereal and other plant rusts
	Uromyces	Bean rusts
	Crinipellis	'Witches broom' of cocoa
	Rhizoctonia	Damping-off diseases/sharp eyespot
	Sclerotinium	Root and stem rot of many plants
Phylum:	CHYTRIDIOMYCOTA	
	Olpidium	Root parasite that transmits plant viruses
Phylum:	ZYGOMYCOTA	
	Rhizopus	Moulds on soft fruits and vegetables
	Mucor	Storage rots on fruits and vegetables

See Myconet 2002 at http://www.umu.se/myconet/Myconet.html for Ascomycota taxonomy.

Table 1.2 Common plant pathogens in the Oomycota phylum

Bremia lactucae	Lettuce downy mildew
Peronospora parasitica	Downy mildew of crucifers and *Arabidopsis*
Peronospora tabacina	Downy mildew/blue mould of tobacco
Pythium spp.	Damping-off of seedlings/numerous hosts
Phytophthora cinnamomi	Root rot of many woody trees
Phytophthora infestans	Late blight of potatoes and tomatoes
Phytophthora parasitica	Black shank of tobacco
Plasmopara viticola	Downy mildew of grapes
Albugo candida	White blister rust of crucifers

such as cabbages and cauliflowers. These organisms are transmitted as zoospores, which may survive as resting structures in the soil for many years. The zoospores are motile in water in the soil and produce a plasmodium on contact with root hairs that invades the roots. The plasmodia then enlarge and multiply inside the roots, producing the characteristic club-root symptoms. Some other protozoa, such as the *Polymyxa* species, are important for their ability to transmit plant viruses (see Section 8.4), and colonise the roots of plants such as wheat and barley, although the protozoa themselves do not appear to have a detrimental effect on the plant.

1.2.4 Bacteria

Whilst it is the case for fungi that many more species are parasitic on plants than on animals, the opposite is true for bacteria. Of the 1600 bacterial species known, which include many serious and life-threatening disease-causing agents of humans, only about 100 species are known to cause disease on plants. However, these can be particularly significant and devastating especially in warmer more humid climates such as the tropics. There is also the potential for human pathogenic bacteria to become internalised into plants, although whether they survive long term inside plants and are a significant threat to consumers is unclear.

Most plant-pathogenic bacteria are rod-shaped (an exception being the filamentous *Streptomyces scabies* that causes potato scab disease), and *Table 1.3* details the taxonomic classification of the major plant-pathogenic bacteria. Many of these species can exist as saprophytes, living off organic matter in the soil or on the outer surface of plants. However, particular isolates have developed the capacity to colonise plants internally as well. In so doing they have developed the production of pathogenicity factors such as toxins, cell-wall-degrading enzymes or plant hormone production (see Chapter 5) to obtain nutrients for their growth and reproduction from living plants. Often the pathogenicity factors are encoded in pathogenicity islands (see Section 6.9), or on extrachromosomal plasmids (see Section 6.10).

The human concept of what is a plant pathogen and what is mutualistic is particularly blurred for some plant–bacterial interactions. The root-nodulating *Rhizobia* spp., *Bradyrhizobia* spp. and *Azorhizobia* spp. are considered as mutualistic because they can improve plant growth and yield. However, from the bacterial perspective the colonisation process is to enhance survival and growth. There are some strains that can form nodules but fail to fix nitrogen, so should these be considered as plant pathogens? In this book we will not be considering the molecular interactions between rhizobia and plants apart from where analogies and comparisons have been demonstrated

Table 1.3 Significant plant pathogenic bacteria

Taxon	Disease
BACTERIA	
Division: GRACILICUTES (Gram-negative)	
Family: Enterobacteria	
Erwinia amylovora	Fireblight/necrosis on apples, pears and ornamentals
Erwinia carotovora	Soft rots on potatoes and other vegetables
Erwinia herbicola	Galls on *Gypsophila*
Pantoea stewartii	Stewart's wilt on corn
Family: Pseudomonadaceae	
Pseudomonas syringae	Leaf spots, blights, cankers, knots on many plants
Xanthomonas campestris	Blights on peppers, cotton and other hosts
Xanthomonas oryzae	Blight on rice
Ralstonia solanacearum	Bacterial wilt on bananas, tomatoes, potatoes and tobacco
Family: Rhizobiaceae	
Agrobacterium tumefaciens	Crown gall disease
Agrobacterium rhizogenes	Hairy root disease
Rhizobia spp.	Mutualistic root-nodulating bacteria
Family: Un-named	
Xylella fastidiosa	Citrus variegated chlorosis/Pierce's disease of grapevine
Division: FIRMICUTES (Gram-positive)	
Streptomyces scabies	Potato scab
Rhodococcus neoformans	Fasciation
Clavibacter michiganensis	Ring rot of potato
MOLLICUTES	
Family: Spiroplasmataceae	
Spiroplasma citri	Corn stunt, citrus stubborn
Family: Un-defined	
Phytoplasmas	Cause numerous diseases (see Section 1.2.5)

between these and plant pathogen interactions. However, there is undoubtedly much about molecular signalling between plants and microbes that has been learnt from the extensive research into plant–rhizobial interactions.

1.2.5 Phytoplasmas and Spiroplasmas

One class of bacteria, referred to as the mollicutes, are characterised by their lack of cell walls and relatively small genome size (530–1350 kb), making them particularly fastidious microbes. The 530 kb genome of the Bermuda grass white leaf phytoplasma represents the smallest known genome of a living organism, and phytoplasmas also have remarkably low guanine plus cytosine (G+C) content in their genomes (23–29%), which is at the lowest levels theoretically possible for coding DNA. Divided into two families, the spiroplasmas (which have a helical structure) and the phytoplasmas (which are pleiomorphic, ranging from 200–800 μm in diameter), these organisms have not been well characterised at the molecular level because they are generally (apart from some of the spiroplasmas) impossible to grow in axenic culture, and it is presumed that this reflects the absence of many genes required for amino acid, vitamin and fatty acid biosynthetic pathways in their genomes.

They are normally transmitted by sap-sucking insects, in particular by leaf, plant and treehoppers, vectors in which the bacterium is able to multiply and infect internal organs including the salivary glands. In many cases, the specific vectors for these diseases have not been identified although advanced molecular diagnostic techniques are aiding with such investigations. When infected insects feed, the bacterium enters the phloem where it multiplies. There is no evidence that these organisms progress into other parts of plants, and they are generally referred to as phloem-limited. Diseases caused by these organisms are particularly prevalent in tropical and sub-tropical regions, and include aster yellows in many ornamentals and vegetables (lettuce, carrots, celery, etc.), coconut lethal and lethal yellowing diseases, rice yellow dwarf, X-disease of peach and cherry, corn stunt and grape yellows. The means by which these organisms produce their often devastating symptoms, such as tissue proliferation, phyllody (development of floral parts into leafy structures), abnormal elongation of internodes, stunting, yellowing and death may suggest the production of toxins and/or modifications to plant hormone levels, but there are as yet little molecular data to support this.

1.2.6 Viruses

Viruses are common in plants, and many species have been identified and characterised at the molecular level. Many plant viruses can infect a wide range of plant species, producing different symptoms in the different hosts, and this has often led to classification of the same virus under more than one name. However, the extensive work of viral taxonomists using molecular and other diagnostic techniques has resulted in a relatively comprehensive classification system (*Table 1.4*). Despite this, there are many examples of latent and asymptomatic viruses where the taxonomy is less clear, and just as new virus and virus-like entities are constantly being discovered in humans and animals, particularly as new strains evolve, so new virus-like entities are regularly being identified in plants.

The majority of plant viruses possess single-strand RNA genomes encapsidated in simple protein coats comprising multiple copies of one or a very few coat protein subunits. These particles may be isometric or rod-shaped, and there are only a few examples of more complex structures, for example containing lipids and/or bacilliform structures. There are also some plant viruses with dsRNA, ssDNA and dsDNA genomes (see Section 7.1). Transmission of viruses between plants is mainly by vectors, in particular arthropods such as aphids, whiteflies, leaf and other hoppers. Some viruses are transmitted by biting insects such as grasshoppers, by nematodes, by fungi, mechanically by larger animals, vegetatively or in seed and pollen. Once inside the plant, the virus replicates in individual cells by modifying and utilising the plant's replication machinery, and spreads between cells progressively before entering the vascular systems for long-distance systemic spread. Unlike animal viruses, plant viruses can progress in plants without crossing cell membranes, by using cell-to-cell plasmodesmata connections. A further group of plant pathogens, the viroids, consist of single-strand circular RNA molecules with no protein coats and with no capacity to encode proteins (see Section 7.8).

1.2.7 Other agents of plant disease

Plants are susceptible to many other biotic and abiotic factors that cause disease-like symptoms or make the plants more susceptible to secondary invasion by pathogens. Diagnostics can be used to determine whether

Table 1.4 Taxonomy of the main plant virus genera

Family	Genus	Type species	Genome	Shape	Transmission
Geminiviridae	*Mastrevirus*	*Maize streak virus*	ssDNA	Geminate	Leafhoppers
	Curtovirus	*Beet curly top virus*	ssDNA	Geminate	Leafhoppers
	Begomovirus	*Bean golden mosaic virus*	ssDNA	Geminate	Whiteflies
(unnamed)	*Nanovirus*	*Subterranean clover stunt virus*	ssDNA	Isometric	Aphids
Caulimoviridae	*Caulimovirus*	*Cauliflower mosaic virus*	dsDNA	Isometric	Aphids
	Badnavirus	*Commelina yellow mottle virus*	dsDNA	Bacilliform	Leafhoppers
Reoviridae	*Phytoreovirus*	*Rice dwarf virus*	dsRNA	Isometric	Leaf/planthoppers
	Fijivirus	*Fiji disease virus*	dsRNA	Isometric	Leaf/planthoppers
	Oryzavirus	*Rice ragged stunt virus*	dsRNA	Isometric	Leaf/planthoppers
Partitiviridae	*Alphacryptovirus*	*White clover cryptic virus 1*	dsRNA	Isometric	Mechanical/seed
	Betacryptovirus	*White clover cryptic virus 2*	dsRNA	Isometric	Mechanical/seed
(unnamed)	*Varicosavirus*	*Lettuce big-vein virus*	dsRNA	Isometric	*Olpidium* (fungus)
Rhabdoviridae	*Cytorhabdovirus*	*Lettuce necrotic yellows virus*	−ssRNA	Bacilliform	Aphids
	Nucleorhabdovirus	*Potato yellow dwarf virus*	−ssRNA	Bacilliform	Leafhoppers
Bunyaviridae	*Tospovirus*	*Tomato spotted wilt virus*	−ssRNA	Spherical	Thrips
(unnamed)	*Tenuivirus*	*Rice stripe virus*	−ssRNA	Thin filamentous	Planthoppers
Sequiviridae	*Sequivirus*	*Parsnip yellow fleck virus*	+ssRNA	Isometric	Aphids
	Waikavirus	*Rice tungro spherical virus*	+ssRNA	Isometric	Leafhoppers
Comoviridae	*Comovirus*	*Cowpea mosaic virus*	+ssRNA	Isometric	Beetles
	Nepovirus	*Tobacco ringspot virus*	+ssRNA	Isometric	Nematodes
	Fabavirus	*Broad bean wilt virus 1*	+ssRNA	Isometric	Aphids
Potyviridae	*Potyvirus*	*Potato virus Y*	+ssRNA	Flexuous rod	Aphids
	Rymovirus	*Ryegrass mosaic virus*	+ssRNA	Flexuous rod	Mites
	Bymovirus	*Barley yellow mosaic virus*	+ssRNA	Flexuous rod	*Polymyxa* (protozoa)
Luteoviridae	*Luteovirus*	*Barley yellow dwarf virus – PAV*	+ssRNA	Isometric	Aphids
	Polerovirus	*Potato leafroll virus 2*	+ssRNA	Isometric	Aphids
	Enamovirus	*Pea enation mosaic virus 1*	+ssRNA	Isometric	Aphids

(Continued)

Table 1.4 (Continued)

Family	Genus	Type species	Genome	Shape	Transmission
Tombusviridae	*Tombusvirus*	*Tomato bushy stunt virus*	+ssRNA	Isometric	Mechanical
	Carmovirus	*Carnation mottle virus*	+ssRNA	Isometric	Mechanical
	Necrovirus	*Tobacco necrosis virus A*	+ssRNA	Isometric	*Olpidium* (fungus)
	Machlomovirus	*Maize chlorotic mottle virus*	+ssRNA	Isometric	Thrips
	Dianthovirus	*Carnation ringspot virus*	+ssRNA	Isometric	Nematodes
Bromoviridae	*Bromovirus*	*Brome mosaic virus*	+ssRNA	Isometric	Beetles
	Cucumovirus	*Cucumber mosaic virus*	+ssRNA	Isometric	Aphids
	Alfamovirus	*Alfalfa mosaic virus*	+ssRNA	Isometric	Aphids
	Ilarvirus	*Tobacco streak virus*	+ssRNA	Isometric	Seed/pollen/thrips
Closteroviridae	*Closterovirus*	*Beet yellows virus*	+ssRNA	Flexuous rod	Aphids
	Crinivirus	*Lettuce infectious yellows virus*	+ssRNA	Flexuous rod	Whiteflies
(unnamed)	*Sobemovirus*	*Southern bean mosaic virus*	+ssRNA	Isometric	Beetles
(unnamed)	*Marafivirus*	*Maize rayado fino virus*	+ssRNA	Isometric	Leafhoppers
(unnamed)	*Umbravirus*	*Carrot mottle virus*	+ssRNA	Isometric	Aphids
(unnamed)	*Tobamovirus*	*Tobacco mosaic virus*	+ssRNA	Rod	Mechanical
(unnamed)	*Tobravirus*	*Tobacco rattle virus*	+ssRNA	Rod	Nematodes
(unnamed)	*Hordeivirus*	*Barley stripe mosaic virus*	+ssRNA	Rod	Seed/pollen
(unnamed)	*Furovirus*	*Soil-borne wheat mosaic virus*	+ssRNA	Rod	*Polymyxa* (protozoa)
(unnamed)	*Pomovirus*	*Potato mop-top virus*	+ssRNA	Rod	*Spongospora* (protozoa)
(unnamed)	*Pecluvirus*	*Peanut clump virus*	+ssRNA	Rod	*Polymyxa* (protozoa)
(unnamed)	*Benyvirus*	*Beet necrotic yellow vein virus*	+ssRNA	Rod	*Polymyxa* (protozoa)
(unnamed)	*Capillovirus*	*Apple stem grooving virus*	+ssRNA	Flexuous rod	Mechanical
(unnamed)	*Trichovirus*	*Apple chlorotic leaf spot virus*	+ssRNA	Flexuous rod	Mechanical
(unnamed)	*Tymovirus*	*Turnip yellow mosaic virus*	+ssRNA	Isometric	Beetles
(unnamed)	*Carlavirus*	*Carnation latent virus*	+ssRNA	Flexuous rod	Aphids
(unnamed)	*Potexvirus*	*Potato virus X*	+ssRNA	Flexuous rod	Mechanical

See http://life.bio2.edu/index.htm for details. N.B. The means of transmission shown does not mean that all viruses in the genus are transmitted that way.

particular disorders, for example yellowing, have been caused by pathogens or are the result of nutritional deficiencies, temperature or chemical stresses on the plant. The symptoms may often be similar, but it is not within the scope of this book to detail the ways in which abiotic factors can affect plant health. However, specific examples will be mentioned where similarities have been shown to occur in the responses of plants to pathogen attack and abiotic factors. Similarly, infestation of plants by nematodes and insects will not be covered except to mention specific examples where there are significant similarities and comparisons to be made in the plant defence responses.

1.3 The significance of plant diseases

1.3.1 Historically important diseases

There are numerous historical examples where plant pathogens have had a major impact on human populations and health, and *Table 1.5* lists some of these. In the past, disease epidemics have resulted in crop failures and

Table 1.5 Historically significant plant diseases

Agent	Where	When
Famines		
Potato blight (*Phytophthora infestans*)	Europe	1845–1846
Brown spot of rice (*Bipolaris oryzae*)	Bengal	1942–1943
Human toxicity		
Ergot (*Claviceps purpurea*)	Europe	Middle Ages
Wheat ear/head blight and root rots (*Fusarium* spp.)	Europe and USA	Current
Economic		
Cereal rust fungi (*Puccinia* spp.)	Worldwide	Annually
Cereal smut fungi (*Ustilago* spp.)	Worldwide	Annually
Cereal leaf blotch (*Mycosphaerella graminicola*)	Worldwide	Annually
Southern Corn Leaf Blight (*Cochliobolus heterostrophus*)	USA	1970
Downy Mildew of Grapes (*Plasmopara viticola*)	Europe	1870–1880
Coffee rust (*Hemiliea vastatrix*)	SE Asia/Africa	1870–1880
	South America	1970–present
Sigatoka on bananas (*Cercospora* spp.)	Worldwide	Current
Monilia pod rot of cocoa (*Moniliophthora roreri*)	S. America	Current
Soybean rust (*Phakopsora pachyrhizi*)	SE Asia	Current
Sugarcane rust (*Puccinia kuehnii*)	America	Current
Rice blast (*Magnaporthe grisea*)	Asia	Current
Citrus canker (*Xanthomonas campestris*)	USA/S America	Current
Bacterial wilt of bananas (*Ralstonia solanacearum*)	America	Current
Rice leaf blight (*Xanthomonas oryzae*)	Japan/India	Current
Soft rot of vegetables (*Erwinia* spp.)	Worldwide	Current
Coconut lethal/lethal yellows phytoplasmas	Worldwide	Current
Citrus tristeza virus	Worldwide	Current
Rice tungro virus	SE Asia	Current
Sugarcane mosaic virus	Worldwide	Current
African cassava mosaic virus	Africa, S Asia	Current
Banana bunchy top virus	Asia, Australia, Africa	Current
Environmental		
Chestnut blight (*Cryphonectria parasitica*)	USA	1904–1940
Dutch elm (*Ophiostoma novo-ulmi*)	USA/Europe	1930–present
Jarrah Dieback (*Phytophthora cinnamomi*)	Australia	Current
Oak decline (*Phytophthora* spp.)	USA	Current

starvation, or the presence of mycotoxins in foods has resulted directly in death. For example, the Romans noted the importance of rusts and mildews and created two Gods, Robigo and Robigus to whom gifts and prayers were offered as appeasement to prevent epidemics occurring. Potato blight in Ireland in the 1840s resulted in the complete failure of the potato crop in 1846, and more than 1 million people died and a further 1.5 million emigrated, as the Irish population dropped from 8 to 5 million. Ergotism, resulting from the consumption of ergot-infected rye and other grains, caused many deaths throughout Europe during the Middle Ages when rye was cultivated by the peasantry as a source of grain. Symptoms of convulsive ergotism and gangrenous ergotism are often described in historical literature as St Anthony's Fire, and the hallucinogenic affects of ergot alkaloids may have been a contributing factor to the phenomenon of witchcraft.

In the modern era and particularly in the developed world, the effects of pathogen epidemics are more often economic, causing price fluctuations as supply and demand dictate. Destruction of coffee plantations, initially in Sri Lanka and then in Africa, by the coffee rust fungus *Hemileia vastatrix*, resulted in many of these countries turning to cultivation of other crops such as tea, and societies (e.g. the British) responded by turning from coffee to tea consumption. More recently, coffee rust epidemics in South America have significantly affected global coffee prices. The southern corn leaf blight epidemic in the southern states of the USA in 1970, in which a new race T of *Bipolaris maydis* developed that was specifically toxic against T-cms (Texas cytoplasmic male sterility factor) maize, caused a 50% yield reduction in the USA. The estimated cost of the epidemic was $1 billion, and the price of corn rose dramatically.

Some diseases can still cause famines in less-developed countries or if particular social and human factors occur. A famine in Bengal in 1943 was caused in part by the rice brown spot fungus (*Bipolaris oryzae*), but was compounded by factors related to the occurrence of World War II, to result in more than 4 million deaths in India. Other plant diseases may have significant environmental consequences. The Dutch elm disease epidemics of the 20th century in the USA and Europe have resulted in the death of millions of elm trees in hedgerows, woods and residential areas, and affected native flora and fauna. Similarly, the Jarrah Dieback disease caused by *Phytophthora cinnamomi* in Western Australia has been responsible for the death of many native eucalyptus trees and has had devastating environmental affects.

A few plant diseases can result in added value to crops. Grapes in parts of France such as the Bordeaux region are deliberately infected with *Botrytis cinerea* to result in 'noble' rot and the sweet and distinctly flavoured Sauterne premium wines. Similarly, in parts of Mexico and Central America, maize is infected with the smut fungus *Ustilago maydis* and the infected galls have been used as food for centuries and are considered as a delicacy that confers certain health benefits. Deliberate infection of cereals with ergot has also been performed by some pharmaceutical companies as a source for the ergot alkaloids that have specific applications in human medicine such as in treatment of migraines and some heart disorders.

1.3.2 Emerging diseases

Whilst the effects of disease epidemics on human populations may have changed, the potential for new epidemics remains. In human medicine, there is increasing evidence of new diseases emerging as bacteria, fungi and viruses become resistant to current control measures (the superbugs), or due to the migration and movement of people carrying diseases and their vectors between countries, or to environmental factors such as global warming increasing the chances for diseases and their vectors to survive in different parts of the world. Exactly the same is true with plant diseases. The extensive use of plant monocultures in agriculture with limited genetic diversity, and the increased reliance on agrochemicals for disease control can put pressure on pathogen populations to evolve and new strains to develop. The southern corn leaf blight epidemic was a particularly salient reminder of this. In addition, changes in agronomic practices, such as increased use of irrigation, can aid spread of soil-borne pathogens, and *Phytophthora* and *Pythia* diseases are becoming an increasing problem in irrigated crops.

The movement of plants and plant products between countries is also an important source of new diseases. The Jarrah Dieback epidemic in Australia was believed to be the result of transmission of *Phytophthora* in soil carried on the wheels of vehicles and by humans. The wheat yellow (stripe) rust fungus *Puccinia striiformis,* that arrived in Australia for the first time in 1979, and has had a major impact on wheat production as it has evolved and spread throughout Australasia, was believed to have entered the country as a contaminant on the clothing of a person travelling from western Europe. The increase in world trade of plant products, as foods for supermarket shelves in the developed world are sourced from different countries to satisfy demand, increases the potential for pathogens to be transferred between countries unless strict quarantine and testing procedures are in place. For example, the drought that occurred in parts of western Europe in 1976 resulted in reduced yields of potato, and these were sourced from other countries such as Mexico. The result was introduction of the A2 mating type of potato blight into Europe for the first time. Prior to this, only the A1 mating type had been present, and introduction of the second mating type has increased the potential for the pathogen to evolve more rapidly as it can now go through its sexual cycle.

Climatic changes are also likely to influence the spread of plant diseases. As temperate regions warm, they are more likely to suffer bacterial diseases that prefer warmer more humid environments. Vectors for bacterial and viral diseases will survive more readily and/or occur earlier in the crop-growing season, increasing the chances that they will spread disease. A further reason for the emergence of new diseases is the increase in our ability to identify the causal agents of plant abnormalities. A number of characteristics that in the past have been put down to abiotic factors such as nutrient deficiencies or environmental factors have been found to be caused by microbes, as diagnostic techniques that can identify these organisms have been developed. Thus the potential for new and emerging epidemics remains high.

1.4 The control of plant diseases

The control of plant diseases and use of molecular biology for this is discussed in Chapters 12, 13 and 14. In general terms, control is normally through combinations of both prevention and cure. Preventative techniques include the use of diagnostics for identification of pathogens in seed, propagative material and soils, and the use of cultural practices to remove potential inoculum for fields and soils, or fumigation and sterilisation of soil for glasshouse crops. Removal of material on which diseases occur and of plants that may be harbouring diseases from the vicinity of crop plants may also be important. Chemical treatments of seed prior to planting with protectant chemicals, or use of biological control agents to out-compete potential pathogens are further preventative controls, as is the use of resistant varieties of plants. Curative approaches may also involve the use of agrochemicals, particularly as many of the newer chemicals are being designed to be both preventative and curative.

With the increasing range of control methods that are available, the possibilities of integrating these into effective, durable and multifaceted control programmes has increased. Such programmes should involve disease-forecasting components to ensure control methods are used more effectively and economically and that chemical inputs are not used unnecessarily, and should involve as many alternative controls as possible. The reliance on a single chemical or single resistance gene to prevent disease should be discouraged because of the relative ease with which pathogen populations can overcome such controls, and formulated mixtures of agrochemicals and effective pyramiding and deployment of multiple resistance genes in cultivars, or durable resistance mechanisms should be encouraged. Integrated control programmes and the ability to communicate these through developments in information technology will all aid the control of plant diseases.

1.5 Molecular biology in plant pathology

1.5.1 A historical perspective

As new technologies have been developed through the years, these have been applied to biological systems to solve problems and further human understanding, and this has been the case in plant pathology. The concept of Mendelian genetics led to breakthroughs in the understanding of how plants resist disease and how this can be used in crop protection. The advent of electron microscopy led to the ability to visualise more microbes and viruses, and to determine where and how these enter and develop inside plants. Wendell Stanley was awarded the Nobel Prize for Chemistry in 1946 for his work on *Tobacco mosaic virus* (TMV) purification in the 1930s that culminated in electron microscope observations of the virus in 1936. As protein-sequencing techniques developed, the TMV coat protein became the third protein to be completely sequenced in 1960, following insulin and pancreatic ribonuclease, and the development of immunological techniques has led to the use of these in diagnostics of plant pathogens and also in localisation of microbes *in planta*. The emergence of nucleic acid sequencing techniques resulted initially in their use for determining the sequences of small DNA and RNA viruses, such as *Cauliflower mosaic virus* (CaMV) in 1980, and TMV in 1982. As sequencing techniques have become more sophisticated, with the use of sequencing machines, robotics and increased software programs and resources to support the processing of sequence data, so larger and larger genomes have been completed.

1.5.2 The use of model organisms

As molecular biology techniques have become more sophisticated and expensive to use, there has been a drive to use them more efficiently. This has often meant the scientific community focusing on model organisms for their studies. For plants, the model has been *Arabidopsis thaliana,* because of its relatively small genome (125 megabases (Mb), approximately 25 500 genes), rapid life cycle (8–10 weeks), and because it is amenable to mutational analyses, gene manipulation and transformation. The genomic sequence of *Arabidopsis* is complete and those of some of the major crop species such as rice (362–389 Mb) and maize (2500 Mb) will follow soon.

With pathogens, it has been less easy to reach a consensus on models, since there are many unique and diverse infection strategies. Many viral genomes have been sequenced and the roles of the genes they encode elucidated. To investigate interactions with plants, attention is now turning to detailed studies of those that infect *Arabidopsis*. The same is true for bacterial pathogens, where *Pseudomonas syringae* pv. *tomato,* particularly strain DC3000, has been widely used for molecular studies. The sequences of other plant-pathogenic bacteria such as *Xylella fastidiosa, Agrobacterium tumefaciens* (5.67 Mb) and *Ralstonia solanacearum* (5.8 Mb) are complete. With oomycetes, *Peronospora parasitica* (downy mildew), which infects *Arabidopsis,* has been well studied, although the lack of a transformation system and its obligate nature have impeded such studies. Potato blight, *Phytophthora infestans* has also been studied as a more economically important oomycetes, but its large and polyploid genome has limited molecular studies, and amongst the fungal pathogens, *Magnaporthe grisea* (rice blast), *Botrytis cinerea* (grey mould) and *Blumeria graminis* (powdery mildew) are organisms for which extensive expressed sequence tag (EST) libraries are becoming established, although many others have also been studied in detail as discussed in Chapters 2–4. A draft sequence of *M. grisea* is now available, that of the cotton pathogen *Ashbia gossypii* will be available soon and complete genome sequences for model fungi such as *Saccharomyces cerevisiae* and *Neurospora crassa* are already available. *Table 1.6* gives important genomic information websites for a number of plants and plant pathogens.

Table 1.6 Some useful genomics websites

Arabidopsis	http://www.arabidopsis.org/
Wheat	http://www.cerealsdb.uk.net/library.htm
Rice	http://www.irri.org/genomics/
Maize	http://www.maizemap.org/
Phytophthora infestans	http://www.ncgr.org/programs/genomics/
Blumeria graminis	http://www.crc.dk/phys/blumeria/
Botrytis cinerea	http://www.genoscope.cns.fr
Cladosporium fulvum	http://www.ncbi.nlm.nih.gov/dbEST
Magnaporthe grisea	http://www.ncbi.nlm.nih.gov/dbEST http://www-genome.wi.mit.edu/
Mycosphaerella graminicola	http://www.ncbi.nlm.nih.gov/dbEST
Pseudomonas syringae	http://pseudomonas-syringae.org
Xanthomonas campestris	http://genoma4.iq.usp.br/xanthomonas/
Ralstonia solanacearum	http://sequence.toulouse.inra.fr/R.solanacearum
General	http://microbialgenome.org/links/fungus.html http://www.oardc.ohio-state.edu/phytophthora/genome_links.htm http://cogeme.ex.ac.uk/

1.5.3 Transformation techniques

A key procedure for examining the role of genes in organisms is that of transformation, for example to reinsert cloned genes into mutants to complement the mutation and confirm function, or to make mutant lines through antisense and RNA interference techniques. Plant transformation, using *Agrobacterium tumefaciens* and/or biolistics is routine for many species. Recently, systems such as BIBAC (binary bacterial artificial chromosome) vectors and TAC (transformation-competent artificial chromosome) vectors have been developed into which genomic DNA fragments of 80–150 kb can be cloned and used directly for *Agrobacterium*-mediated transformation of the insert DNA into plants. Attempts are also being made to develop systems for targeted gene replacement, so that specific mutations could be made in a gene *in vitro* and the mutated version of the gene then used to directly replace the endogenous copy in the genome. These systems are available for some fungal species (though with only modest efficiency for some plant-pathogenic fungi), and homologous recombination has been reported at low efficiency in *Arabidopsis*. In addition viral vectors based on *Potato virus X* (PVX) have been developed for rapid transient transformation assays. This technique is particularly useful for testing gene silencing in reverse genetics experiments as discussed in Section 1.5.5.

Bacterial transformation has been routine for a number of decades and reliable transformation of many fungi and oomycetes has also become more routine. Approaches include electroporation of DNA into fungal protoplasts using selectable markers such as fungicide resistance genes, and recently, *A. tumefaciens* has been used to transform fungi such as *Colletotrichum gloeosporioides* (anthracnose) and *Mycosphaerella graminicola* (wheat leaf blotch), and biolistics for many others. New selectable markers have also been developed, particularly for obligate biotrophic fungi that cannot be grown axenically. The recent transformation of *B. graminis* used the bialaphos (BASTAR) resistance *bar* gene to confer resistance to this herbicide. Because the barley plants used also contained this resistance gene, they were not killed by the bialaphos selection, and the only surviving fungal colonies on the plants were the transformants.

1.5.4 Forward genetics

The classical way to identify the specific function of a gene is to use forward genetics. Mutants with altered phenotypes are identified by screening mutagenised populations, and by carefully choosing the mutant phenotype and designing elegant screens, mutations can be targeted to a specific pathway, or even to specific points within a pathway. The most common methods used to clone genes in forward genetics are insertion mutagenesis, map-based cloning or by PCR using primers designed from heterologous genes in other organisms. Insertion mutagenesis refers to the insertion of a known piece of DNA into a gene to modify expression of that gene (*Figure 1.1*). For example, the *Tn5* transposon has been widely used to 'tag' genes in bacteria and endogenous transposons are being developed for use in fungi. Restriction enzyme-mediated insertion (REMI) has also been widely used in fungi, where a plasmid containing a specific restriction enzyme site is transformed into the fungus along with the restriction enzyme, which then cleaves the DNA to allow it to insert into the fungal genome. In plants, endogenous transposons

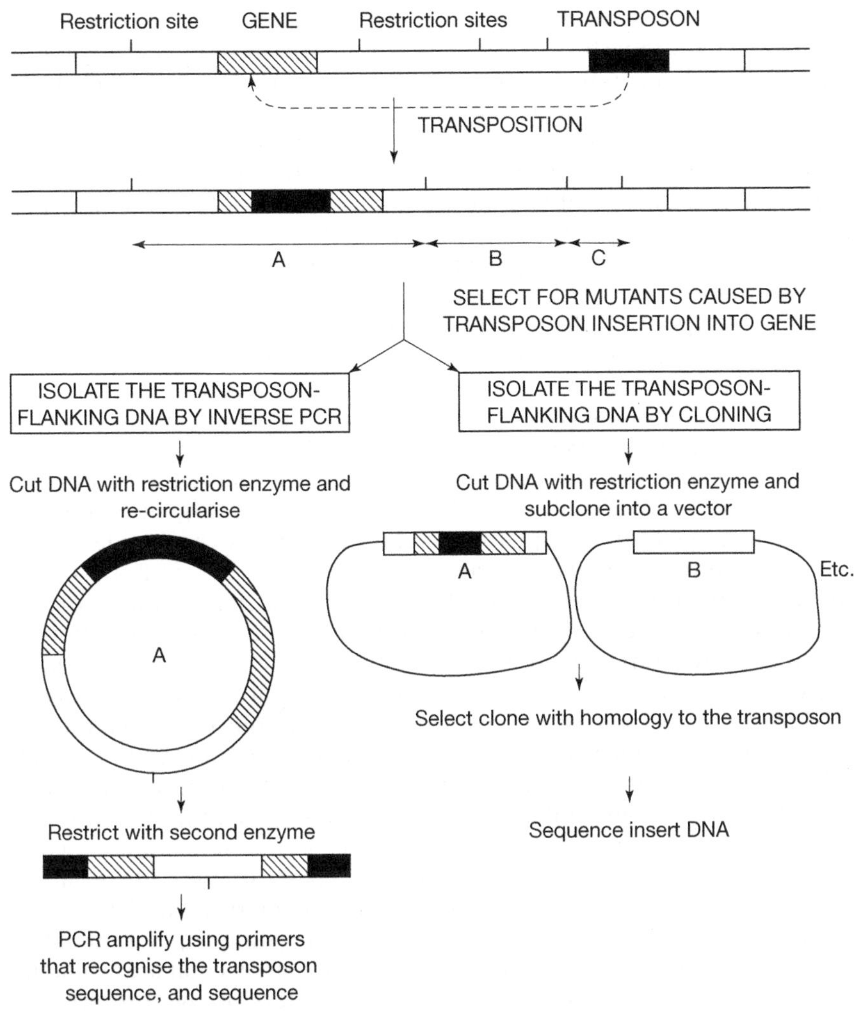
Restriction site
GENE
Restriction sites
TRANSPOSON
TRANSPOSITION
A
B
C
SELECT FOR MUTANTS CAUSED BY TRANSPOSON INSERTION INTO GENE
ISOLATE THE TRANSPOSON-FLANKING DNA BY INVERSE PCR
ISOLATE THE TRANSPOSON-FLANKING DNA BY CLONING
Cut DNA with restriction enzyme and re-circularise
Cut DNA with restriction enzyme and subclone into a vector
A
B
Etc.
A
Select clone with homology to the transposon
Restrict with second enzyme
Sequence insert DNA
PCR amplify using primers that recognise the transposon sequence, and sequence

Figure 1.1

Insertion mutagenesis by transposon tagging. Transposon insertion inactivates the gene resulting in a mutant phenotype. Genomic DNA is then isolated, restriction digested, and the DNA flanking the transposon can be isolated by inverse PCR or subcloning as shown.

such as *Mu*, *Ac/Ds*, *En/Spm* and *Tam* have been widely used in maize and *Antirrhinum*, and some of these have been transferred into and used in other species such as *Arabidopsis*, flax and tomato. Random T-DNA insertions have also been used in plants, and T-DNA mutagenised populations of *Arabidopsis* are publicly available through the *Arabidopsis* Stock Centres.

Transposon insertions usually result in null alleles by disrupting the open reading frame of the gene, although insertions into regulatory elements can alter expression levels. The main advantage of insertion mutagenesis is that the inserted DNA serves as a convenient tag to facilitate cloning of the affected gene, and a number of techniques have been devised to clone the genomic DNA flanking the insertion element, including plasmid rescue and inverse PCR. A disadvantage is that if the gene insertion is lethal, no mutant line can be recovered. In addition, mutations generated by transposable elements are often unstable because the element can transpose out of the gene as well as into it. To circumvent this and create stable insertions, two element systems have been developed in plants in which one element contains the transposase enzyme but lacks the target excision sequences, whilst the second element has these excision sequences but no transposase. Only the second element is capable of transposition but does so only in the presence of the transposase-expressing element. Mutants are generated by combining the two elements in a single plant by crossing the two lines to initiate transposition. When a plant with the desired phenotype is identified, the transposase-expressing element is segregated out, stabilising the mutation. Insertion elements have also been developed to identify insertions into or near regulatory elements in plant genomes (gene traps). A reporter gene (e.g. beta-glucuronidase (GUS)) is transformed into plants without a promoter so that it is expressed only if it inserts adjacent to a functional promoter. The expression pattern of the reporter will reflect the expression pattern of the gene normally controlled by the promoter, so genes expressed at particular stages during plant microbe interactions can be identified. However, the procedure is inefficient, since the majority of insertions are not into regulatory elements and therefore only a small number of the transgenic lines generated show any detectable reporter gene expression.

While insertion mutagenesis is generally the fastest way to mutagenise and identify a gene, it is not useful for studying any pathways that are essential during the life cycle of an organism, where it results in a lethal phenotype. Chemical mutagens such as EMS (ethylmethane sulphonate), which induces point mutations, or fast neutron ionising radiation, which results in small deletions, can produce more subtle non-lethal mutants. For these, map-based cloning (*Figure 1.2*) or PCR Tilling (see Section 1.5.5) can be used to identify the mutant gene. Map-based cloning is based on the co-segregation of the mutant phenotype and known, mapped markers, and the resolution of mapping depends on the size of the segregating population that is scored in the mapping cross, and the number of markers that are available. Molecular markers are now routinely used for developing chromosome maps, and a number of techniques have been developed to generate these. Among the first DNA-based markers were restriction fragment length polymorphisms (RFLPs). More recently, PCR-based techniques have been developed that can rapidly screen smaller samples of genomic DNA, for example RAPDs (random amplified polymorphic DNA sequences) generated using random primers. Markers are usually the result of base differences in the priming sites that result in the presence of the PCR product in one parent and absence in the other. Because RAPDs often result in presence/absence polymorphisms, heterozygous individuals cannot be distinguished from one of the homozygotes in diploid organisms without testing in the next generation, making them less useful than in haploid organisms. As a result, RAPDs have been replaced by co-dominant markers such as CAPS (cleaved amplified polymorphic sequences) in diploid organisms, where the basis of allelic variation is the presence of a restriction site in one parent but not in the other. PCR primers are used to generate a genomic fragment spanning the polymorphic restriction site, and the PCR products are digested with the diagnostic restriction enzyme and the digests are run on a gel for scoring. Normally, both of the parental genotypes and the heterozygote are all clearly distinguishable.

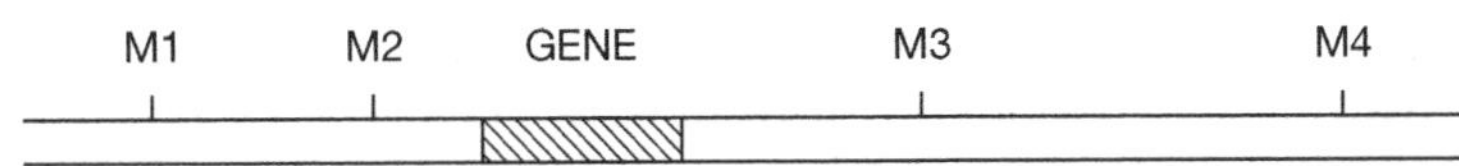

1. Marker sequences (M1–M4) (molecular markers or phenotypic markers) are mapped relative to the gene of interest by genetic crosses and measuring segregation ratios. The closest flanking markers are identified.

2. The genomic DNA is cloned into large capacity YAC (yeast artificial chromosome) or BAC (bacterial artificial chromosome) libraries (bacterial and small fungal genomes may be sequenced directly without the need for subcloning).

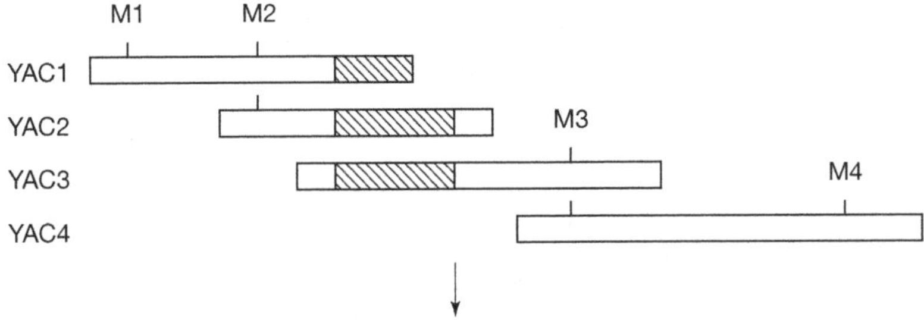

3. The order and overlap between YACs (BACs) that possess the markers is established and the DNA in the relevant YAC (BAC) clones can then be sequenced.

Figure 1.2

MAP-based cloning. The gene to be cloned is flanked by markers M2 and M3, and a strategy for cloning the gene based on this information is shown.

Simple sequence length polymorphisms (SSLPs) and microsatellites have also been developed as mapping markers in plants and fungi. The polymorphisms are based on differences in lengths of short repeat sequences, and the PCR products are simply run on a gel to check for differences in length. Markers based on single nucleotide polymorphisms (SNPs) are also being developed and have great potential for use in mapping, marker-assisted breeding and molecular diagnostics as discussed in Chapters 12 and 13. The real-time PCR techniques used for detection of SNPs and for quantitative PCR are based on ways of monitoring for the accumulation of PCR products over time through the labelling of primers, probes or amplification products with fluorogenic compounds. The most commonly used fluorogenic compounds rely on fluorescence resonance energy transfer (FRET) between fluorogenic labels or between a fluorophore and a quencher. In the simplest approach, compounds that fluoresce when associated with dsDNA (e.g. SYBR-green 1) are incorporated into the PCR reaction and fluorescence is monitored over time. A more elaborate approach used in the LightCycler (*Figure 1.3a*) uses two primers labelled with different fluorophores, one at the 3′ terminus and the other at the 5′ terminus, so that when the two primers hybridise to the template DNA, they are located within ten nucleotides of each other, and fluorescence occurs and is detected. This approach has the advantage of being able to monitor fluorescence changes during both annealing and denaturation in PCR, giving more reliable results.

The Taqman approach (*Figure 1.3b*) is an alternative strategy, and when used to quantify two alleles in mapping, the primers for the two alleles are labelled with a quencher dye (e.g. 6-carboxy-N,N,N′,5N′-tetrachlorofluorescein (TAMRA)) at the 3′ end and with a different reporter dye at the 5′ end, for example TET (6-carboxy-4,7,2′,7′-tetrachlorofluorescein) could be used on allele 1, and FAM (6-carboxyfluorescein) on allele 2. PCR is then directed by flanking primers and the 5′-nuclease activity of the Taq DNA polymerase causes the primer that binds most efficiently to the polymorphic sites in the sample to degrade during extension. This can be detected in automated DNA-sequencing-type machines. The molecular beacons approach (*Figure 1.3c*) involves a hybridisation probe labelled at one end with a fluorescent dye and at the other with a quencher. When the probe hybridises to the target sequence (during the annealing step), separation of fluorescence from the quencher results in a detectable signal. Other primer designs include Sunrise (*Figure 1.3d*) and Scorpion primers (*Figure 1.3e*).

(a) The LightCycler primers – two primers are used, one labelled at the 3′ end with a donor (D) the other at the 5′ end with an acceptor (A). When donor and acceptor are close enough in the PCR reaction, fluorescence occurs.

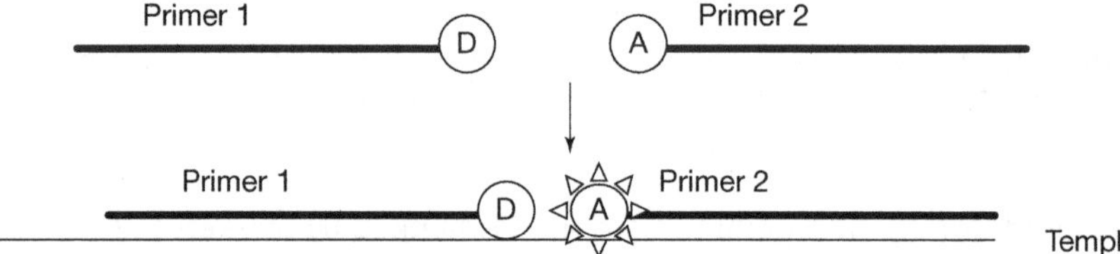

(b) The Taqman assay. One primer is labelled with a fluorescent reporter (F) at one end and a quencher (Q) at the other that stops it fluorescing. If PCR is primed upstream by a second primer, the endonuclease activity of the Taq polymerase removes the reporter away from the quencher so that fluorescence occurs.

(c) Molecular beacons. The primer forms a hairpin loop when not associated with the template DNA, such that the fluorescence and quencher parts are tightly associated and no fluorescence occurs. When the primer associates with template DNA, this association is lost and fluorescence occurs.

Figure 1.3

Quantitative/real-time PCR techniques. The technique depends on monitoring the accumulation of PCR products over real-time by detecting fluorescence. Various ways of designing primers that fluoresce upon PCR amplification have been devised as shown.

(d) Sunrise primers. When used for first-strand PCR, the fluorescence in the primer remains quenched. However, when the opposite strand is duplicated, the hairpin structure is relaxed and quenching is lost.

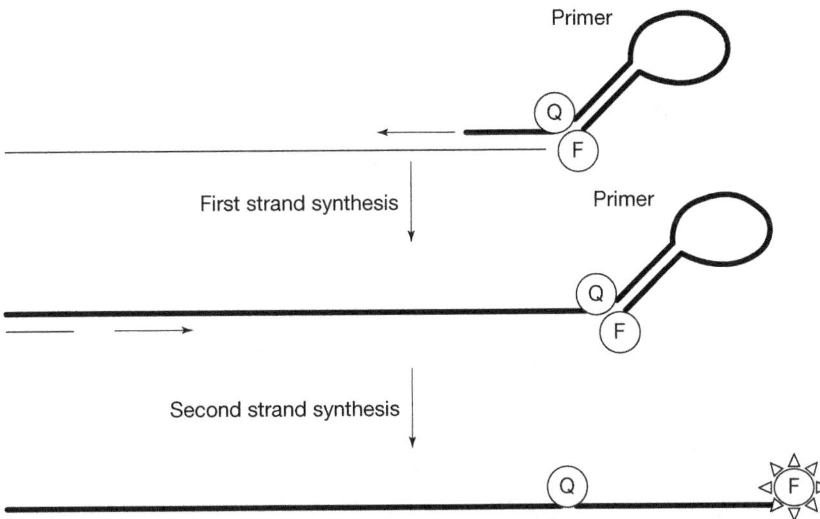

(e) Scorpion primers. Fluorescence is quenched in the primer by the hairpin structure. However, once the first strand has been synthesised, the primer is designed to open up and anneal to a downstream sequence on the nascent strand allowing fluorescence to occur.

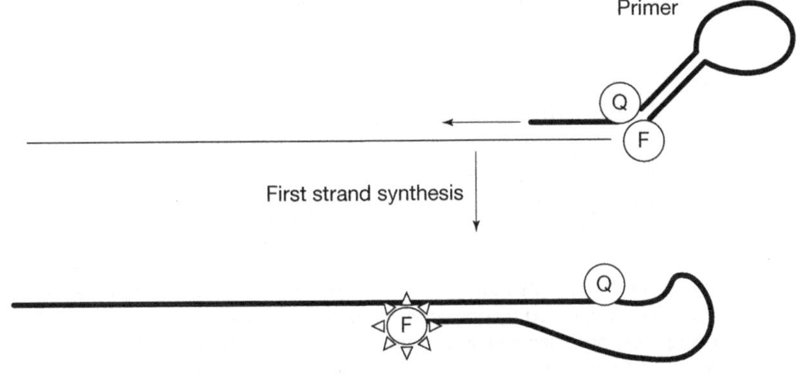

Figure 1.3

(Continued).

Methods are also being developed in mammalian systems in which SNPs form the basis for microarray-based mapping and genotyping, and these techniques are readily transferable to plants and microbes.

A further method used in map-based cloning is amplified fragment length polymorphism (AFLP) (*Figure 1.4*). This is a PCR-based technique in which genomic DNA is cut with two restriction enzymes, and adapters specific to each type of cohesive end are ligated on to the fragments. Primers complementary to the two adapters are used to amplify the restriction fragments.

1. Restrict the genomic DNA with two enzymes that generate cohesive ends.

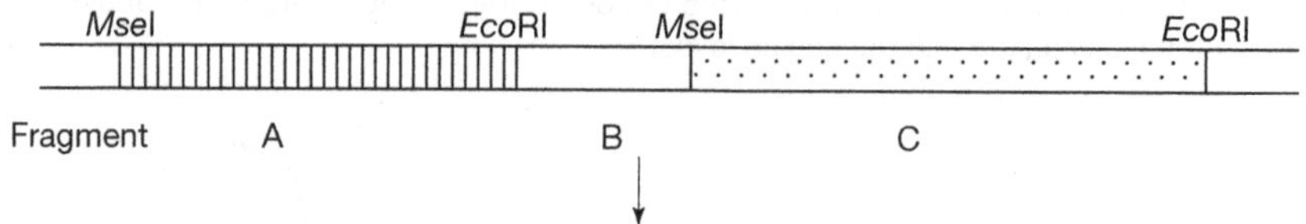

2. Ligate MseI and EcoRI adapters.

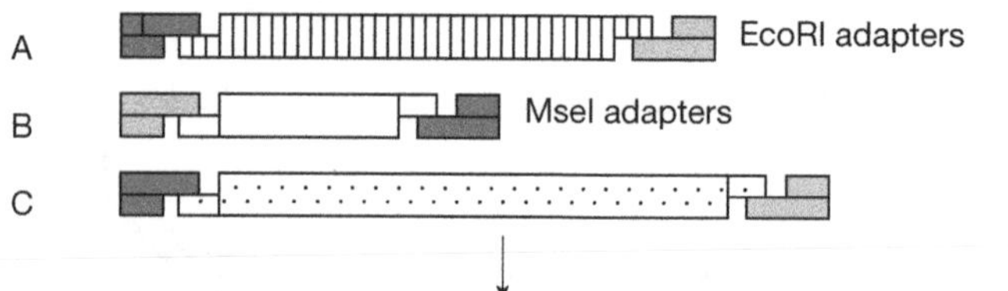

3. Amplify all the fragments using EcoRI and MseI universal primers.

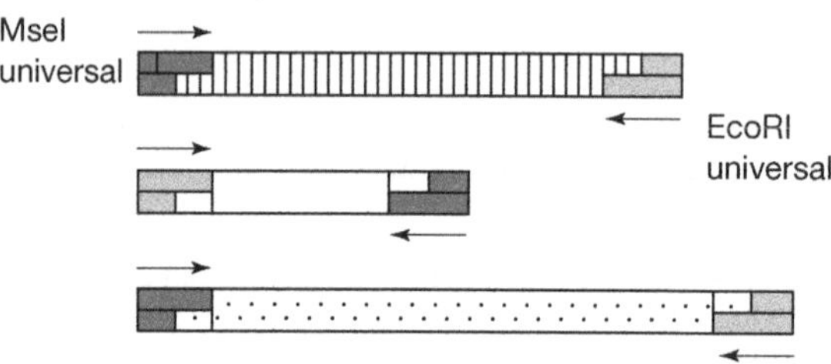

4. Amplify using selective primers that are the same as the universal primers but extended by 1, 2 or 3 randomly selected additional bases. Only the subset of fragments that have these bases adjacent to the restriction site will be amplified.

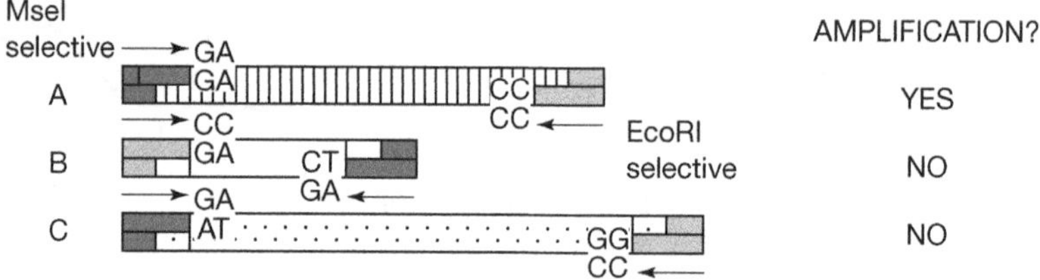

5. By using one of the primers radioactively or fluorescently labelled, amplification products can be visualised by polyacrylamide gel electrophoresis or automated sequencing. Polymorphisms in and around the restriction enzyme sites between genotypes can be detected as the presence/absence of bands.

Figure 1.4

Amplified fragment length polymorphisms (AFLPs). Genomic DNA is cut and specific adapters are ligated onto all the restriction fragments. These are then amplified as shown and polymorphisms in and around the restriction sites can be detected.

Two to three extra bases are added at the 3′ end of the primers, so that only a subset of the restriction fragments serve as templates for the PCR reaction. This reduces the complexity of the mixture of PCR products to around 50–100 products per reaction, so that when run on a polyacrylamide gel, the majority of bands represent a single PCR product, and differences in band presence or absence can be detected. By using various combinations of primers with different extensions, vast numbers of genomic fragments can be screened for polymorphisms. Map-based cloning using AFLP markers was used to clone the barley *mlo* gene, which encodes resistance to *Blumeria graminis* (see Section 10.10).

1.5.5 Reverse genetics

With so much sequence information now available, reverse genetics has become a way of connecting sequence to function. One approach that has been used is PCR Tilling (Targeting Induced Local Lesions in Genomes). Essentially, this involves producing a population of plants with random mutations, for example through chemical mutagenesis. DNA is prepared from the individual plants (normally the M2 progeny from the initial M1 mutagenised plants). This DNA is then mixed in pools, for example for ten individuals in each pool, and the gene of interest is then PCR amplified from each pool. If a mutation is present in one of the plants in the pool, the PCR products will be of two types, wild-type and mutant, and when this DNA is heated and cooled, it will form heteroduplexes that can be detected by denaturing high-performance liquid chromatography (DHPLC). The individual plant within the pool can then be identified and, following appropriate control experiments, the phenotype associated with the mutation in the gene of interest determined. The technique has been widely used in *Arabidopsis*, where mutant lines can be held as a community resource, and automation of the PCR and DHPLC techniques facilitates rapid screening of these populations.

A second approach that can be used for increasing or decreasing the expression of a specific gene and then analysing the phenotype of the resultant organism is that of gene silencing, such as sense or antisense suppression, which work through post-transcriptional gene silencing (PTGS) mechanisms (see Section 8.8.3). In plants, *Agrobacterium* and *Potato virus X* (PVX)-based transient expression systems have been developed to facilitate these approaches. In nematodes such as *C. elegans*, dsRNA sequences are generated and used directly for silencing but these methods are not yet available in plants or their pathogens. Other techniques have also been developed for targeting insertions or deletions into specific genes in organisms, such as homologous recombination, but as discussed above (see Section 1.5.3), these techniques have yet to be developed for effective use in plants and most plant-pathogenic fungi.

Techniques have also been developed for assigning functions to particular domains of genes and for producing proteins with modified traits, such as disease resistance genes with altered specificities (see Section 13.4). DNA shuffling is an *in vitro* recombination technique in which fragments of closely homologous genes or variants of the same gene are exchanged to generate genes of different base compositions. PCR and techniques such as incremental digestion of fragments with exonucleases create fragments of genes which are reassembled into recombinant genes, and the phenotype associated with these can be determined.

1.5.6 Dissection of signalling pathways

Once a gene in a desired pathway has been cloned, there are a number of ways to use this information to identify other genes whose products

1. The bait

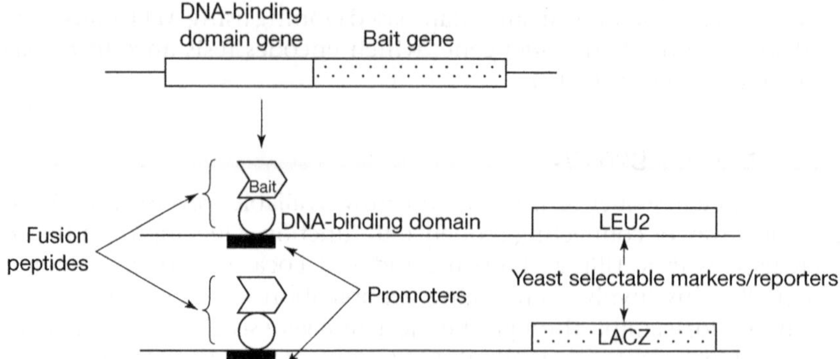

2. The prey – make a cDNA library of genes fused to transcriptional activation domains.

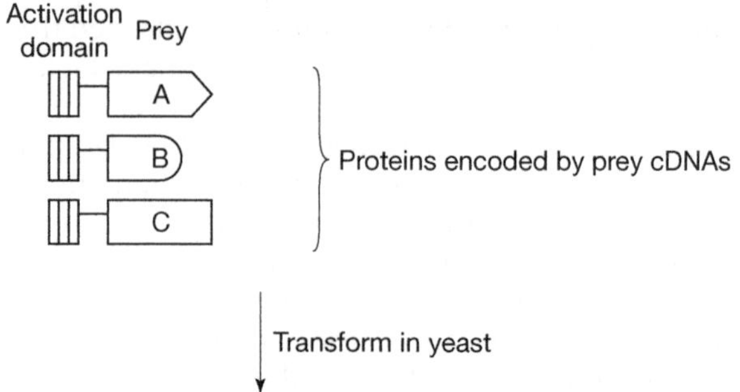

3. If the prey and bait interact, transcription occurs and the yeast grows on selective media and expresses the reporter gene.

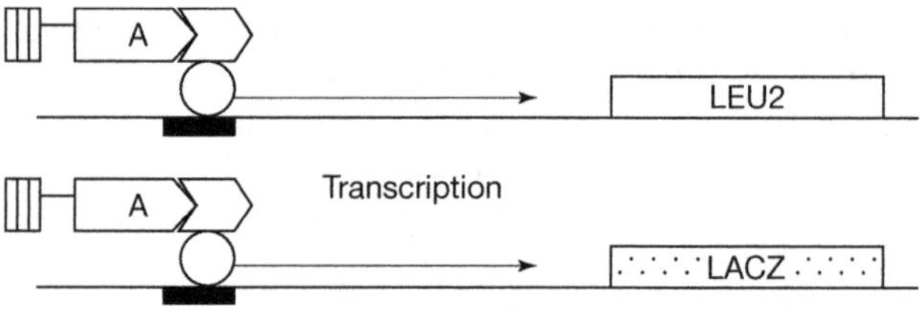

Figure 1.5

The yeast two-hybrid system. The gene of interest (bait) is fused to a DNA-binding domain, and the prey is made from tissue in which interacting proteins are likely to be expressed. When the prey and bait interact, the reporter and selectable markers are expressed.

function in the pathway. These include the yeast-based two- and three-hybrid screens, which are methods to identify proteins that interact with known proteins as shown in *Figure 1.5*. In addition, if the downstream genes in a pathway are already known, and it is important to identify the regulatory elements and transcription factors that control their expression, the yeast one-hybrid screen may be used, in which the yeast reporter gene is placed under the control of the promoter element of interest. The promoter element becomes the bait to trap the DNA-binding protein. A cDNA library is expressed as a protein fusion with a transcriptional activation domain in yeast cells with the reporter construct. Any fusion with the activation domain that results in binding to the element will lead to expression of the reporter.

1.5.7 Gene expression profiling

Gene expression profiling refers to the analysis of the differential expression of large numbers of genes. By comparing the expression profiles, an indication of which genes may play a role in the particular response can be identified. For example, one could identify all genes that are up- or down-regulated in a plant at a particular time following pathogen infection. Microarrays are emerging as the method of choice for large-scale analysis of gene expression, and the small size of microarrays means that the expression of thousands of genes can be examined simultaneously in a single experiment. Ideally, a unique gene set (the unigenes) from the target organisms should be arrayed for maximal information, but where these are not available, EST (expressed sequence tag) collections are often used. Microarrays are essentially a means of performing northern hybridisations in reverse. Instead of putting mRNA on a blot and probing with a defined sequence to examine expression, the cDNA or genomic sequences are arrayed on the solid support and probed with labelled RNA from different timepoints. Because the DNA on the support is in excess, the hybridisation signal is proportional to the amount of the specific RNA in the total RNA sample that is homologous to the DNA in that spot.

There are two types of microarrays currently in use, those in which oligonucleotides of specific sequences are synthesised *in situ* on a glass substrate (Genechip), and those in which DNA fragments are spotted on a glass substrate. The Genechip is sometimes referred to as a synthesis array or Affymetrix (the company who manufacture the arrays). With existing technologies, around 300 000 different oligonucleotides, equivalent to 8000 different genes, can be synthesised in an area of approximately 1.6 cm^2. The normal procedure is to synthesise a small number of oligonucleotides (up to 20) spanning different regions of each cDNA onto the array and incorporate a number of internal controls. Hybridisation to the arrays is done using fluorescently labelled cDNA probes generated from mRNA of the samples to be tested, and with synthesis arrays, single probes are used on each array and the results compared using appropriate software packages.

The second type of microarray is sometimes referred to as a deposition or a cDNA microarray. The deposition type is analogous to a dot blot in that DNA fragments are spotted directly onto the substrate, but since a non-porous glass substrate is used for the microarray instead of a membrane, the density of the spots is much greater. The direct spotting of the DNA fragments means that full-length cDNAs or genomic fragments can be used as well as short oligonucleotides. With current technology, deposition arrays can be generated at a density of approximately 10 000 genes per 3.24 cm^2. With the deposition arrays, two separate RNA samples are labelled with different fluorescent labels and combined in a hybridisation mix on a single array. After hybridisation and washing, the microarray is scanned sequentially at the two emission wavelengths of the probes. The relative expression of each gene under two different conditions is therefore determined in a single

1. Prepare the array

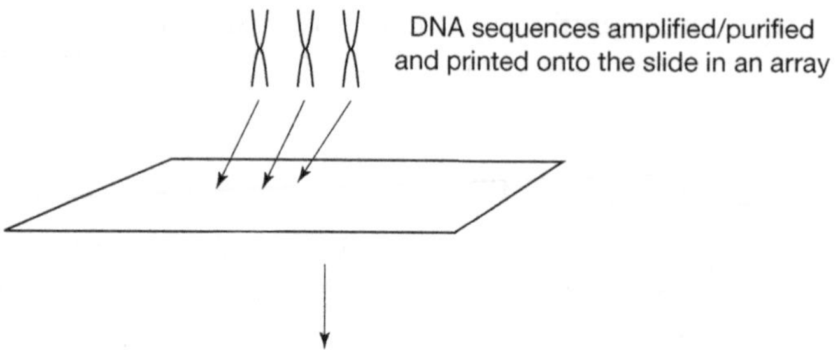

2. Hybridisation

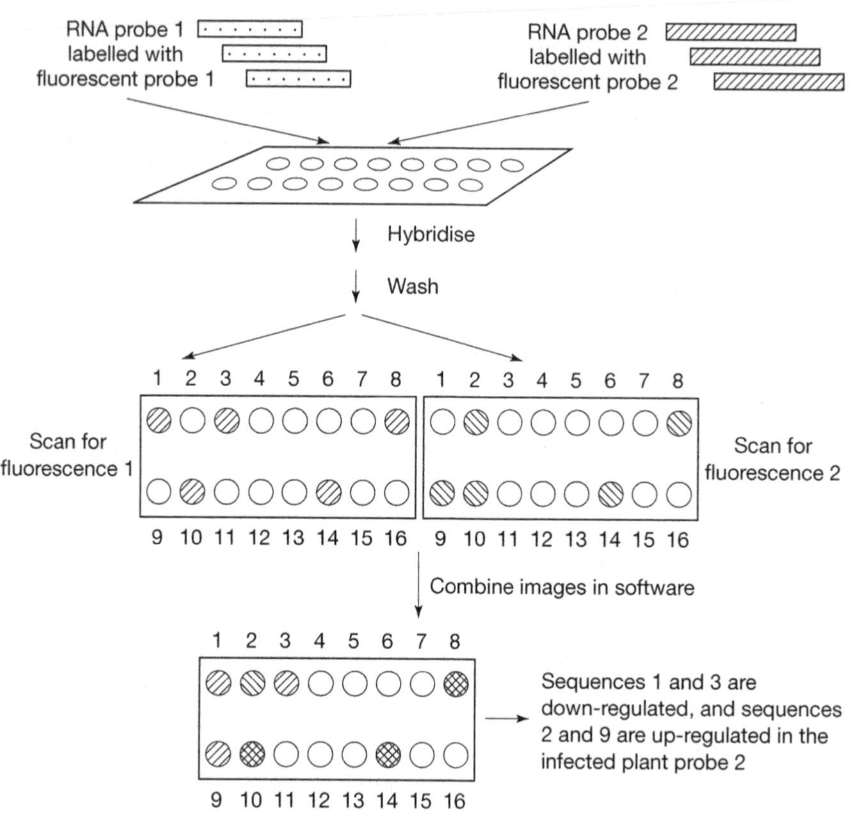

Figure 1.6

Spotted (deposition) DNA microarrays. DNA sequences from the plant (e.g. an EST library) are spotted in a specific array. In the example shown, probe 1 will be made from the uninfected plant and probe 2 from the infected plant.

hybridisation experiment. *Figure 1.6* indicates how this might be used to determine which genes in a plant are induced by a particular pathogen. A large amount of information on microarray technology is available on the Internet, and some useful websites are given at the end of this chapter.

Other approaches developed for analysing differential gene expression include differential display, cDNA-AFLP, suppression subtractive hybridisation

(SSH) and serial analysis of gene expression (SAGE). Of these cDNA-AFLP, SSH and SAGE have proved to be the more reliable and robust techniques. cDNA-AFLP is essentially the same as AFLP (*Figure 1.4*) except that double-stranded cDNA is made by reverse transcription of the mRNA and the double-stranded cDNA is then used for the restriction enzyme digestions. Linkers are ligated on to the ends of the restriction fragments and PCR is done using primers based on the linker sequences as described in Section 1.5.4. The PCR products from the samples to be compared are run on a polyacrylamide gel and differential bands are eluted and cloned. Since long primers based on known sequence are used for cDNA-AFLP, it is more reproducible and generates fewer false positives than differential display. By using various combinations of two base pair extension primers, it is possible to systematically screen cDNA populations from multiple mRNA samples for any differences. For example, a range of timepoints or conditions can be compared simultaneously alongside each other on the same gel and because band intensities are quantitative, the expression profile of particular fragments over time can be analysed, as opposed to microarrays and SSH where only two timepoints are compared against each other at a time.

SSH generates a library of cDNA clones that is enriched in differentially expressed genes. While not all cDNAs obtained with SSH represent differentially expressed genes, the method is fast and the cDNA population produced can be screened by using each clone individually to probe RNA blots, or by making arrays of the library on filters and probing these arrays with the labelled cDNA from the two RNA samples used to generate the subtracted library. One advantage of SSH is that low abundance messages can be more readily identified than by other methods.

SAGE is another PCR-based technique that can be used especially when extensive sequence databases are available for the organisms. Essentially, short (10–14 bp) sequences of cDNAs (that represent the mRNAs that are being expressed in the sample) are concatamerised with 4 base-pair tags between them. These sequences are generated by digesting the cDNAs with restriction enzymes and ligating adapters that contain the recognition sequences for type IIS restriction enzymes. The type IIS restriction enzyme then cleaves the cDNA 20 base pairs from the adapter, and these sequences are ligated to form the concatamers. The concatamers are sequenced and the sequences compared to databases to identify the genes that are being expressed, and the frequency of a particular fragment will be proportional to its relative expression level in the corresponding mRNA.

1.5.8 Proteomics

Proteomics refers to characterisation of all the proteins expressed by the genome of an organism. While gene expression profiling provides information at the level of transcript accumulation, proteomics provides information on the proteins. Depending on how the experiment is set up, information such as where and when each protein accumulates, to what level it accumulates, and post-translational modifications to the proteins can be obtained. As new sub-cellular techniques are developed, proteomics should provide more precise information on gene expression since the functional product of most genes is not the RNA but the protein.

The basis of most separations for protein analysis is 2-dimensional gel electrophoresis, and a number of new techniques have been developed to identify proteins from the small amount of sample that can be obtained from a 2-D protein gel. In addition, identification of proteins has been aided by the vast amounts of sequence available in databases. A protein spot is eluted from the gel and the protein is subjected to proteolytic cleavage into peptide fragments. These fragments are analysed by matrix-assisted laser desorption/ionization-time of flight (MALDI TOF) mass spectrometric analysis, and this generates a list of peptide masses for the peptide fragments. The combination of fragment sizes is characteristic for a specific protein and can be predicted from the gene sequence, so the results for a protein can be compared to a database of calculated peptide masses for each open reading frame in the genome. If no match is found in the sequence database, proteins can be analysed by peptide sequencing. In the future, array technology will be adapted for use in proteomics. For example, it may be possible to array antibodies for a large number of proteins on a chip to analyse for changes in protein levels in an analogous way to how mRNA changes are currently measured. Alternatively, protein arrays could be used to probe for interacting proteins.

1.5.9 Metabolite profiling

A further global approach being developed to study plant pathogen interactions is at the level of metabolites, using metabolic profiling analogous to mRNA or protein profiling. Changes occur in the chemical constituents of plants as part of the response to pathogen attack. For example, phytoalexins are an important component of the defence response (see Section 9.3.4), and there are probably many other equally important metabolites that are as yet unknown. One of the difficulties in metabolite analysis is their diversity. As a result, global analysis of all metabolites is a much greater technical challenge than analysis of mRNAs or proteins, but techniques for extraction, purification and detection are being developed.

1.5.10 Bioinformatics

Genomic sequencing along with genomics-based and proteomic-based arrays are generating vast amounts of data, and an important challenge is to make effective use of this information. To cope with all these data requires new technologies, and for this the field of bioinformatics has emerged. Bioinformatics refers to the use of computers to sort, analyse and provide relevant and appropriate biological information, and covers topics such as electronic publishing, complicated database analysis and making large sets of data publicly available. An important component of the information explosion is the Internet, and some relevant Internet resources for molecular plant pathology are given below.

The plant pathology internet guidebook:
  http://www.pk.uni-bonn.de/ppigb/ppigb.htm

Publicly accessible analysis programs and databases:
  http://srs.ebi.ac.uk/
  http://www.infobiogen.fr/services/dbcat/
  http://www3.oup.co.uk/nar/Volume_27/Issue_01/summary/gkc105_gml.html
  http://rana.stanford.edu

References and further reading

Agrios, G.N. (1997) *Plant Pathology*, 4th Edn. Academic Press, San Diego.

Berbee, M.L. (2001) The phylogeny of plant and animal pathogens in the Ascomycota. *Physiological and Molecular Plant Pathology* **59**: 165–187.

Dickinson, M.J. and Beynon, J. (2000) *Annual Plant Reviews, Volume 4 – Molecular Plant Pathology*. Sheffield Academic Press, Sheffield.

Dong, S., Wang, E., Hsie, L., Cao, Y., Chen, X. and Gingeras, T.R. (2001) Flexible use of high-density oligonucleotide array for single-nucleotide polymorphism discovery and validation. *Genome Research* **11**: 1418–1424.

Harrington, C.A., Rosenow, C. and Retief, J. (2000) Monitoring gene expression using DNA microarrays. *Current Opinion in Microbiology* **3**: 285–291.

Ingold, C.T. and Hudson, H.J. (1993) *The Biology of Fungi*, 6th Edn. Chapman and Hall, London.

Judson, N. and Mekalanos, J.J. (2000) Transposon-based approaches to identify essential bacterial genes. *Trends in Microbiology* **8**: 521–526.

Kahmann, R. and Basse, C. (1999) REMI (restriction enzyme mediated integration) and its impact on the isolation of pathogenicity genes in fungi attacking plants. *European Journal of Plant Pathology* **105**: 221–229.

Kuhn, E. (2001) From library screening to microarray technology: strategies to determine gene expression profiles and to identify differentially regulated genes in plants. *Annals of Botany* **87**: 139–155.

Lee, I-M., Davis, R.E. and Gunderson-Rindal, D.E. (2000) Phytoplasma: phytopathogenic mollicutes. *Annual Review of Microbiology* **54**: 221–255.

Lucas, J.A. (1998) *Plant Pathology and Plant Pathogens*, 3rd Edn. Blackwell Science, Oxford.

McCallum, C.M., Comai, L., Greene, E.A. and Henikoff, S. (2000) Targeting induced local lesions in genomes (TILLING) for plant functional genomics. *Plant Physiology* **123**: 439–442.

Ramonell, K.M. and Somerville, S. (2002) The genomics parade of defense responses: to infinity and beyond. *Current Opinion in Plant Biology* **5**: 291–294.

Soanes, D.M., Skinner, W., Keon, J., Hargreaves, J. and Talbot, N.J. (2002) Genomics of phytopathogenic fungi and the development of bioinformatic resources. *Molecular Plant Microbe Interactions* **15**: 421–427.

Sweigard, J.A. and Ebbole, D.J. (2001) Functional analysis of pathogenicity genes in a genomics world. *Current Opinion in Microbiology* **4**: 387–392.

Tillib, S.V. and Mirzabekov, A.D. (2001) Advances in the analysis of DNA sequence variations using oligonucleotide microchip technology. *Current Opinion in Biotechnology* **12**: 53–58.

Waller, J.M., Lenné, J.M. and Waller, S.J. (2002) *Plant Pathologist's Pocketbook*, 3rd Edn. CABI Publishing, Wallingford.

Wang, M-B. and Waterhouse, P.M. (2001) Applications of gene silencing in plants. *Current Opinion in Plant Biology* **5**: 146–150.

Fungal and oomycete diseases – establishing infection

2

To cause disease, a pathogen must find a suitable host plant, pass through the external protective layers of the host, and gain access to the nutrients that it requires for its own growth and development. So how does a potential fungal pathogen enhance its chances of finding a suitable host plant, and how does it penetrate the host surface? A key environmental factor in dispersal, spore germination and plant penetration is moisture. A second major factor is the ability of many fungi to detect and respond to host cues, such as chemical signals, electrical stimuli, pH and host surface chemistry, hardness and topography. It is the nature of these signals and the molecular mechanisms through which they are detected by fungi and oomycetes and translated into responses that we focus on in this chapter.

2.1 Dispersal of spores

Fungi and oomycetes are generally dispersed as spores that are either deposited in the soil or transmitted through the air. There are many unique and different mechanisms for release of spores from a parent colony. In some cases, the spores are held on structures above the boundary layer of still air adjacent to the leaf surface to enhance their chances of being lifted away in air currents, upon stem vibration, leaf flutter, or in rain splash. In other cases the spores are actively discharged by specific dispersal mechanisms that eject them away from the plant.

2.2 Finding a suitable host

Once dispersed, the spores must eventually come into contact with a suitable host plant if they are going to cause disease. They must do this using pre-programmed mechanisms and endogenous energy reserves until such time as they gain access to the host cell nutrients. For air-borne and splash-borne spores, the secret to success lies in producing vast quantities of spores that can remain viable for a long time. It has been estimated that a single maize smut (*U. maydis*) gall may contain up to 2×10^{11} spores and a hectare of wheat heavily infected with stem rust (*Puccinia graminis*) may produce up to 5 kg or 2×10^{12} urediospores per day. These rust urediospores, which may be transmitted in air currents at altitudes in excess of 3000 metres, are heavily pigmented with carotinoids and able to survive exposure to UV damage. As they are deposited in rain or through the direct action of wind currents, some of these spores will land on suitable host plants to initiate another disease cycle and ensure the success of the pathogen.

Whilst a passive process for finding a host is appropriate for many fungi, it is not suitable for soil-borne spores, which do not germinate simply because the soil is moist. The ability of spores to remain dormant is referred to as fungistasis, and whilst there is no fully satisfactory explanation for this phenomenon, it is clear that many spores require nutrients, particularly glucose and other sugars, to stimulate germination. The significance of this lies in the fact that developing roots release carbohydrates and other nutrients. This mechanism ensures that spores only germinate when plants are in the vicinity. In many cases, non-specific chemical exudates are enough to trigger germination. However, some fungi only respond to specific chemicals released from their particular host plant species. For example, *Phytophthora sojae* is attracted to isoflavones such as daidzein and genistein that are exuded from soybean roots, chemicals to which other *Phytophthora* species are not attracted. Similarly, sclerotia of *Sclerotium cepivorum* (onion white rot), only germinate in response to n-propyl and alkyl sulphides that are produced through the conversion (by soil bacteria) of non-volatile sulphide-containing compounds released specifically from the roots of host onion, garlic and a few other *Allium* species.

Chemotactic responses to general and specific chemical signals, and also to electric currents generated around plant roots, are particularly apparent in the oomycetes *Phytophthora* and *Pythia* and the protozoan *Plasmodiophora*. Here, the dormant zoospores respond to the signals by germinating and producing flagella, which enable them to swim through moisture in the soil toward the source of the signal. Many zoospores also possess a default mechanism such that if they fail to reach a suitable host after a few hours swimming, the spores encyst and return to dormancy, releasing a further zoospore at a later stage. Chemotaxis and germination in response to chemical signals is also a feature of some air-borne fungi and oomycetes. The zoospores of *P. infestans* swim on the leaf surface in water droplets before encysting and germinating. *B. cinerea* also requires nutrients, especially sugars, that are present on the surface of leaves to stimulate germination, and it has been suggested that gaseous compounds such as ethylene, released during the ripening process in climacteric fruit and during senescence stimulate the germination of certain post-harvest pathogens such as *B. cinerea* and *Colletotrichum* species (see Section 2.7.2).

2.3 Spore attachment to the plant

There are considerable chemical and physical differences between the surfaces of different plant species, between different parts of a plant, and also depending on the age and maturity of the plant. The waxy cuticle is generally hydrophobic and repels water droplets and the microbes that they contain. This hydrophobicity can be measured by contact angles in water droplets, and that of some plants such as rice (170° for a 10 μl drop) is higher than for artificial non-stick surfaces such as Teflon (90–130°). A fungal spore that is splashed or deposited onto a plant surface in rain must attach firmly and rapidly to the plant surface to provide time to go through the penetration process, and similarly a spore that has been blown onto a plant under dry conditions must be able to attach, and remain attached when the water required for germination arrives.

In most cases this attachment is a two-stage process. There is an initial rapid non-specific adhesion to the plant surface, stabilised by hydrophobic interactions between the spore and the cuticular surface. For rust urediospores for example, it is the spines on the surface of the spores that are involved in this early contact. In the presence of moisture, a strong and reversible adhesion, mediated through active metabolism in the spores and molecules referred to as adhesins released from the spores occurs. In the rice pathogen *M. grisea,* a glyco-conjugated adhesive mucilage is released from a compartment in the spore apex upon wetting of the conidium. Similarly, the necrotrophic, broad host range pathogen *B. cinerea* appears to go through an initial weak adhesion prior to formation of a fibrillar matrix that develops at the germ-tube and causes the strong adhesion prior to penetration. In rust fungi, adhesive pads can be formed experimentally between spores and hydrophobic surfaces such as Teflon. However, when formed on a leaf surface, additional adhesive materials released from the host plant through the action of esterases and cutinases are contained in the pads.

Soil-borne fungi and oomycetes must also attach and adhere to roots prior to penetration. In the case of *Phytophthora* species, the approach is similar whether they are foliar-penetrating pathogens such as *P. infestans* or root-penetrating such as *P. sojae.* The motile zoospores either swim on the leaf surface or in the soil towards the source of chemical signals, and then encyst. This process occurs within 30 minutes of the spore reaching a host, and in the process the wall-less zoospore sheds its flagella, secretes an adhesive and produces a cell wall.

2.4 The germination process

Whilst the molecular mechanisms for induction of spore germination by environmental factors such as water and low-molecular-weight nutrients have not been studied in detail in plant-pathogenic fungi, there have been extensive studies in model fungi such as *Aspergillus nidulans* and *Saccharomyces cerevisiae.* Here it has been shown that the sensing of extracellular carbon sources controls the breakdown of trehalose by regulating the activity of the enzyme trehalase. Trehalose (α-D-glucopyranosyl-α-D-glucopyranoside) accumulates to form up to 15% of the dry mass of many fungal spores, and in *S. cerevisiae,* the cytoplasmic neutral trehalase Nth1 responsible for trehalose breakdown is tightly regulated by a phosphorylation mechanism that involves a cAMP-dependent protein kinase (PKA) (*Figure 2.1*). A similar PKA gene has been identified in the plant pathogen *M. grisea,* which significantly contains the amino-acid motif that is proposed to be the target of PKA-mediated phosphorylation of Nth1. Furthermore, mutations in this gene result in a reduction of pathogenicity. However, a similar PKA-dependent pathway does not appear to mediate glucose responses in all fungi, and *Neurospora crassa* and *Schizosaccharomyces pombe,* for example, use alternative signalling pathways. A second general role for PKA-dependent pathways during germination is to co-ordinate changes in the expression of genes encoding components of the protein synthesis machinery during periods of growth resumption or in response to nutritional shifts, as has been shown to occur in yeast for transcriptional activation of ribosomal protein genes. A third role for PKA-dependent pathways is in the response to more specific stimuli of germination. For example,

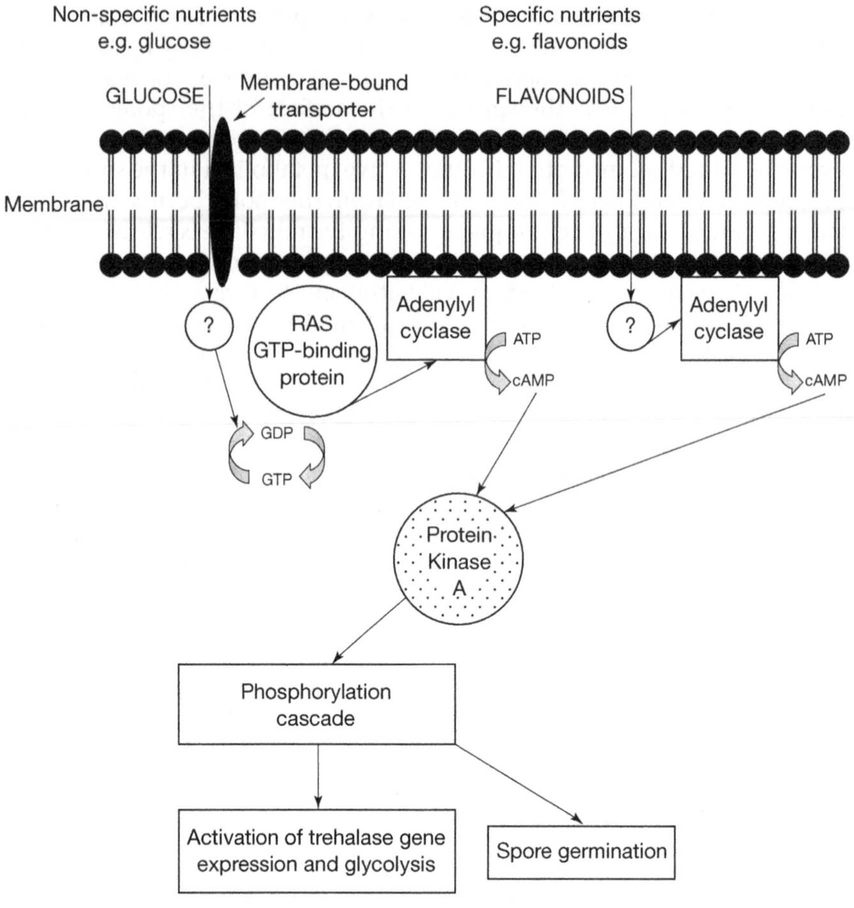

Figure 2.1

cAMP signalling pathways in response to external signals. The glucose pathway (left-hand side) is initiated by glucose transporters, whilst the separate pathway induced by flavonoids in *Fusarium solani* is shown down the right-hand side.

a pathway has been identified in the stimulation of *Fusarium solani* germination in response to host flavonoids. This foot and root rot pathogen has a number of strains or *forma speciales* that are pathogenic on different plant species. *Nectria haematococca* (*F. solani* f. sp. *pisi*) is stimulated to germinate in the presence of specific flavonoids such as pisatin in root exudates from pea, whilst the related bean pathogen *F. solani* f. sp. *phaseoli* is stimulated by the flavan genictein and the flavanone naringenin. Interestingly, these specific root exudates are the ones used by pea and bean symbiotic strains of *Rhizobium* species for induction of *nod* (nodulation) genes that are required for the production of Nod factors, that in turn signal developmental changes in plants to allow nodulation (see Section 3.3.6). In *Rhizobia* the signalling occurs through a NodD transactivator protein that binds to regulatory motifs (nod boxes) upstream of the other *nod* genes. However, there is no evidence for a similar signalling mechanism in *F. solani*. Instead, the use of inhibitors of adenylate cyclase and cAMP-dependent protein kinase, along with measurements of cAMP levels indicate that a cAMP-dependent

PKA is involved, and that this pathway is separate from the pathway(s) used by non-specific nutrients such as sugars and amino acids (*Figure 2.1*).

2.5 Penetration methods

Once the fungus has found a suitable plant, adhered and germinated, it must grow into the plant to obtain its nutrients. Fungi have the capacity to do this either directly through the cuticle, or to grow towards natural opening such as stomata or wound sites and enter through these. Some fungi can use a variety of methods, such as *B. cinerea,* which can access plants through wounds or direct penetration. Others will use only one method, such as *M. grisea* which uses direct penetration, whilst rusts use stomatal penetration. However, care must be taken in making these generalisations, since whilst the repeating asexual urediospore stage of most rusts penetrate exclusively through stomata, basidiosporelings of most rusts use direct penetration as their mode of entry (*Figure 2.2*).

To directly penetrate the cuticle of a plant requires entry through layers of wax, cutin, pectin and cellulose fibrils, whilst to go through natural openings,

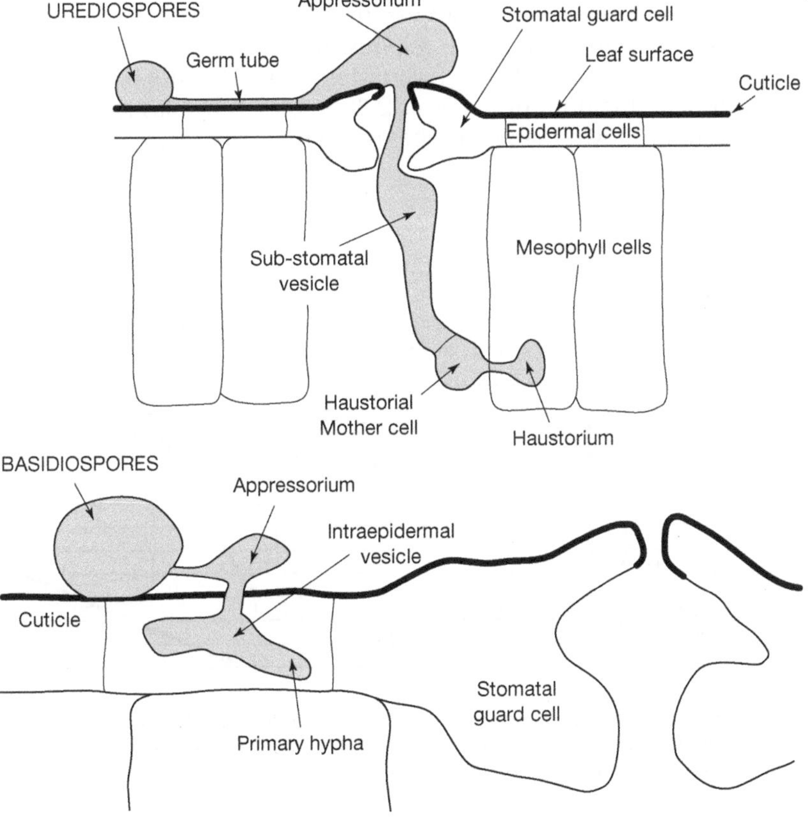

Figure 2.2

Penetration mechanisms. The alternative penetration mechanisms of rust urediospores (through natural openings) and rust basidiospores (direct penetration).

germ-tubes must grow towards the entrance points. Many fungi go through a sequence of events following germination that involves a short period of germ-tube growth along the leaf surface followed by differentiation of the hyphal tip to form a swelling known as the appressorium. As the germ-tube grows along the leaf surface, more mucilage is synthesised to attach it to the surface, although at this stage, many fungi will go through this process whether they are on a host plant, a non-host plant, or even an artificial surface. Following maturation of the appressorium, a narrow hyphal thread or penetration peg grows downwards from the lower surface and through the cuticle of the plant or the natural opening. Entry is believed to involve the combined effects of mechanical force and enzymic dissolution of the epidermis, with the relative importance of these varying between different plant–pathogen interactions.

2.6 Germ-tube elongation

Once a spore has anchored itself to the plant and the germ tube emerges, the germ tube will grow along the surface on the plant until it receives signals to penetrate. This is a pre-programmed process, fuelled by energy reserves from the spore, and in most cases, the direction of germ-tube growth appears to be an essentially random process. However, there are cases where growth has been shown to be directional and in response to host factors, the best example being the response of rust fungal germ-tubes to surface topography in a process known as thigmotropism (*Figure 2.3*). Studies using plastic replicas of leaf surfaces have shown that germtubes re-orientate themselves when they

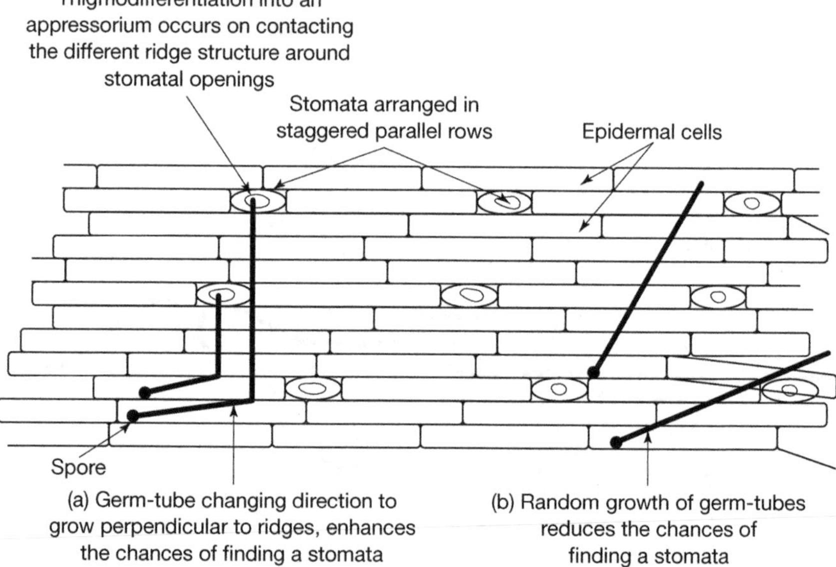

Figure 2.3

Thigmotropism – a schematic diagram of a cereal leaf surface. In (a) the germ-tubes re-orientate upon contact with epidermal ridges, compared to (b) random directional growth. On contacting the stomata the germ-tube differentiates to form an appressorium.

contact the ridges between epidermal cells to grow perpendicular to such ridges. This is believed to be a particular advantage for those rusts that grow on cereal leaves, in which the epidermal cells and stomata are arranged in ordered rows, as a means of enhancing the chances of the germ-tube locating a stomata. The molecular mechanisms involved are unknown, but vesicles accumulate at the apices of germ-tubes in close association with actin filaments and there is a rapid and continuous supply of cell wall precursors to the advancing germ-tube tip. In addition, rust germ-tubes differentiate into appressoria in response to the spacing and height of ridges surrounding stomatal lips, a process referred to as thigmodifferentiation (see Section 2.7.1).

2.7 Induction of appressorial development

There is increasing evidence that particular physical and chemical characteristics of plant surfaces induce appressorium development, including hydrophobicity, surface hardness, chemical composition, cutin monomers and wax polar lipids. Together, these cues relay the presence of a conducive environment and cause developmental changes in the fungus.

2.7.1 Physical factors

Restriction enzyme-mediated insertion (REMI) (see Section 1.5.4) has been undertaken in a number of fungi to tag and isolate genes involved in development and pathogenicity. Amongst the genes identified have been some that are required for appressorium differentiation, for example *PTH11* from *M. grisea,* which encodes a putative transmembrane protein located in the plasmalemma. Models have been proposed based on analogies with mammalian systems, in which transmembrane protein receptors known as integrins (which bind to actin microfilaments of the cytoskeleton on their cytoplasmic side and matrix glycoproteins such as vitronectin and fibronectin on their extracellular side), could be involved in sensing surface topography and hardness and transmitting the signal through the cell wall. In the bean rust fungus *Uromyces appendiculatus,* which like other rust fungi shows thigmodifferentiation into appressoria in response to contact with stomatal ridges on host plants, a fibronectin-like gene has been cloned that is up-regulated during infection structure differentiation, supporting this model for surface sensing and signalling.

Perception of environmental signals through transmembrane receptors generally results in conformational changes in the protein and interactions with heterotrimeric G proteins inside the cell (*Figure 2.4*). G proteins are known to regulate many fundamental events in fungi such as mating, pathogenicity and virulence, and G protein subunits have been isolated from a wide range of plant pathogens. In the case of the chestnut blight fungus *Cryphonectria parasitica,* targeted gene disruption in one of these has been shown to result in a reduction in virulence. In the human pathogenic fungus *Cryptococcus neoformans,* disruption of G protein subunits results in the inability to form melanin, and since melanin formation is a key component in appressorial function for a number of plant-pathogenic fungi (see Section 2.8.3), environmental signals may function through a similar mechanism in these. In *M. grisea,* three genes encoding G protein alpha subunits have been cloned, one from each of the three major G alpha protein

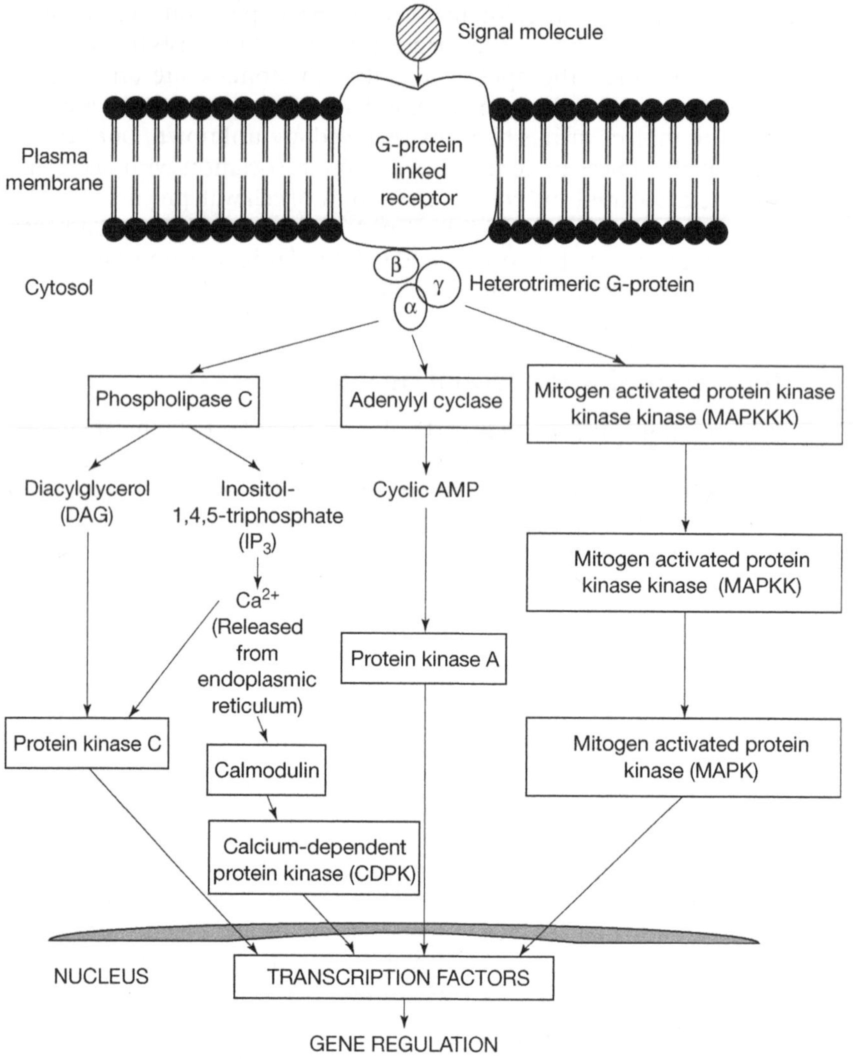

Figure 2.4

Signalling cascades emanating from G-protein-linked receptors. Following signal perception by G-protein linked receptors, the signal can be transduced through a complex, interlinked network of pathways. See text for examples in plant pathogens.

subgroups. However, mutation analysis suggests that it is only MAGB, (which belongs to the inhibitor of adenylyl cyclase type), that is associated with pathogenicity, and that mutations in the gene encoding this protein can be complemented by exogenous addition of cAMP.

G-proteins are important intermediates in a number of signal pathways involving protein kinases. In the phospholipase C pathway, the result is calmodulin-mediated activation of a Ca^{2+}-dependent protein kinase (*Figure 2.4*), and calmodulin genes have been shown to be induced in *Colletotrichum* species when conidia contact a hard surface. Inhibitors of calcium channels and calmodulin have also been found to block induction of thigmodifferentiation in rusts, suggesting that this may also involve the phospholipase C pathway.

In a second signal transduction pathway, the G-protein interacts with adenylyl cyclase to result in cAMP generation and protein kinase A activation. Genes encoding the catalytic and regulatory subunits of PKAs have been isolated from *M. grisea* and targeted deletions of these have shown a clear delay in appressorial formation and the ability to cause disease. Deletions in similar genes from *C. trifolii* and *C. lagenarium* resulted in delayed appressorium penetration and inability to penetrate host surfaces, and PKA and cAMP are also believed to be involved in appressorial formation in barley powdery mildew (*Blumeria graminis*), based on studies using PKA inhibitors and exogenous cAMP applications, although gene disruption experiments have yet to confirm this.

A third G-protein-mediated pathway involves mitogen-activated protein kinase (MAPK) cascades, which are serine/threonine kinases that differentially alter the phosphorylation status of transcription factors in the nucleus and alter gene expression. Three MAPK genes have been isolated from *M. grisea,* although only two of these have a confirmed role in pathogenicity. *PMK1* (pathogenicity MAP kinase) has been shown to be functional through its ability to complement mutations in the yeast FUS3/KSS1 MAPK. *pmk1* mutants in *M. grisea* are unable to differentiate appressoria although they do show normal vegetative growth. These germ-tubes are however responsive to cAMP for germ-tube tip deformation, suggesting that this kinase acts downstream from the surface recognition and infection structure formation pathway. Conversely, mutants in a second MAPK gene *MPS1* (MAP kinase for penetration and sporulation) are able to form appressoria but cannot penetrate and grow invasively in the host plant. This kinase is similar to the yeast SLT2 MAP kinase, which is activated by membrane stretching and regulates cell integrity and cytoskeleton reorganisation. *MPS1* may have a similar role in organisation of the cytoskeleton during appressorial formation in *M. grisea.*

MAPK genes have been isolated from other phytopathogenic fungi, including *Colleotrichum* species, *Botrytis cinerea, Cochliobolus heterostrophus, Pyrenophora teres* and *Fusarium oxysporum.* Targeted disruptions of these genes, such as *CMK1* in *C. lagenarium, BMP1* in *B. cinerea, PTK1* in *P. teres* and *CHK1* in *C. heterostrophus* have resulted in the inability of the fungi to form appressoria and penetrate plants, although mutants could generally grow normally *in vitro.* In the case of *CMK1,* appressorial melanisation was inhibited and there was evidence that these pathways have other roles, since *cmk1* mutants had reduced conidiation and produced spores that failed to germinate. Similarly, disruption of *fmk1* in *F. oxysporum* f. sp. *lycopersici* (vascular wilt) resulted in loss of pathogenicity, but the mutants grew in axenic culture. The mutants had defects in surface hydrophobicity and reduced levels of cell wall degrading enzymes such as pectate lyases (see Section 2.9).

2.7.2 Chemical factors

As well as physical signals, there are a number of naturally occurring chemical components of plant surfaces such as waxes and cutin-derived fatty acids that are known to induce appressorial formation. In anthracnose (*C. gloeosporioides),* waxes isolated from avocado host plants stimulate appressorial formation, whilst waxes from non-host plants do not. However, it appears that these waxes contain both inducers and inhibitors of germination and appressorial formation, and that it is the balance between these that is important. Ethylene, the volatile plant hormone released during ripening in climacteric fruit such as mangoes, tomatoes and bananas has been shown to induce spore germination and multiple appressorial formation in *C. gloeosporioides* and *C. musae*. Although ethylene receptors have not yet been identified in fungi, they have been isolated from plants where they appear to be histidine kinase-type receptors, related to bacterial two-component sensing mechanisms (see Section 5.4), that induce gene expression through kinase signalling cascades. Other volatiles emanating from plants may also have important roles in stimulating gene expression in different fungi, for example trans-2-hexen-1-ol released through cereal plant stomata stimulates appressorial formation in the stem rust *P. graminis*.

A further factor that induces gene expression in many fungi, and for which there is increasing evidence in plant pathogens, is nutrient starvation. In *N. crassa* and *A. nidulans*, starvation for certain amino acids activates general amino acid control pathways and leads to an increase in transcription of amino acid biosynthetic genes such as those encoding ornithine carbamoyltransferase (OCT) and glutamine amidotransferase/indoleglycerolphosphate synthase/phosphoribosylanthranilate (GIP). Similarly, in chestnut blight (*Cryphonectria parasitica*), starvation for arginine, histidine or tryptophan results in increased *Oct* and *Gip* transcription. In *A. nidulans* and *N. crassa*, positively acting regulatory genes (*areaA* in *A. nidulans*, *nit-2* in *N. crassa*), encode zinc-finger DNA-binding transcription factors that bind promoter domains containing a GATA sequence. Similar genes have been identified in *C. fulvum* (*Nrf-1*) and *M. grisea* (*Nut-1*). Nitrogen starvation has been shown to stimulate expression of the *Avr9* avirulence gene in the tomato leaf mould fungus *C. fulvum* (see Section 4.2.3) in which 12 putative AREA-binding sites are present in the promoter region, and the *MPG1* hydrophobin gene in *M. grisea* (see Section 2.8.2), indicating that global regulatory circuits may be essential in many plant-pathogenic fungi to regulate metabolic pathways required for pathogenicity. In *M. grisea*, two additional nitrogen-regulatory genes have been identified by mutation screens, *npr1* and *npr2*, that may be alternative global nitrogen regulators.

It is likely that other signalling pathways involving the induction of gene expression in response to detection of external stimuli will be discovered over the coming years. One of the major challenges facing fungal developmental biologists is to establish how these different signalling pathways cross-talk, and how the detection of a specific signal results in a developmental response and change in the pathogen.

2.8 Appressorial development

2.8.1 Morphology

Although the morphology of appressoria differ considerably between species, their main function appears to be to anchor the fungus firmly to the plant and penetrate the surface either directly through the cuticle or through natural openings. Physical force and cell wall-degrading enzymes are the key factors involved in penetration. For generating physical force, the structure of the appressoria, along with factors such as hydrophobins, melanisation and build-up of turgor pressure appear to be the key to success.

In *M. grisea* the mechanism of appressorium differentiation has been well documented through cytological studies (*Figure 2.5*). By 2 to 4 h after initial germination, the tip of the germ-tube stops elongating and forms a hook. Mitosis occurs during germ-tube elongation and one of the resultant daughter nuclei migrates into the hook, which becomes separated by a septum, and specialised outer wall layers are synthesised around this differentiating appressorium. The appressorium wall remaining in contact with the plant possesses only a single cell wall layer, lacks chitin and differentiates into the appressorium pore during maturation of the cell. After 4 to 8 h of development, dihydroxynaphthalene (DHN)-melanin (see Section 2.8.3) is deposited in a continuous fibrillar layer between the plasma membrane and the appressorium cell wall. This melanin binds to chitin in all but the appressorium pore which itself becomes surrounded by a ring that seals the appressorium to the plant surface. A similar structure has been observed in

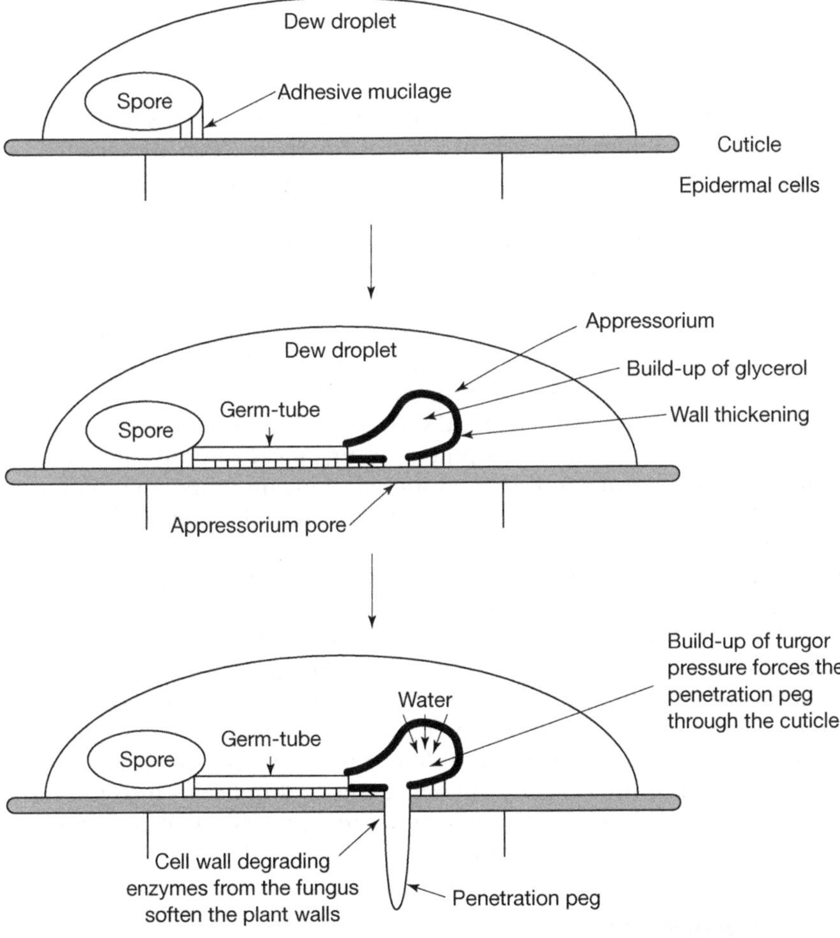

Figure 2.5

The penetration process of *Magnaporthe grisea.*

the melanised appressoria of *Colletotrichum* species. A period of quiescence then occurs, during which enzymes may be released to soften the plant surface below the infection peg. By 24–31 h, appressorium maturation and turgor generation occur, accompanied by loss of the glycogen rosettes that are present during melanisation. Simultaneously, a penetration peg emerges and extends from the pore into the cuticle. During this, actin localises in the penetration peg, possibly to control germ-tube growth and penetration peg extension through integrin and extracellular matrix protein signalling systems (see Section 2.7.1). Evidence for this comes from punchless mutants, which are unable to form a penetration peg, and are therefore non-pathogenic. These have a mutation in an integral membrane protein, structurally related to the tetraspanin family, which in animals form part of a membrane signalling complex controlling cell differentiation and adhesion.

The sequence of appressorium formation in *M. grisea* is similar to that in *Colletotrichum* species, which probably reflects a similar infection strategy in both. However, not all fungi produce appressoria by this method. In *B. cinerea*, the hyphal tip swells, but does not form an appressorial structure and there is no evidence for turgor pressure generation, and in *C. fulvum* the germ-tube simply grows through the stomata of the host plant.

2.8.2 Hydrophobins

Fungal hydrophobins are small proteins of 96–187 amino acids that contain eight cysteine residues arranged in a defined pattern. They are secreted by fungi and undergo polymerisation in response to air–water, or hydrophobic surface interfaces. Once polymerised, they form rodlet layers as structural components of spores and hyphal walls that provide a hydrophobic surface presumably as protection against desiccation. Hydrophobins may also play a role in spore dispersal and their abundance and conservation among fungal species indicates that they perform a number of other functions in development. One of the best-studied hydrophobins is that encoded by the *MPG1* gene in *M. grisea*. During appressorial formation, the MPG1 hydrophobin self-assembles at the rice leaf surface, providing a layer upon which subsequent appressorium development occurs. *mpg1* mutants are reduced in their ability to form appressoria, although these mutants can be complemented by hydrophobins from other fungi expressed under control of the *MPG1* promoter in *M. grisea*. Models have recently been proposed in which the MPG1 protein acts as part of a mechanosensory pathway in the developing appressorium that regulates the developmental pathway. However, such a mechanism is not universal to fungi, since there is no evidence for hydrophobins or rodlet formation in *B. cinerea,* whilst a number of hydrophobins have been found in *C. fulvum,* which does not form appressoria. In the fungus causing Dutch elm disease, *Ophiostoma novo-ulmi,* the hydrophobin cerato-ulmin has a different function as a phytotoxin, and transformation of the *Cu* gene encoding this hydrophobin into the non-pathogenic fungus *O. quercus,* results in its conversion into a virulent pathogen because of this phytotoxicity.

2.8.3 Melanisation

DHN-melanin is a dark fungal pigment produced by polymerisation of 1,8-dihydroxynaphthalene. This monomer is obtained through a series of reactions involving polyketide synthesis in which joining and cyclisation of five acetate molecules occurs, followed by four subsequent steps alternating reduction and dehydration reactions (*Figure 2.6*). In *M. grisea,* three melanin biosynthetic genes have been identified, *ALB1, RSY1* and *BUF1,* which encode a polyketide synthase, a scytalone dehydratase and a polyhydroxynaphthalene reductase respectively, and a second napthol reductase gene has also been identified. The role of melanin in the appressorium cell wall is to retard the efflux of glycerol from the appressorium, allowing hydrostatic turgor to build up. Melanin reduces the pore size in the appressorial wall to less than 1 nm, which allows water but no larger molecules to pass. Therefore water influx occurs as osmotically active solutes accumulate. Melanin biosynthesis appears to play a similar role in appressorium-mediated penetration by *C. lagenarium,* and genes encoding polyketide synthase (PKS1), polyhydroxynaphthalene reductase (THR1) and scytalone dehydratase (SCD1) have been isolated and shown to be required for appressorium-mediated penetration. However, similar genes identified in *Alternaria*

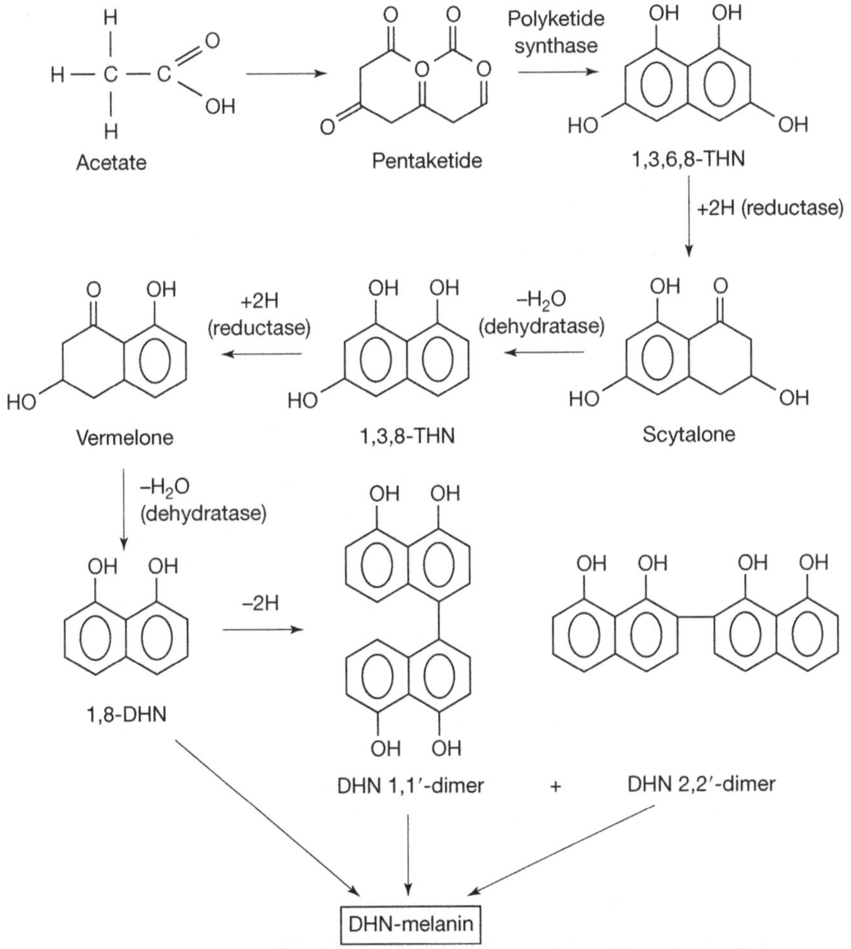

Figure 2.6

The biosynthetic pathway for dihydroxynaphthalene (DHN)-melanin.

alternata for regulating melanin production are not involved in appressorial melanisation and penetration, indicating that this is not a universal strategy.

2.8.4 Turgor pressure

The metabolic processes that occur in fungi prior to penetration are poorly understood. Spores of many fungi can germinate in water with no requirements for exogenous nutrients during appressorium morphogenesis. Energy for the process is therefore derived entirely from storage compounds within the spores. In *M. grisea,* trehalose and mannitol are the major carbohydrates stored in dormant conidia and are the most likely candidates to support the energy requirements for germ-tube emergence and appressorium development. Trehalose is rapidly broken down during germination (see Section 2.4), and during germ-tube elongation internal glycerol levels rise dramatically, probably linked to membrane biosynthesis. After a sharp decline during appressorial differentiation, glycerol levels rise within appressoria to a maximum at 48 h after conidial germination, and concentrations of up to 3.2 M have been estimated. This concentration of glycerol is responsible for generating the large hydrostatic turgor produced by appressoria. Theoretically 3.2 M glycerol is sufficient to generate a turgor pressure of 8.7 MPa, and experimental observations have detected pressures of at least 5.8 MPa. Pressure generated by appressoria is then translated into mechanical force allowing the penetration peg to rupture the plant cuticle.

Glycerol is well known as a compatible solute in fungi (an osmolyte) and is known to accumulate to very high concentrations in certain species in response to hyperosmotic stress. In *S. cerevisiae* for example, glycerol is formed from glycolysis-derived dihydroxyacetone-3-phosphate by reduction to glycerol-3-phosphate and subsequent dephosphorylation by a specific phosphatase. There are a number of possible sources for glycerol in plant-pathogenic fungi. Glycogen granules have been observed in *M. grisea,* which disappear in the appressorium cytoplasm during turgor generation, and these represent one potential source for glycerol biosynthesis. Trehalose, which is found in *M. grisea* conidia in large concentrations, is likely to constitute an alternative precursor for glycerol and recent evidence also indicates that lipid bodies, which are numerous in early appressorium development, may be another potential source. In particular, triacylglycerol lipase activity has been shown to increase during appressorial maturation, and it is believed that the mass transfer of storage lipids and carbohydrate reserves is controlled by the *PMK1* MAP kinase pathway (see Section 2.7.1), and that degradation of these is controlled by the CPKA protein kinase A signalling pathway. The importance of glycerol accumulation and turgor pressure for penetration of other plant pathogens has not been proven, and not all fungi produce this amount of turgor. For example, that in rusts has been calculated as 0.35 MPa, which is presumably sufficient for these fungi to penetrate through stomatal openings.

2.9 Cell-wall degrading enzymes (CWDEs)

Whilst physical force has a major role to play in penetration and sound explanations have been presented for the development of turgor pressure

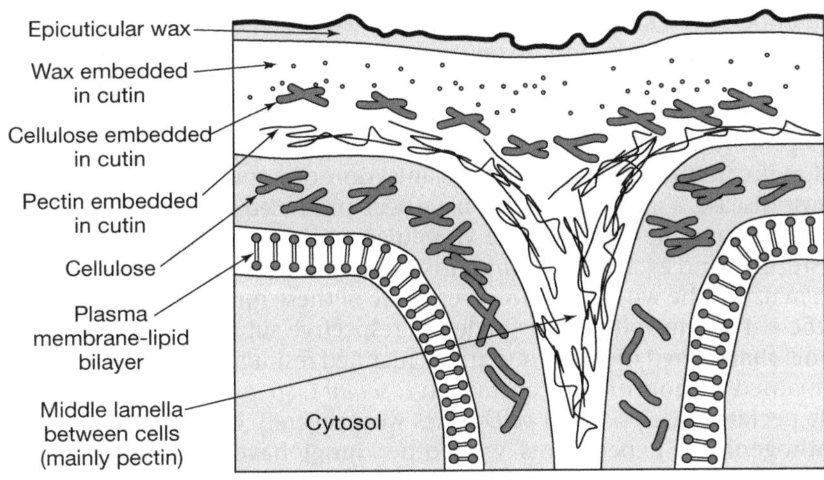

Figure 2.7

Composition of a typical plant epidermis, indicating the various polymers that a fungus or oomycete needs to degrade/penetrate to access the cytosol.

combined with ways to channel this into a penetration peg, there are many fungi which do not form melanised appressoria and may not be able to generate such high turgor pressures, yet are still capable of directly penetrating the cuticle of plants. *Figure 2.7* shows the composition of a typical plant epidermis indicating the different polymers through which a fungus must pass to penetrate. Despite numerous attempts to purify and characterise cell wall degrading enzymes (CWDEs) from fungi over the years, there have been no definite conclusions about their roles in cuticle penetration. In an experiment in which the cutinase gene from *Nectria haematococca* was introduced into a pathogen of papaya, *Mycosphaerella* spp., that normally requires wounds for invasion, the fungus gained the ability to directly penetrate fruits, indicating that in some cases, CWDEs can have an important role. More recently, cloning and gene disruption experiments have been attempted to shed more light on their possible involvement.

Pectinases are amongst the first CWDEs secreted by plant pathogens upon contact with plants, and in the largest amounts. The pectin matrix, consisting of homogalacturonan and rhamnogalacturonan with side chains of arabinans, xylano and/or arabinogalactans, is in both the cell wall and middle lamella between cells in plants. The main pectinase enzymes produced by fungi are endo- and exo-polygalacturonases (PGs) that use hydrolytic cleavage, and the pectate lyases (PLs) that use β-elimination cleavage and formation of a double bond in one of the resultant galacturonate residues. Pectin lyases, polymethylgalacturonases, pectic methylesterases and rhamnogalacturonases may also be produced by some fungi.

In many cases, particularly for the pectinases, cellulases and cutinases, there are multiple genes encoding enzymes with the same function (multigene families) indicating significant redundancy. For example, in *N. haematococca,* there are four functional pectate lyases, *B. cinerea* has five endo-PGs, and *Cochliobolus carbonum* and *M. grisea* have at least four xylanases each. Whilst this may allow the fungus greater flexibility in its pathogenicity, it hampers gene knockout experiments aimed at determining a role in penetration. For example, in *F. oxysporum* f. sp. *lycopersici,* the endo-PG PG5 is expressed during the early stages of infection and is induced by pectin and D-galacturonic acid, but repressed by glucose. Targeted inactivation of the *Pg5* gene however had no detectable effect on pathogenicity. In *C. carbonum,* a fungus that does not require melanisation or appressorial formation to cause disease, a series of mutants have been produced in which single copies of genes for endo-PG, xylanase 1, cellulase and β-1,3-glucanase have been disrupted, and each of these mutants was found to have the same virulence on maize as the wild-type. However, none of these mutants was completely deficient in the ability to degrade the respective substrate. Even strains of fungi that carried multiple mutations retained residual enzyme activity and remained pathogenic. In contrast, in *F. solani* f. sp. *pisi,* double mutants in the pectate lyase *pelA* and *pelD* genes were severely compromised in their pathogenicity. Experiments with other fungi have yielded contrasting results, whilst in the case of *B. cinerea,* the importance of CWDEs has been confirmed by an alternative approach in which antibodies to a 60 kDa lipase protein (that has cutinolytic activity) were applied to plants prior to inoculation resulting in the inability of the fungus to penetrate.

Because of the lack of success using knockouts of individual genes encoding CWDEs, an alternative approach used in *C. carbonum* has been to knockout a regulatory gene. In culture, the expression of most CWDEs in most fungi is inhibited by glucose and other simple sugars by a process known as catabolite or glucose repression. In yeast, release from this repression requires a protein kinase SNF1p, which is responsible for phosphorylation of Mig1, the equivalent of CREA in *Aspergillus.* Mig1 and CREA are zinc finger DNA-binding proteins that bind to promoter regions of CWDEs and repress activity. Phosphorylation of these proteins inhibits this binding. The equivalent of the *SNF1p* gene (*ccSNF1*) has been cloned from *C. carbonum* and targeted gene disruption mutants created. These mutants produce reduced levels of CWDEs and are significantly less virulent on susceptible maize, producing fewer spreading lesions although the morphology of these was normal. However, it has not yet been confirmed whether this is due to reduced penetration of the plant or the inability to degrade walls between cells resulting in a reduced ability to access nutrients.

References and further reading

Annis, S.L. and Goodwin, P.H. (1997) Recent advances in the molecular genetics of plant cell wall-degrading enzymes produced by plant-pathogenic fungi. *European Journal of Plant Pathology* **103**: 1–14.

Deising, H.B., Werner, S. and Wernitz, M. (2000) The role of fungal appressoria in plant infection. *Microbes and Infection* **2**: 1631–1641.

D'Souza, C.A. and Heitman, J. (2001) Conserved cAMP signalling cascades regulate fungal development and virulence. *FEMS Microbiology Reviews* **25**: 349–364.

Forsberg, H. and Ljungdahl, P.O. (2001) Sensors of extracellular nutrients in *Saccharomyces cerevisiae. Current Genetics* **40**: 91–109.

Henson, J.M., Butler, M.J. and Day, A.W. (1999) The dark side of the mycelium: melanins of phytopathogenic fungi. *Annual Review of Phytopathology* **37**: 447–471.

Idnurm, A. and Howlett, B.J. (2001) Pathogenicity genes of phytopathogenic fungi. *Molecular Plant Pathology* **2**: 241–255.

Kahmann, R. and Basse, C. (2001) Fungal gene expression during pathogenesis-related development and host plant colonisation. *Current Opinion in Microbiology* **4**: 374–380.

Knogge, W. (1996) Fungal infection of plants. *The Plant Cell* **8**: 1711–1722.

Kronstad, J.W. (ed.) (2000) *Fungal Pathology*. Kluwer Academic Press, Dordrecht.

Snoeijers, S.S., Pérez-García, A., Joosten, M.H.A.J. and de Wit, P.J.G.M. (2000) The effect of nitrogen on disease development and gene expression in bacterial and fungal plant pathogens. *European Journal of Plant Pathology* **106**: 493–506.

Xu, J-R. (2000) MAP kinases in fungal pathogens. *Fungal Genetics and Biology* **31**: 137–152.

Fungal and oomycete diseases – development of disease

3

Having entered the plant, the fungal or oomycete pathogen must access nutrients for its own growth and development. The strategies that have evolved for doing this are diverse, and studies into the molecular signalling and biochemical processes involved are still in their infancy. This is partly because each pathogen uses its own unique strategy for obtaining nutrients. Indeed it is crucial to remember that when considering a plant disease, we are looking at the outcome of an intimate interaction between the plant and the pathogen, so studies on one fungus can not always be extrapolated to others.

The main criteria for determining which plant pathogen interactions to study at the molecular level have been the ease with which the system can be examined. Can the fungus be grown in culture? Is it possible to perform genetic studies, mutagenesis experiments and the analysis of mutants? And perhaps most importantly, can the fungus be transformed efficiently? As a result, non-obligate, easily culturable fungi such as rice blast and tomato leaf mould have been better studied, although recent advances in the ability to transform obligate biotrophs such as the barley powdery mildew fungus should result in progress in this area as well over the coming years. Despite the technical difficulties and diversity in strategies, there has been impressive progress in our understanding of fungal pathogenicity, and comparative genomics and use of bioinformatics will increase this in the future.

3.1 The basic concepts – necrotrophy versus biotrophy

There are two main classifications of fungal pathogens. Necrotrophs are those that kill the host cells as they invade, and then utilise the nutrients that are released for their growth and development. These can be considered as the 'thug' pathogens, and they generally use toxins and/or cell-wall-degrading enzymes as their weapons. The opposite extreme is referred to as biotrophy, in which the pathogen needs to tap into living cells to obtain its nutrients. These are the 'conmen' pathogens, that appear to go unrecognised by the plant as they divert nutrients and photoassimilates for their own growth and development.

Of course, not all pathogens fit into one or other category, and many (the hemi-biotrophs) go through an initial biotrophic phase in plants before causing cell damage and essentially becoming necrotrophic. A further category, such as *Botrytis* and *Colletotrichum* species may enter the plant and then go through a quiescent phase, remaining latent until conditions are favourable for them to continue development. There are also fungi (e.g. *Aspergillus flavus*) that are not *per se* pathogenic to plants, but are considered as pathogens because the mycotoxins they produce cause human health problems and their presence reduces the value of the crop. Some mycotoxin-producing endophytes such as ergot (*Claviceps purpurea*) and *Epichloë* spp. may even be considered as mutualistic on plants because of the protection they can give them against pests, although the alkaloids they produce are potentially lethal to humans. Having summarised the different ways in which pathogens may cause disease on plants, we can now examine the molecular changes that they invoke in infected plants, the signalling that takes place between pathogen and plant, and the genes that are important in the pathogens for causing disease.

3.2 Host barriers

Having breached the outer defences of plants, the fungus may need to overcome chemical barriers before it becomes established. Preformed antimicrobial compounds (phytoanticipins) and inhibitors of cell-wall-degrading enzymes such as polygalacturonase inhibitor proteins (PGIPs) are often present in healthy plants, and other compounds are released when plants are stressed or challenged by pathogens (phytoalexins and the oxidative burst) that present a formidable defence against invasion. These are discussed in detail in Sections 9.2 and 9.3.

3.3 Overcoming host barriers

3.3.1 Quiescence

One approach for evading plant defences is that adopted by a number of post-harvest pathogens, in particular some *Colletotrichum* and *Botrytis* species. The strategy here is to germinate, penetrate, and upon contacting the chemical defences, enter a quiescent phase. In most cases, dark, thick-walled appressoria represent the quiescent structures although the means by which the fungus detects the inhospitable nature of the host and responds to form these structures is unclear. In addition, the factors that cause quiescent appressoria to re-establish aggressive infections have not been unequivocally demonstrated. Many of these infections occur on fruits in which the levels of phytoanticipins and PGIPs are known to decrease upon ripening, and it may be that the fungus detects these changes and becomes aggressive once the levels are sufficiently low. A further possibility that has been suggested from studies of *C. gloeosporioides* on climacteric fruit such as mangoes and tomatoes, is that the fungus detects increases in the fruit-ripening hormone ethylene to trigger aggressive infection as discussed in Section 2.7.2.

3.3.2 Detoxification of phytoanticipins

Some fungi have enzymes to degrade antimicrobial compounds, and the ability to produce such enzymes can determine the host range of the pathogen. One of the best-studied examples is the degradation of the phytoanticipin avenacine, a triterpenoid saponin present in oats, by the enzyme avenacinase. Avenacinase is present in oat-colonising isolates of *Gaeumannomyces graminis* (take-all), but not in those that can only colonise wheat. The avenacinase enzyme has been cloned and gene-disruption mutants of oat-attacking isolates created, and these lose the ability to attack oats, but retain their ability to attack wheat. Furthermore, varieties of oats that lack avenacin are susceptible to these mutants and to normal wheat-attacking isolates of the fungus.

Studies on other phytoanticipin-detoxifying enzymes have suggested an additional role for these enzymes. Addition of the tomatinase gene to *N. haematococca* enhanced the ability of the fungus to grow on green tomatoes, rich in tomatine, and addition to *C. fulvum* also increased fungal virulence. However, disruption of the tomatinase gene in *Septoria lycopersici* (tomato leaf spot) had no effect on pathogen growth in the presence of physiological levels of α-tomatine, although there was increased sensitivity to higher levels. These mutants failed to cause disease on *Nicotiana benthamiana,* but they grew normally on tomato plants. However, there was enhanced plant cell death and elevated expression of defence genes in the tomatoes, and subsequent experiments indicate that this is because tomatinase has a dual role in pathogenicity. In addition to detoxifying the phytoanticipin through enzymic hydrolysis, the resultant breakdown products induce a signal transduction mechanism in the plants that culminates in suppression of plant defence responses.

3.3.3 Detoxification of phytoalexins

One of the best-studied isoflavonoid phytoalexins is pisatin from the garden pea, but most fungal pathogens of peas are able to detoxify this compound through a pisatin demethylase enzyme (pda/cyp57). This enzyme, which has been best studied from *N. haematococca* is a cytochrome P450 mono-oxygenase, and therefore belongs to a group of enzymes found in most organisms involved in degradative and biosynthetic reactions. These enzymes are encoded by *Cyp* genes, and the primary stimulatory signal for induction during pea pathogenesis has been shown to be pisatin itself, acting through a 35 bp DNA element (pisatin-responsive element) present in the transcription control region of the gene. This strongly implies that the *Pda1* gene has evolved specifically in the fungus for the role of phytoalexin detoxification. When this gene was introduced into fungi that were normally sensitive to pisatin, they were able to detoxify pisatin. However, this was not necessarily correlated with the ability to infect pea, indicating that it is not the only factor involved in pathogenicity. *N. haematococca* is also able to detoxify the chickpea phytoalexins medicarpin and maackiain through a mono-oxidase gene. The *MAK1* gene responsible for this has been transformed into *Colletotrichum destructivum,* and the transformants were able to detoxify medicarpin in alfalfa more rapidly than non-transformed strains.

Other phytoalexin-detoxifying enzymes have also been identified. *B. cinerea* produces a laccase (stilbene oxidase) that is able to detoxify the grape stilbene phytoalexin resveratrol. It also produces a glutathione S-transferase enzyme that may be involved in conjugating toxic compounds to glutathione, thereby removing potentially fungitoxic compounds. However, disruption mutagenesis of this gene did not affect virulence of the pathogen on tomato.

3.3.4 ATP-binding cassette (ABC) transporters

An alternative to detoxification of antimicrobial compounds is to transport them out of the cell, and the main class of transporter proteins in fungi involved in protection against plant defence compounds are of the ATP-binding cassette (ABC) type (*Figure 3.1a*). In *M. grisea* an ABC transporter, ABC1 has been identified through an insertional mutagenesis screen and selection for mutants that have lost pathogenicity. Whilst the *ABC1* mutants are able to penetrate cells normally, they die during initial colonisation of host tissue, suggesting that they are susceptible to plant defence products. The rice phytoalexin sakuranetin can induce expression of the *ABC1* gene, although it is not certain whether it is this or other antifungal compounds that are the main substrates for transport.

A number of ABC transporters have been identified in other fungi such as *Gibberella pulicaris* (a necrotrophic potato pathogen), and *B. cinerea,* where one of these, *BcatrB* is induced by the grapevine phytoalexin resveratrol. Disruption of this gene results in isolates with a lower virulence on

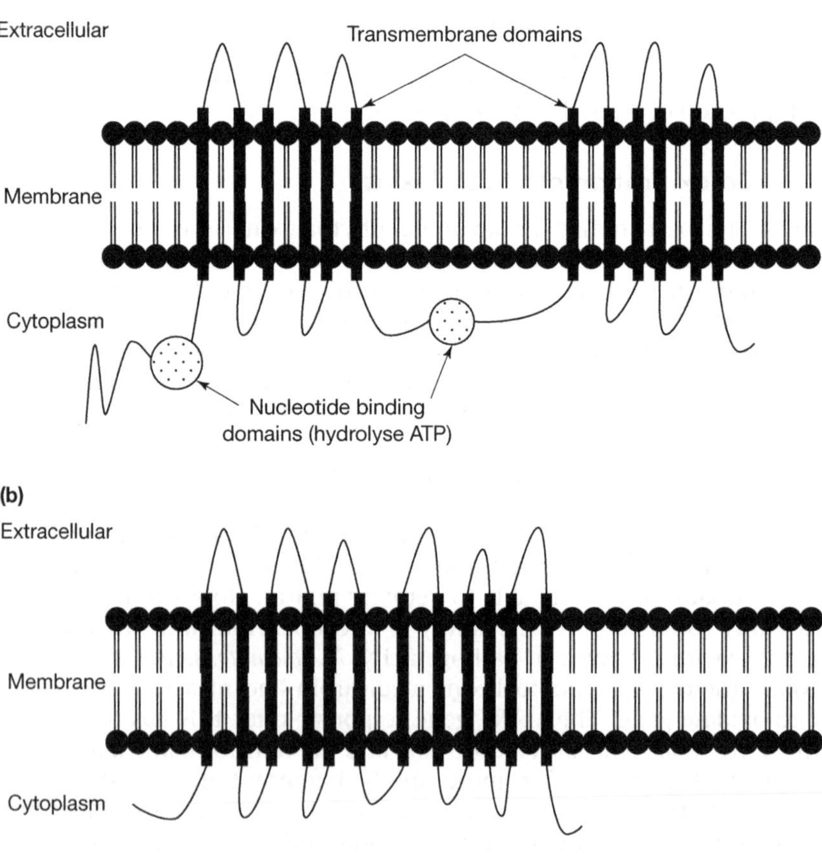

Figure 3.1

(a) ABC transporters and (b) MFS transporters. ABC transporters hydrolyse ATP to transport solutes across membranes, whilst MFS transporters use a proton-motive force composed of membrane potential and electrochemical proton gradients.

grapevines. Interestingly, both this and the ABC1 transporter from *M. grisea* are strongly induced by azole and phenylpyrrole fungicides, and, as in other fungi, such as *A. nidulans, Candida albicans* and *S. cerevisiae,* it is becoming clear that plant-pathogenic fungi can utilise their ABC transporters to export fungicides and other toxic compounds as part of fungicide and multidrug resistance.

3.3.5 Suppression of active oxygen species

Evidence that fungi produce suppressors of plant defence responses initially came from work on the pea pathogen, *Mycosphaerella pinodes,* where infection resulted in increased susceptibility of the plants to some normally non-pathogenic fungi such as *Alternaria alternata*. The *M. pinodes* produces specific glycopeptides, termed suppprescins, that suppress defence responses in plants, and it is believed that this is through an effect on the signal transduction pathways in plants that would normally result in defence gene activation. There is delayed production of pathogenesis-related proteins such as phenylalanine ammonia lyase (PAL) and chalcone synthase for example.

One of the induced defences in plants upon pathogen invasion is the production of reactive oxygen species and programmed cell death (see Section 9.3), and in some fungi there appears to be active export of metabolites and proteins from the fungus into the plant to suppress these defences. Superoxide dismutase (SOD) and catalase enzymes may be secreted to scavenge reactive oxygen species, and convert these to less active products. *Sclerotinia sclerotiorum* secretes oxalic acid (which is presumed to interfere with signal transduction pathways in plants) as a means of suppressing the oxidative burst. *A. alternata* has been shown to produce and secrete mannitol in response to host factors, to suppress reactive oxygen species-mediated defences analogous to the suppression of animal defence responses by the human pathogenic fungus *Cryptococcus neoformans*. Production of mannitol is also necessary for pathogenicity of the tomato leaf mould fungus, *C. fulvum*. In addition, it has been shown that mannitol dehydrogenase, which catalyses conversion of mannitol to mannose, is induced in some plants such as tobacco in response to pathogen attack, suggesting that plants may in turn have evolved counter–counter-defence mechanisms.

3.3.6 Avoidance of recognition

Many biotrophic fungi possess a more subtle approach to plant invasion, and the key to their success is often to avoid recognition by the plant for as long as possible and thus avoid triggering any defence responses. In *C. gloeosporioides,* a gene *CgDN3* has been identified that encodes a small nitrogen-starvation-induced secreted peptide. Isolates in which this gene is mutated from normal appressoria, but cannot infect and reproduce *in planta,* because a hypersensitive defence response is induced. Thus CgDN3, which has some homology to a family of plant cell wall receptor protein kinases, appears to be required to avert induction of the hypersensitive response in the compatible host, the tropical pasture legume *Stylosanthes guianensis*. An alternative approach that has been identified for avoiding recognition is to mask the fungal structures that are formed *in planta*. A gene

Clh1 has been cloned from bean anthracnose, *C. lindemuthianum* which, as shown by immunolabelling, is switched on during the initial biotrophic phase of infection and then switched off during the necrotrophic phase. This fungus forms intracellular hyphae to obtain nutrients rather than haustoria, and the protein encoded by this gene, which is a proline-rich glycoprotein that resembles plant cell wall proline- and hydroxyproline-rich proteins, is believed to coat the hyphae to produce a 'pseudo' plant cell wall that the plant is unable to recognise as being foreign.

A similar evasive strategy has been proposed for the rust fungi. In this case, the germ-tubes and appressoria of the fungi have been shown to be coated in chitin, which is a well-recognised elicitor of plant defence responses. However, the chitin is deacylated on the structures that form within the plant, such that they are coated in chitosan, and it has been postulated that this is a means of evading recognition by the plant. Chitin is also lacking from the infection hyphae of *C. lagenarium* and *C. graminicola,* possibly indicating that this is a general strategy for evading plant defences. It may also be that the modified chitin-related compounds that are produced act in an analogous fashion to nodulation factors produced by mutualistic rhizobial bacteria, as signals for cytological changes in plants. Here, lipochitin oligos of 3–6 unit length (Nod factors) that are N-acylated at the non-reducing end, are the signals secreted by the bacteria that determine host specificity and that induce morphological changes in the plant including changes in root hair curling and activation of cortical cell division (*Figure 3.2*). *Nod* genes in the bacterium encode enzymes such as N-acyl transferases and chitin synthases that are required for the biosynthesis and secretion of these Nod factors, which appear to activate changes in gene expression in plants through rapid changes in ion fluxes at the root hair plasma membrane, and membrane depolarisation. Furthermore, it has been shown that Nod factors can stimulate associations between mycorrhizal fungi and plants, and it has been postulated that there is clear conservation in the signalling pathways of the rhizobial and mycorrhizal mutualistic interactions with plants. Whether the same is true for biotrophic plant pathogens has yet to be determined.

3.4 Establishing infection

Having penetrated the plant and evaded the defences, the invading pathogen must obtain nutrients from the host to enable it to grow, develop and ultimately produce spores. Necrotrophic organisms will do this using cell-wall-degrading enzymes and toxins to break down the cells, whilst biotrophs often produce specialist feeding structures (haustoria) to tap into the host cells.

3.5 Cell-wall-degrading enzymes

The nature of cell-wall-degrading enzymes produced by fungi has already been discussed (see Section 2.9), and it is clear that these are often important for ramification through the host and that they are regulated by a catabolite repression system. In some *Colletotrichum* species, it has been suggested that the fungus deliberately secretes ammonia into the host to raise the pH, which allows the fungus to secrete its *pelB*-encoded pectate lyase (which is

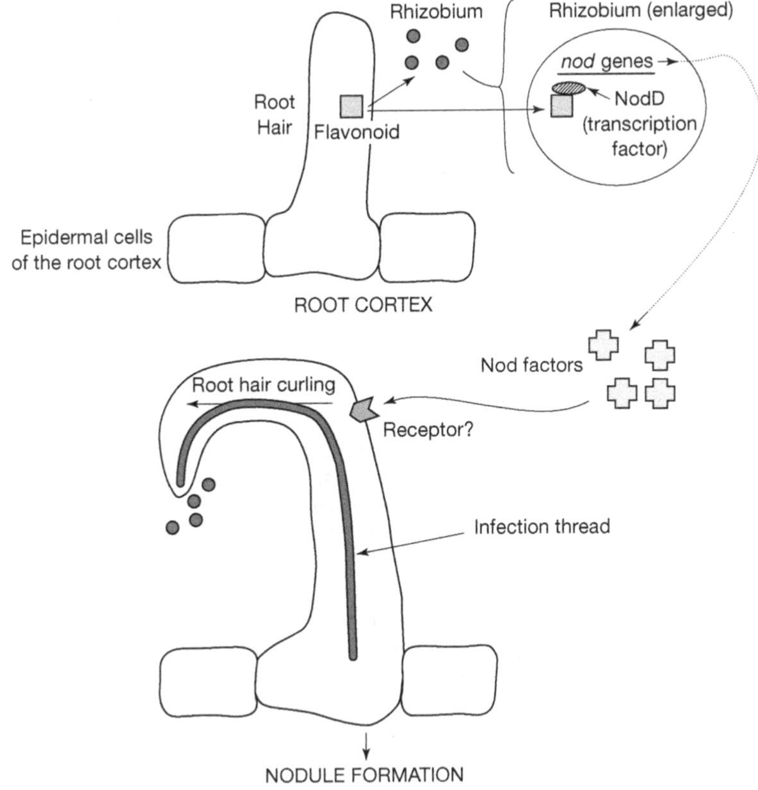

Figure 3.2

The infection of root hairs by *Rhizobium*. The root secretes specific flavonoids, which activate the NodD transcription factor in the *Rhizobium*. This results in biosynthesis of the Nod factors, which are detected by the plant to induce root-hair curling and nodulation.

not secreted below pHs of 5.7). How the production of ammonia is regulated is yet to be determined.

3.6 The role of toxins

Plant-pathogenic fungi produce a wide range of toxins that may cause whole plant death or merely subtle changes in gene expression. Fungal toxins can be divided into three groups (*Table 3.1*). The host-selective toxins that are produced by only a few fungal species such as *Alternaria* and *Cochliobolus* are only toxic to hosts of these pathogens. By contrast, the non-selective toxins can cause damage to both the host plant and other plant species not normally attacked by that pathogen. There are also toxins produced by plant-pathogenic fungi that are not toxic to the plants, but that are often fatal to animals and humans that may consume the infected plant material, and we shall consider this group of mycotoxins here as well.

3.6.1 Host-selective toxins

Amongst the best-characterised host-selective toxins are three that are produced by species of *Cochliobolus* (T-toxin, Victorin-C and HC-toxin) and the

Table 3.1 Toxins produced by plant-pathogenic fungi. Modes of action are discussed in the text and the chemical structures are shown in *Figure 3.3*.

Host-selective toxins	
Victorin	*Cochliobolus victoriae*/oats
T-toxin	*C. heterostrophus*/maize
HC-toxin	*C. carbonum*/maize
HS-toxin	*C. sacchari*/sugarcane
Ptr	*Pyrenophora tritici-repentis*/wheat
AK	*Alternaria alternata*/Japanese pear
ACT	*A. alternata*/tangerines
AF	*A. alternata*/strawberries
AM	*A. alternata*/apples
AAL	*A. alternata* f. sp. *lycopersici*/tomato
Host non-selective toxins	
Fusicoccin	*Fusicoccum amygdali*/almond
Cercosporin	*Cercospora* spp./various
Tentoxin	*Alternaria* spp./various
Naphthazarins	*Nectria haematococca*/peas
Dothistromin	*Dothistroma septospora*/pines
Mycotoxins	
Aflatoxins	*Aspergillus flavus* and *A. parasiticus*/grains
Fumonisin	*Fusarium moniliforme*/maize
Ergot alkaloids	*Claviceps purpurea*/grasses
Trichothecines	*Fusarium* spp./cereals

Alternaria toxins (*Figure 3.3*). T-toxin, produced by the maize pathogen *C. heterostrophus* (Southern corn leaf blight) is amongst the most infamous of plant pathogens in recent history, following the devastating epidemic that swept across North America in 1970. The race T of the fungus, which was first detected in 1969 and caused the epidemic, contains two genes, a polyketide synthase (*PKS1*) and a decarboxylase (*DEC1*) that are required for toxin biosynthesis. The original race O that was present prior to 1969 lacks these genes. The T-toxin itself is a mixture of several linear polyketols ranging from C_{35} to C_{41}, that binds to a 13 kDa inner mitochondrial membrane protein, the product of the *T-urf13* gene (*Figure 3.4*). This creates a pore in the mitochondria which causes leakage of small molecules, resulting in cessation of ATP synthesis and cell death. Since energy production is blocked in the plant cells, they are unable to mount any defence responses and the fungus invades. The URF-13 protein also confers male sterility on the maize, and it was this trait (the Texas male sterility cytoplasm, T-cms) that was bred into much of the US maize in the 1960s for ease of hybrid production, and which was responsible for so much of the maize that was planted in 1970 being susceptible to race T.

Like the T-toxin, Victorin, produced by *C. victoriae* on oats acts by inhibiting mitochondrial function. In this case, the toxin is an unusual chlorinated peptide compound that targets a 100 kDa victorin-binding protein, which is a subunit of the glycine decarboxylase enzyme. This is essential for photorespiration and plant cells appear to go through an induced senescence and programmed cell death with cleavage of RUBISCO (ribulose bisphosphate carboxylaisa/oxygenase). The HC-toxin produced by *C. carbonum* (northern leaf spot) however does not directly cause plant cell death, and is believed to

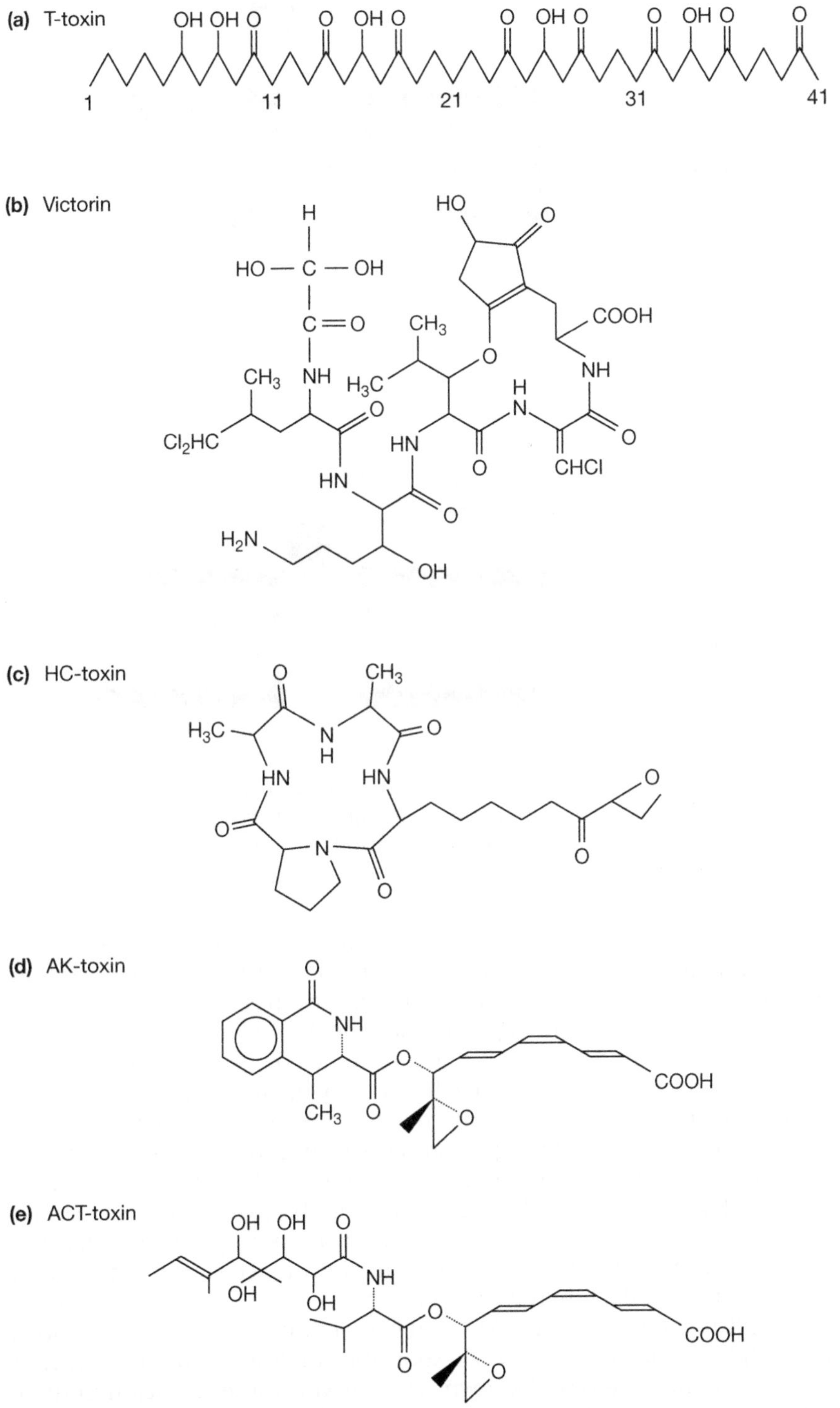

(a) T-toxin

(b) Victorin

(c) HC-toxin

(d) AK-toxin

(e) ACT-toxin

Figure 3.3

The chemical structures of some host selective toxins from *Cochliobolus* spp. (a, b and c) and *Alternaria alternata* (d and e).

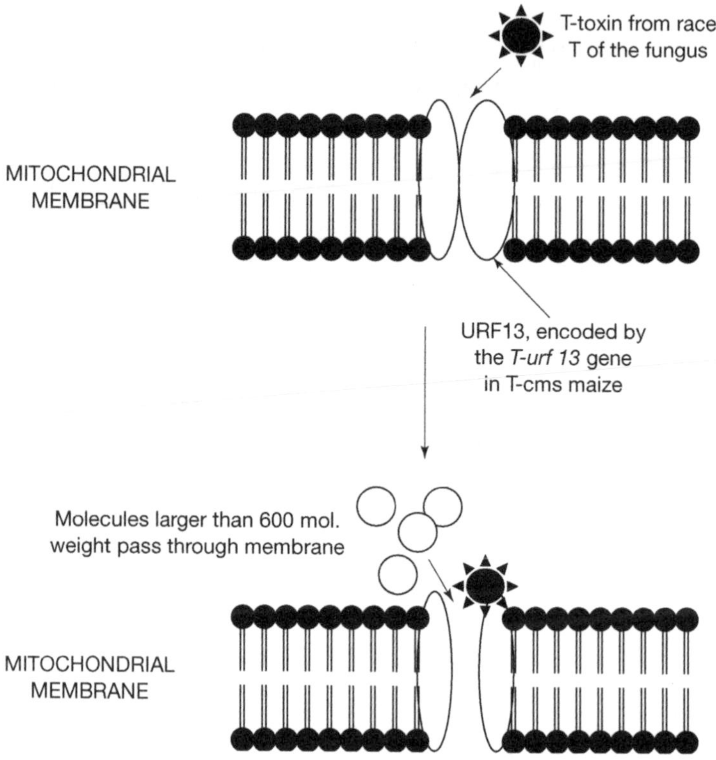

Figure 3.4

The effect of T-toxin from *C. heterostrophus* on T-cms maize. The T-toxin binds to an inner mitochondrial membrane protein URF-13, creating a pore through which small molecules leak. This results in a cessation of ATP synthesis culminating in cell death.

work through suppression of plant defence responses. This toxin is a cyclic tetrapeptide containing D-amino acids and inhibits histone deacetylases in the plant nucleus, causing hyperacylation of histones and changes in gene expression. One interesting feature of the genes for HC-toxin biosynthesis is that they are clustered in the genome surrounded by two copies of the *ToxA* gene encoding an MFS (major facilitator superfamily) transporter (*Figure 3.1b*). These transporters are like ABC transporters, but do not hydrolyse ATP, and the transport is driven by a proton-motive force composed of membrane potential and electrochemical proton gradients. Despite a number of attempts, knockouts of the *ToxA* gene have not been generated, suggesting that it is essential for fungal survival, and it has been postulated that the genes are involved in export of the toxin, not only to secrete it into the plant, but also to protect the fungus from its effects.

In *A. alternata* there are at least seven known host–parasite combinations in which host-selective toxins are responsible for disease, and the chemical composition of these is well known. These toxins cause cell death in three different ways, affecting host plasma membranes to cause potassium and other electrolytes to leak, affecting mitochondria and uncoupling photophosphorylation (light-dependent conversion of ADP to ATP), or decreasing photosynthetic carbon dioxide fixation. The AAL toxin from

A. alternata f. sp. *lycopersici* is amongst the best studied and is structurally related to sphinganine and the fumonisin mycotoxins produced by *Fusarium moniliforme* (the sphinganine-analogue mycotoxins or SAMs). These are inhibitors of sphinganine *N*-acyltransferase, blocking ceramide biosynthesis, and in mammalian systems, ceramides and ceramide-containing lipids are important in diverse intracellular signalling pathways. Disruption of these pathways often results in programmed cell death, and the classical symptoms of programmed cell death have been shown to occur in tomato plants treated with these toxins.

In recent years, work has focused on cloning the genes involved in toxin biosynthesis, through combinations of REMI (see Section 1.5.4) and mutant screening, and a number of these have been identified. The *AMT* gene, crucial for production of the AM toxin (a four-member cyclic depsipeptide) in apple pathotypes of *A. alternata,* has been identified as a peptide synthetase, and two genes for AK biosynthesis, *AKT1* and *AKT2* have also been identified and cloned by insertion mutagenesis. Interestingly, it appears that there are homologues of the genes in strains of *Alternaria* that produce similar toxins, but not in those that produce different toxins, suggesting horizontal transfer of biosynthetic gene clusters (see Section 4.7).

3.6.2 Host non-selective toxins

The symptoms caused in plants by host non-selective toxins are often very similar, being chlorotic lesions and/or wilt. This generally results from the toxins inducing reactive oxygen species (ROS) in the plant, that cause membrane breakdown and nutrient leakage. Amongst the toxins that work this way is cercosporin, synthesised by several phytopathogenic *Cercospora* species, causing such diseases as brown eye spot of coffee, frog-eye of tobacco and grey leaf spot of corn. Similar perylenequinone toxins are produced by a number of other phytopathogens including *Cladosporium* spp. and *A. alternata.* These toxins are photodynamically active pigments that induce lipid peroxidation in plant cells (*Figure 3.5*). They are non-toxic to the fungi, which reduce the compounds to inactive forms that lack the photodynamic effect, but in plants, a spontaneous oxidation occurs to the active form. They are secreted from the fungus via MFS transporters, and the gene encoding one of these, *Cfp* (cercosporin facilitator protein), has been cloned. Mutations in this gene result in the inability to produce cercosporin, and reduced virulence on soybean.

Nectria haematococca also produces non-selective toxins, the napthazarins, that can work through light-dependent pathways to form ROS resulting in lipid and membrane damage. However, these toxins also work through light-independent pathways. Dothistromin, a cebetin toxin produced in pine needles by *Dothistroma septospora* (blight in pines), is also involved in ROS formation in plant cells by light-independent pathways. Because it is present in the peanut pathogen *Cercospora arachidicola,* this toxin is also considered to be a mycotoxin.

3.6.3 Mycotoxins

Mycotoxins are low-molecular-weight fungal metabolites that are toxic to vertebrates, and can have drastic effects on human and animal health.

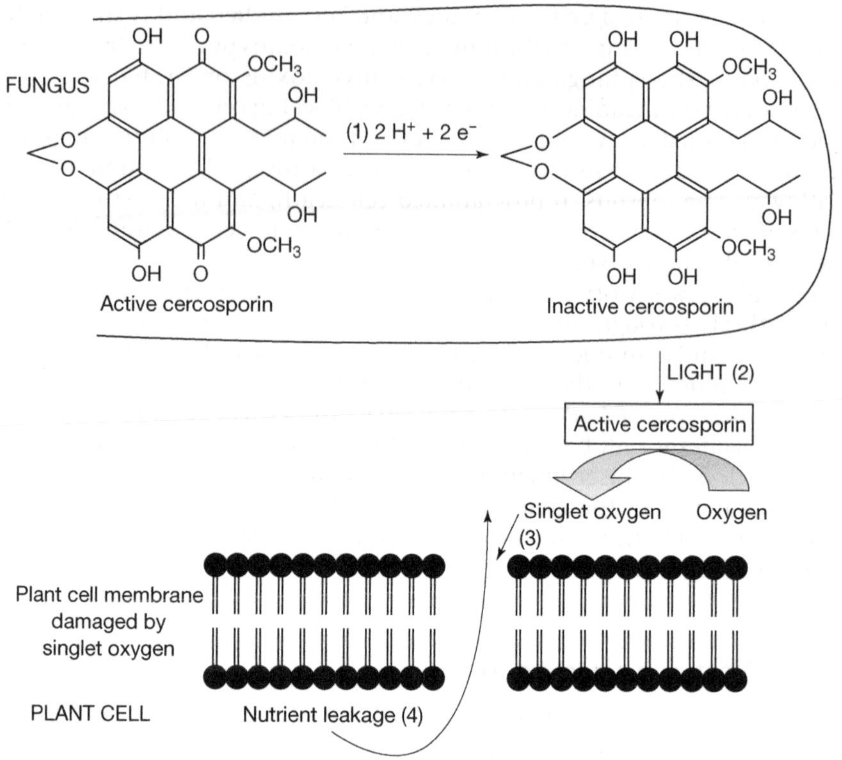

Figure 3.5

Mode of action of cercosporin. In fungi, the cercosporin is reduced to an inactive form (1). When secreted, the toxin is reactivated (2) by light, and generates singlet oxygen, which causes lipid peroxidation in plant cell membranes (3) and leakage of nutrients (4).

There are four important classes of mycotoxins, the ergot alkaloids, aflatoxins, trichothecines and fumonisins (*Table 3.1*).

The ergot alkaloids, produced by *C. purpurea* are amongst the most notorious of mycotoxins historically, causing both gangrenous (e.g. St Anthony's Fire in humans) and convulsive (e.g. Ryegrass Staggers in sheep) ergotism in humans and farm animals. Consumption of infected rye was probably responsible for millions of deaths during the Middle Ages, but outbreaks are now rare, and some of the alkaloids have become useful in medicine, for example as vasoconstrictors. Much of the current interest in these fungi stems from the use of closely related endophytes such as *Epichloë* and *Neotyphodium* on pastures of fescue and ryegrass to protect plants against insects (such as aphids) and diseases, although this has had detrimental effects on mammals and birds because of the alkaloids. The ergot alkaloids are derived from isoprenoid and amino acid precursors and the biosynthetic pathways for a number of them have been elucidated (*Figure 3.6a*).

The aflatoxins, produced by *Aspergillus flavus* and *A. parasiticus* are a group of bisfuranocoumarins that can cause serious liver damage and cancers. Human exposure is often through consumption of contaminated groundnuts, maize and other grains. The genes for aflatoxin biosynthesis

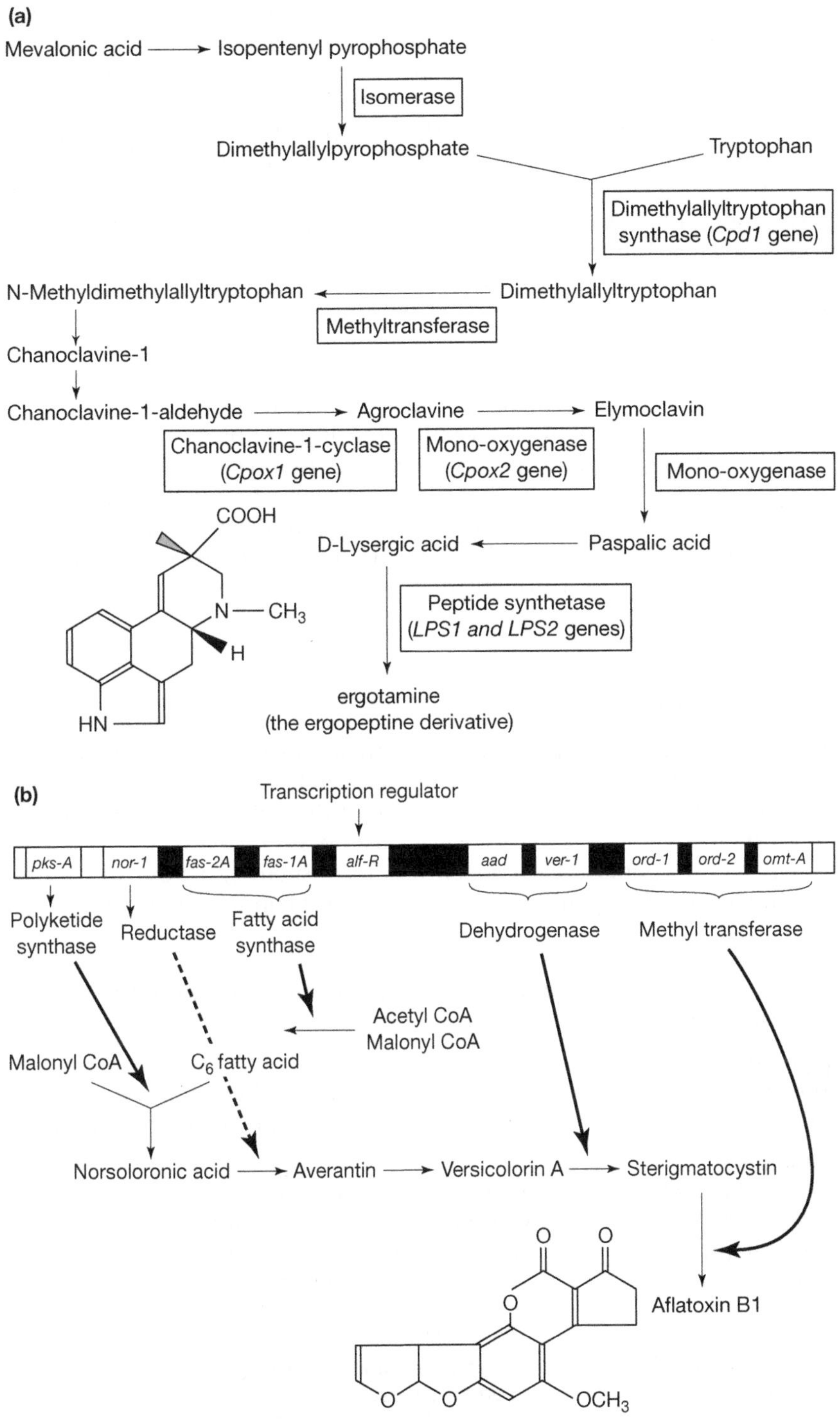

Figure 3.6

Toxin biosynthesis. (a) The ergot alkaloid biosynthetic pathway. The enzymes involved, including putative genes that encode them are boxed. (b) The genes and enzymes involved in aflatoxin B1 biosynthesis.

and regulation of its production are clustered in the fungal genome in a 60 kb region, and have been well studied in *Aspergillus* spp. (*Figure 3.6b*). Interestingly, non-toxin-producing species, such as *A. nidulans*, possess the same cluster of genes and can synthesise the immediate precursor of aflatoxins, sterigmatocystin. It appears that additional cytochrome P450 genes are required in the toxin-producing strains to convert this to aflatoxin.

The trichothecene toxins produced by *Fusarium* spp. are ingested through consumption of cereals such as wheat, barley, rice and maize and are also a serious threat to human health, particularly as these diseases are becoming more widespread on plants. As with aflatoxins, the genes responsible for biosynthesis appear to be clustered in the *Fusarium* genome in a 25 kb region. Disruption of the *Tri5* gene, which controls the first step of trichothecene biosynthesis results in reduced fungal pathogenicity, and other genes controlling the biosynthesis have also been identified. The fourth main group of mycotoxins are the fumonisins, amino polyalcohols produced by several *Fusarium* species. The genes involved in biosynthesis of these are also clustered and appear to encode a cytochrome P450 monooxygenase, an alcohol dehydrogenase, an α-aminotransferase and a dioxygenase amongst others. The mode of action of these toxins in animals and livestock appears to be through their ability to induce programmed cell death (apoptosis).

3.7 Biotrophy

Biotrophic fungi use stealth to obtain their nutrients from plants, and amongst the biotrophic fungi, there are a wide range of growth habits (*Table 3.2*). As well as seeking to avoid recognition by the plant so that they do not incite defence responses, they must obtain nutrients for their own growth. In the apoplast, the concentrations of sucrose, hexose and amino acids are low (typically millimolar), which is not conducive to rapid fungal

Table 3.2 The growth habits of some common biotrophic fungi and oomycetes

Epiphytic	
Non-haustorial	*Myriogenospora* (tangle-top on grasses)
Haustorial	*Erysiphe, Blumeria, Uncinula* (powdery mildews)
Endophytic	
Intercellular hyphae (non-haustorial)	*Taphrina deformans* (peach leaf curl) *Cladosporium fulvum* (tomato leaf mould)
Inter and intra cellular hyphae	*Puccinia* (monkaryotic form) (rusts) *Claviceps purpurea* (ergot) *Ustilago maydis* (maize smut) Mycorrhizal fungi (mutualistic)
Intercellular hyphae/ haustorial	*Puccinia* (dikaryotic form) (rusts) *Uromyces* (rusts) *Bremia* (downy mildews) *Peronospora* (downy mildews) *Phytophthora infestans* (late blight) *Albugo candida* (white blister rust)
Intracellular hyphae	*Colletotrichum* (anthracnose)

growth. Whilst some fungi, such as *C. fulvum* are able to obtain sufficient nutrients through their invading hyphae as they grow in the apoplast, most biotrophic fungi form specialised feeding structures known as haustoria, designed to increase nutrient uptake by tapping into nutrient-rich plant cells. Interestingly, these haustoria have many similarities to the arbuscles of mutualistic vesicular-arbuscular mycorrhiza.

3.7.1 Haustorial structure

The generalised structure of a fungal haustoria (*Figure 3.7a*) essentially consists of a tubular neck connected to a haustorial body that invaginates into

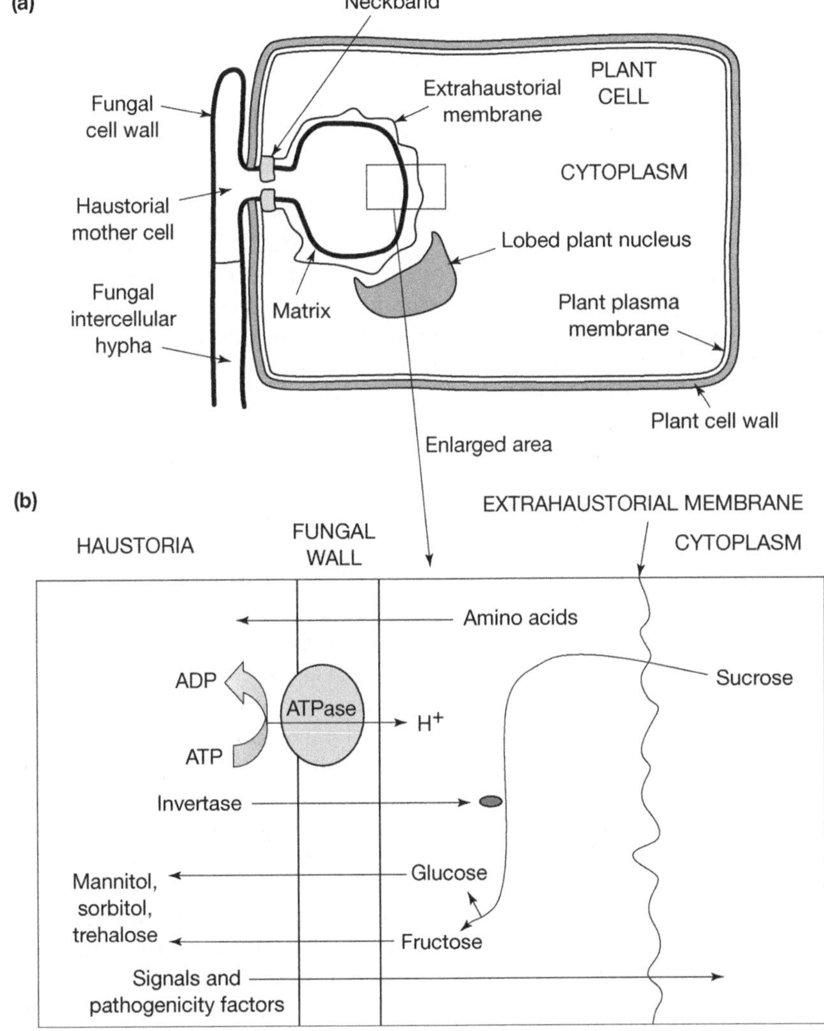

Figure 3.7

General structure of a haustorium. (a) General structure with (b) enlarged cross-section through the wall and matrix to show transport mechanisms. See text for detailed explanations.

the host cell. The neck is surrounded by an electron-dense neckband that bridges the plant and fungal plasma membranes and appears to act as a seal to prevent the flow of solutes from the extrahaustorial matrix to the apoplast. How the fungus triggers neckband formation is unknown. The main body of the haustoria is surrounded by a modified fungal wall, an extrahaustorial matrix, and an extrahaustorial membrane that is derived from the plant plasma membrane, but is thicker than normal membranes, richer in carbohydrates, and appears to lack ATPase activity. It is through these structures that nutrients are transferred and signals between the plant and pathogen are believed to pass. These signals may include some to alter host metabolic pathways along with suppressors of defence responses in plants. There is certainly evidence from cytological studies that major changes occur to plant cells containing haustoria. For example the surface area of the endoplasmic reticulum directly adjacent to the haustoria increases and the plant cell nucleus becomes lobed and heterochromatic as it migrates toward the haustoria. However, the significance of these changes is unknown.

3.7.2 Haustorial function

A number of genes and gene products that are specifically associated with haustoria have been identified by isolating haustoria from infected plants and raising antibodies to their constituent proteins, or making cDNA libraries of the genes that are expressed in them. Amongst these are genes with anti-stress functions such as metallothionins, and others involved in nutrient transfer, in particular amino acid and hexose transporters, and the roles of some of these genes have been confirmed by complementation analyses in yeasts.

Plasma membrane H^+-ATPases are believed to play a crucial role in active nutrient transfer. Following the initial biochemical evidence for the existence of these in haustorial plasma membranes, they have subsequently been cloned. A current model for hexose transport through haustoria in the rust fungi invokes an invertase encoded by the fungus that is secreted into the extrahaustorial matrix to convert sucrose into the hexoses, glucose and fructose (*Figure 3.7b*). Mannitol, sorbitol and trehalose are the main glucose storage compounds in fungi, and ATPase-coupled hexose transporters in the haustorial plasma membrane in combination with hexitol dehydrogenases, transport and convert the hexoses into these. However, in the case of the unrelated oomycete, white blister rust (*Albugo candida*) it has been shown that the increased invertase activity during infection of *Arabidopsis* correlates with increased levels of a host invertase mRNA, so it may be that there are roles for fungal and host-encoded invertases in different interactions.

ATPase-coupled transporters are also present in the membranes to transport essential amino acids for the fungi, possibly those that the fungus is unable to biosynthesise. A further possibility that has been suggested is that whole proteins may be transported into fungi from plants. Some evidence for this comes from work on *Erysiphe pisi* haustoria in which a protein with significant homology to a plant chloroplast protein was found in the fungal haustoria, although it is unclear whether the gene encoding this protein is fungal or plant. Some biotrophic fungi, for example *C. lindemuthianum,* switch to necrotrophic infection after the initial biotrophic phase, and insertion mutagenesis has revealed a zinc-cluster family transcriptional activator, encoded by the *CLTA1* gene that appears to control this switch.

3.8 Prevention of leaf senescence

For a biotrophic fungus to complete its life cycle and sporulate on an infected plant, it must continue to assimilate nutrients. This may take several days or weeks after the initial spore germinated and penetrated the plant. A cereal rust fungus for example, may grow for 2–3 weeks within a wheat plant, perhaps 2–3 cm from the initial point of entry, tapping into more and more mesophyll cells. If one of the benefits of long-term biotrophy is the ability to control transport of nutrients within the plant, the fungus must be able to produce signals for such control. A senescent or dead plant is of no value to such a fungus. Plant hormones and polyamines are known in plants to be involved in the maintenance of juvenility, and there is increasing evidence that biotrophic fungi manipulate the levels of these during infection.

3.8.1 The role of cytokinins

One characteristic symptom exhibited by plants infected with many biotrophic and hemi-biotrophic fungi is the formation of green islands. These are areas of green tissue immediately around infection sites that show delayed senescence, and it is believed that fungi maintain these to ensure the presence of metabolically active cells from which they can obtain nutrients. Fluxes of cytokinins, plant hormones that stimulate metabolism and transport of nutrients, are believed in part to be responsible for this phenomenon. Measurements of cytokinin levels in infected tissues correlate with symptoms in that they are higher in green islands, and there are a number of fungi that are known to produce specific symptoms because of increased levels of auxins and cytokinins, such as peach leaf curl (*Taphrina deformans*) and maize smut (*U. maydis*). In the bacterial gall-forming pathogens, *Pseudomonas syringae* pv. *savastanoi* and *Agrobacterium tumefaciens* (see Section 5.7), genes encoding isopentenyl transferases, the key enzyme in cytokinin biosynthesis from mevalonic acid and isopentenyl pyrophosphate precursors, have been identified. However, similar genes have not been identified in plants or fungi, and it remains unclear whether fungal pathogens modify levels of cytokinins by modifying biosynthetic pathways in plants or producing them *de novo*.

3.8.2 The role of polyamines

Polyamines, such as putrescine, spermidine and spermine (*Figure 3.8*) are derived in plants from arginine, and occur in free forms or bound to negatively charged molecules such as phenolic acids, DNA and RNA. They are necessary for replication and cell division in most cells and have been implicated in the regulation of senescence in plant cells as well as in response to biotic and abiotic stresses. In biotrophic interactions between rusts/mildews and their hosts, increased levels have been found in green islands, along with increased activities of arginine decarboxylase (ADC), ornithine decarboxylase (ODC) and S-adenosylmethionine decarboxylase (AdoMetDC)

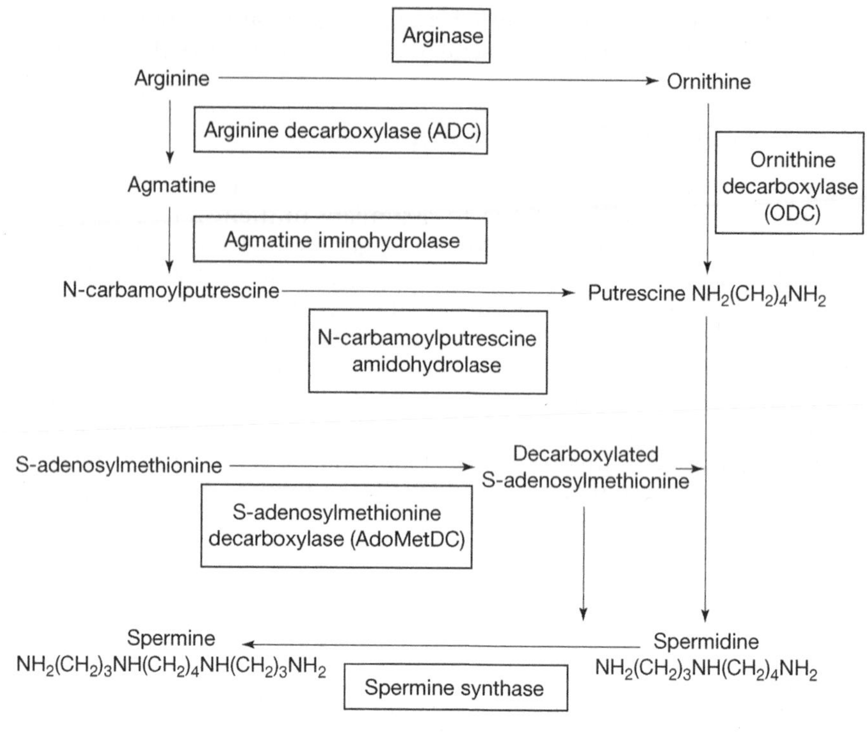

Figure 3.8

Biosynthetic pathways for the polyamines putrescine, spermidine and spermine in plants. The enzymes involved are shown in boxes.

biosynthetic enzymes. It has been suggested that release of free polyamines from their conjugated forms may contribute to these increases. Levels also increase in hyperplastic tumours associated with the protozoa *Plasmodiophora brassicae* (clubroot), whilst there is evidence in some necrotrophic interactions that levels decrease. In *Septoria nodorum,* for example, disruption of the ODC gene resulted in a non-pathogenic strain. How pathogens bring about the changes in polyamine levels and how this relates to cytokinin levels is yet to be determined.

References and further reading

Ashby, A.M. (2000) Biotrophy and the cytokinin conundrum. *Physiological and Molecular Plant Pathology* **57**: 147–158.

Cullimore, J.V., Ranjeva, R. and Bono, J-J. (2001) Perception of lipo-chitooligosaccharidic Nod factors in legumes. *Trends in Plant Science* **6**: 24–30.

Daub, M.E. and Ehrenshaft, M. (2000) The phytoactivated *Cercospora* toxin cercosporin: contributions to plant disease and fundamental biology. *Annual Review of Phytopathology* **38**: 461–490.

Del Sorbo, G., Schoonbeek, H-J. and De Waard, M.A. (2000) Fungal transporters involved in efflux of natural toxic compounds and fungicides. *Fungal Genetics and Biology* **30**: 1–15.

Heath, M.C. (1995) Signal exchange between higher plants and rust fungi. *Canadian Journal of Botany* **73** (Suppl. 1): S616–S623.

Heiser, I., Oßwald, W. and Elstner, E.F. (1998) The formation of reactive oxygen species by fungal and bacterial phytotoxins. *Plant Physiology and Biochemistry* **36**: 703–713.

Idnurm, A. and Howlett, B.J. (2001) Pathogenicity genes of phytopathogenic fungi. *Molecular Plant Pathology* **2**: 241–245.

Markham, J.E. and Hille, J. (2001) Host-selective toxins as agents of cell death in plant–fungus interactions. *Molecular Plant Pathology* **2**: 229–239.

Mendgen, K., Struck, C., Voegele, R.T. and Hahn, M. (2000) Biotrophy and rust haustoria. *Physiological and Molecular Plant Pathology* **56**: 141–145.

Oliver, R.P. and Schweizer, M. (eds) (1999) *Molecular Fungal Biology.* Cambridge University Press, Cambridge.

Osbourn, A.E. (1999) Antimicrobial phytoprotectants and fungal pathogens: a commentary. *Fungal Genetics and Biology* **26**: 163–168.

Perfect, S.E. and Green, J.R. (2001) Infection structures of biotrophic and hemibiotrophic fungal plant pathogens. *Molecular Plant Pathology* **2**: 101–108.

Staples, R.C. (2001) Nutrients for a rust fungus: the role of haustoria. *Trends in Plant Science* **6**: 496–498.

Sweeney, M.J. and Dobson, A.D.W. (1999) Molecular biology of mycotoxin production. *FEMS Microbiology Letters* **175**: 149–163.

Van den Brink, H.J.M., Van Gorcom, R.F.M., Van den Hondel, C.A.M.J.J. and Punt, P.J. (1998) Cytochrome P450 enzyme systems in fungi. *Fungal Genetics and Biology* **23**: 1–17.

Van Etten, H., Temporini, E. and Wasmann, C. (2001) Phytoalexin (and phytoanticipin) tolerance as a virulence trait: why is it not required by all pathogens? *Physiological and Molecular Plant Pathology* **59**: 83–93.

Walters, D.R. (2000) Polyamines in plant–microbe interactions. *Physiological and Molecular Plant Pathology* **57**: 137–146.

Wolpert, T.J., Dunkle, L.D. and Ciuffetti, L.M. (2002) Host-selective toxins and avirulence determinants: what's in a name. *Annual Review of Phytopathology* **40**: 251–285.

Fungal and oomycete genetics

4

The ability of a fungal or oomycete pathogen to cause disease on a plant species defines that species as a host and others as non-hosts. The presence within the fungus of 'pathogenicity factors', i.e. the necessary penetration mechanisms, cell-wall-degrading enzymes, toxins, signalling molecules, biosynthetic genes, nutrient transport mechanisms, etc., will all determine which plant species are hosts. However, whilst the whole population of a particular fungus or oomycete will contain certain morphological and phenotypic features that define it as a species, within that species there will be many genetic variants with more specific characteristics. It is the nature of this genetic variation, how it occurs and how it affects fungal pathogenicity that we focus on in this chapter.

4.1 The concept of race structure

Within a fungal species, there are a number of levels at which genetic variation may affect significant characteristics. For example, the stem rust fungal pathogen *Puccinia graminis* causes disease on cereal plants. However, within this species there are groups of individuals which attack just wheat, others that attack just barley and others, oats. These are referred to as *forma speciales* (e.g. *P. graminis* f.sp. *tritici* on wheat). Within each of these *forma speciales* there are then individuals that can only cause disease on cultivars of wheat that have a particular genotype. These individuals are said to belong to a particular race or pathotype of the fungus. Within a race, the individuals themselves will possess genetic variation affecting many other characteristics, and these individuals will be referred to as isolates. However, it has generally been at the race level that agriculturally significant changes in pathogens have been detected. The appearance of Race T of *C. heterostrophus* that devastated T-cms maize in the US in the 1970s (see Section 3.6.1) is one example, and changes in the race structures of other pathogens are a constant process and threat in the battle between plants and pathogens.

4.2 Avirulence genes

4.2.1 General concepts

The concepts of disease resistance genes and horizontal versus vertical resistance will be discussed in more detail in Chapters 9 and 10, and the presence or absence of resistance genes is a significant factor that determines whether a particular race or pathotype of the fungus will cause disease. In horizontal resistance (*Figure 10.1* in Chapter 10), all plants have a certain but not necessarily the same level of resistance, and this is controlled by many genes in the plant in a complex manner. The resistance offers a reasonable level of defence against all isolates of the pathogen, so it is not possible to define races by their pathogenicity. Vertical resistance in contrast involves some varieties being strongly resistant to some isolates but fully susceptible to others and is generally controlled by gene-for-gene mechanisms (*Figure 10.2* in Chapter 10). In essence, for a fungal pathogen such as *C. fulvum,* there will be a number of resistance genes in tomato (e.g. *Cf-1–Cf-9*) and different cultivars will contain different combinations of these. The fungal population will have a corresponding number of what are termed avirulence genes (*Avr1–Avr9*). If a race of the fungus containing a particular avirulence gene lands on a plant with the corresponding resistance gene, recognition of the elicitor (that results from the avirulence gene) by the product of the resistance gene will result in a defence response and resistance in the plant. If, however, either the avirulence gene is mutated or absent in the fungus, or conversely, the resistance gene is recessive in the plant, disease will ensue (*Figure 10.3* in Chapter 10).

4.2.2 Cloning of avirulence genes

Because avirulence genes and their gene products are central to understanding gene-for-gene resistance, much effort has gone into cloning these genes, characterising the gene products and determining how they elicit a defence response in plants. The best studied of these to date are listed in *Table 4.1*. Much of the pioneering work on cloning fungal avirulence genes was performed by Pierre de Wit in Wageningen, The Netherlands, who was able to identify race-specific elicitors in apoplastic fluids of tomato plants infected with the tomato leaf mould pathogen, *C. fulvum*. Having identified a specific peptide in these fluids that functioned in an AVR9 race-specific manner, he was able to purify and sequence this peptide and use the peptide sequence to identify the gene that encoded it in the fungus. By transforming isolates of the fungus lacking the *Avr9* gene with the cloned sequence, he was able to confirm that this was the gene. Other *Cladosporium* and *Rhynchosporium secalis* avirulence genes have been identified by similar means, whilst those from *M. grisea* have been identified by positional cloning. In *P. infestans, P. sojae* and *P. parasitica,* where detailed genetic maps have been constructed, a number of avirulence genes have recently been positioned as a step towards their cloning. Interestingly, these often appear to be clustered, leading to the suggestion that avirulence determinants may be concentrated in pathogenicity islands (see Section 4.7).

Table 4.1 Cloned fungal and oomycetes avirulence genes, including putative functions of the peptides base on homologies with known genes

Pathogen	AVR protein	Resistance specificity	Putative protein function
Cladosporium fulvum	AVR9	*Cf-9* gene in tomato	Cysteine-knot peptide
C. fulvum	AVR4	*Cf-4* gene in tomato	Chitin-binding protein
C. fulvum	AVR2	*Cf-2* gene in tomato	None
C. fulvum	ECP2	*Cf-ECP2* gene in tomato	None
Magnaporthe grisea	AVR-Pita	*Pi-ta* gene in rice	Zinc metalloprotease
M. grisea	PWL1	Weeping lovegrass	None
M. grisea	PWL2	Weeping lovegrass	None
Rhynchosporium secalis	NIP1	*Rrs1* gene in barley	Toxin/hydrophobin
Phytophthora infestans	INF1	*Nicotiana benthamiana*	None

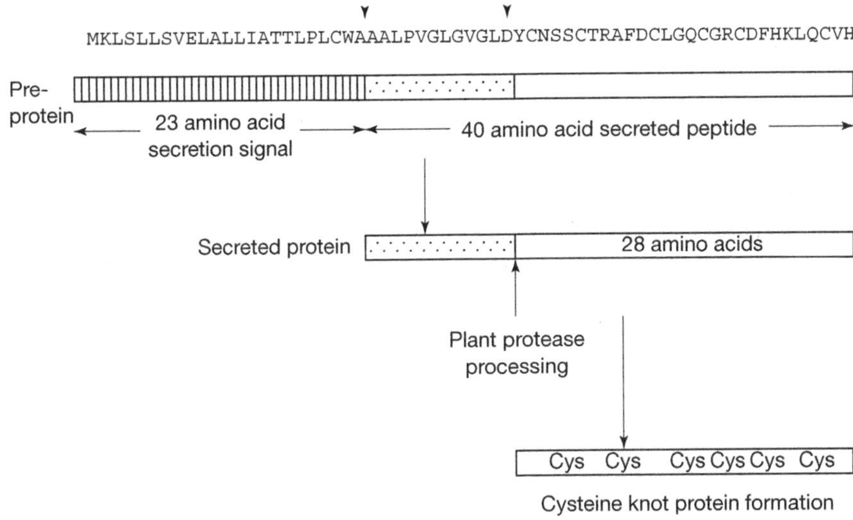

Figure 4.1

Processing of the AVR9 avirulence peptide to produce the secreted elicitor. The secretion signal targets the protein for secretion, and the resultant 40 amino acid peptide is further processed by plant proteases to yield the 28 amino acid cysteine-rich active elicitor.

4.2.3 Avr protein structure and function

Although the fungal avirulence genes identified to date have no specific features common to all of them, they tend to be small peptides with signal peptides suggesting that they are secreted. For example, the structure and processing of the *Avr9* gene from *C. fulvum* is shown in *Figure 4.1*. Similarly, *Avr4* encodes a 135 amino acid precursor, also with a signal peptide for extracellular targeting, and following removal of the 18 amino acid signal and further processing by plant and/or fungal proteases, the resultant elicitor is 86 amino acids long of which 8 are cysteines. AVR2 is also a cysteine-rich protein of 78 amino acids with a predicted signal sequence. *AVR-Pita* from *M. grisea* is a 223 amino acid protein with a signal peptide, and has a small domain with some homology to the active site of neutral zinc proteases, although the significance of this is not known.

The function of avirulence proteins in fungi is unclear, and the very presence of avirulence genes in pathogens has always been an intriguing concept. Why have a gene to encode a product that ensures that you are unable to cause disease on particular cultivars of plant? The logical explanation is that these proteins must have an important function in the fungus, and that there is a cost to losing them in terms of pathogenicity. This role may be as a house-keeping gene or a pathogenicity factor, and their avirulence function is merely a consequence of the fact that the plant has a mechanism for recognising them as foreign and setting off a defensive response. Secreted proteins would be more likely to act as elicitors, because they would be more readily detected by a plant surveillance mechanism. Support for the house-keeping role of avirulence genes comes from the putative involvement of *Avr9* in nitrogen metabolism. The gene is induced by nitrogen starvation and has nitrogen metabolism regulatory sequences analogous to *A. nidulans* AREA sequences in its promoter region. The *NIP1* (*AvrRrs1*) gene from *R. secalis* by contrast appears to be a pathogenicity factor, with the NIP1 gene product (a small cysteine-rich peptide) acting as a necrosis-inducing protein in a non-selective toxin manner. *NIP1* gene disruption mutants have significantly reduced virulence. Support also comes from mutation frequencies amongst avirulence genes. In his original studies on flax rust (*Melampsora lini*), Flor found that the mutation rates of different avirulence genes were very variable, and that the genes most frequently mutated in laboratory experiments were the same as those that were most often lost in the field. This implies that the less readily mutated genes are more critical to pathogen fitness than those for which mutants are common.

4.2.4 The significance of avirulence genes in species specificity

Other avirulence genes have been identified in *M. grisea* and *Phytophthora* species that appear to confer species specificity and/or *forma speciales* status (see *Table 4.1*). The *PWL* (pathogenicity on weeping lovegrass) genes in *M. grisea* are members of a small gene family, in which *PWL1* represents a gene originating from a strain virulent on weeping lovegrass, whilst *PWL2* is from a strain that causes disease on rice. Adding *PWL2* to strains that lack this gene and which are normally virulent on weeping lovegrass makes them avirulent, implying that a defence response is elicited in weeping lovegrass in response to PWL2. The peptides encoded by these genes are small peptides with signal sequences.

In certain species of *Phytophthora* and *Pythia,* elicitins have been identified as determinants of avirulence at the species level. These elicitors appear to be members of multi-gene families. The elicitin encoded by *Inf1* in *P. infestans* ensures that this oomycete is unable to cause disease on *Nicotiana* species, and growth is arrested at an early stage of infection. However strains lacking *Inf1* function, created by gene silencing, are able to infect and sporulate on *N. benthamiana.* The elicitin itself is a 10 kDa protein that is highly expressed and appears to be a secreted structural protein. There is evidence to suggest that it may be a sterol carrier protein, which is of particular significance in *Phytophthora* since oomycetes are unable to synthesise sterols and must obtain them from external sources.

4.3 Fungicide resistance

Treatment with fungicides is one of the main approaches used for maintaining healthy crops and reliable yields, and is an important component of many integrated crop management programmes (see Section 13.5). The main groups of fungicides are listed in *Table 4.2*, along with their mode of action. Within most fungal populations however, and/or in response to fungicide usage, rare mutants resistant to fungicides often occur. As selection pressure is exerted on these populations from repetitive use of a fungicide, these mutants survive and may gradually increase as a proportion of the population. In some cases, this has resulted in the performance of a fungicide being compromised, with either increased dosages or different chemicals being required for disease control. Modern practices of avoiding repetitive use of individual fungicides and favouring mixing and/or alternating fungicides is

Table 4.2 Mode of action of the major groups of fungicides, and nature of resistance in pathogen populations

Fungicide/Fungicide class	Mode of action	Mechanism through which resistance has been acquired (when detected)
Aromatic hydrocarbons	Inhibit amino acid and enzyme production	Unknown (cross-resistance with dicarboximides) (1960 – *Penicillium*/citrus)
Organo-mercurials	Non-specific	Detoxification of binding substances (1964 – *Pyrenophora*/cereals)
Benzimidazoles (MBCs)	β-tubulin/spindle formation	Altered target site – β-tubulin (1970 – many pathogens)
2-aminopyrimidines	Adenosine deaminase	Unknown (1971 – *Blumeria*/barley)
Phosphorothiolates	Unknown	Metabolic detoxification (1974 – *Magnaporthe*/rice)
Phenylamides	RNA polymerase/nucleic acid biosynthesis	Altered target site (RNA polymerase) (1980 – *Phytophthora*/potato)
Dicarboximides	Unknown	Unknown (cross-resistance with aromatic hydrocarbons) (1982 – *Botrytis*/grapes)
Demethylase inhibitors (DMIs)	Sterol biosynthesis	Increased efflux / altered target site (eburicol 14 α-demethylase)/ target site overproduction (1982 – *Blumeria*/barley)
Carboxanilides	Inhibit succinic dehydrogenase/ mitochondrial respiration	Altered target site (succinate ubiquinone oxidoreductase) (1985 – *Ustilago*/cereals)
Dithiocarbamates	Inactivate sulphydryl groups in amino acids and enzymes	
Strobilurins	Mitochondrial electron transport	Unknown (2000 – *Blumeria* and *Septoria*/wheat)
Anilinopyrimidines	Protein secretion/ methionine biosynthesis	

reducing the risk of this occurring. *Table 4.2* lists the suspected mechanism of acquired resistance for a number of fungicides.

Because fungicide resistance generally results from mutations, single-site fungicides are more likely to encounter problems since only a single mutation in the pathogen may be necessary to counter the chemical. With multisite fungicides, mutations in more genes are required simultaneously, making resistance less likely. For example, when the methyl-benzimidazole-carbamate (MBC)-generating fungicides such as benomyl were introduced for the control of a number of diseases in the early 1970s, resistance was detected within 2 years of usage in a wide range of fungi. In most cases this is due to mutations within the target site (β-tubulin) gene so as to reduce the binding affinity of the benzimidazole, with often only a point mutation required to confer resistance. In contrast, the sterol 14 α-demethylase inhibitor (DMI) fungicides (the largest group of systemic fungicides currently in use) have remained highly effective despite many years of use and their single-site mode of action. These fungicides inhibit the enzyme eburicol 14 α-demethylase, encoded by the *Cyp51* gene, which is the rate-limiting step in ergosterol biosynthesis. Inhibition of eburicol 14 α-demethylase in filamentous fungi causes depletion of ergosterol and accumulation of

14-methylsterols, which disrupts chitin and fatty acid metabolism, inhibits membrane-bound proteins and membrane fluidity, and with a high enough dose of fungicide, causes physical membrane damage. However, decreased sensitivity and field resistance to certain DMIs has been reported in at least 13 species of plant pathogens. In the cereal eyespot fungus, *Tapesia yallundae,* genetic studies have indicated that resistance is conferred by a single major gene, with evidence of polygenic components increasing levels of resistance in the most resistant isolates. A number of mechanisms, such as decreased affinity of the target enzyme, overproduction of the target enzyme, active efflux of the fungicide, circumvention of toxic sterol formation, and changes in biosynthetic pathways have been proposed. Point mutations at amino acid residue 136, within the proposed substrate recognition domain of the *Cyp51* gene, have been correlated with DMI resistance in field isolates of *B. graminis* and *Uncinula necator* (grape powdery mildew). Laboratory-generated mutants of *Candida albicans* and selected strains of *Penicillium italicum* resistant to DMIs also show changes in this region. However, in *T. yallundae,* there was no correlation between resistance and mutations within the *cyp51* gene, suggesting that other factors, possibly the level of expression of the gene, may be important.

In studies into metalaxyl (a phenylamide fungicide) resistance in *P. infestans,* genetic crosses have also indicated that resistance is conferred by a single major gene, with variation in the sensitivity influenced by supplementary minor genes. It was also shown that there was no correlation between metalaxyl resistance and fitness of the pathogen, although in studies on other fungi and using other chemicals there is some evidence that loss of pathogen fitness is associated with acquisition of fungicide resistance.

4.4 Mechanisms for generating genetic variation in fungi

Having established that genetic variation in fungi is important to overcome resistance in plants and chemical control measures, what are the factors that control the formation and spread of new genotypes? The five most significant factors are mutation rates in the pathogen, the means of reproduction, the rate of migration of isolates, the population size and the selection pressures caused by agricultural practices. A pathogen with a high mutation rate and large population size will evolve more rapidly, and if the spores are airborne, the rate of gene flow (migration) will be higher. Indeed, of these factors, the rate of migration appears to be the most important for new disease outbreaks.

To generate variation, the means of reproduction is also important. Most fungi will continue to grow and invade fresh plant material for as long as possible and then produce numerous asexual spores (zoospores, conidia, chlamydospores, urediospores, etc.) or other mitotically derived structures such as sclerotia, which are sufficient to re-infect fresh material and repeat the asexual cycle. There are many fungi that appear to reproduce only asexually and for which sexual cycles have not been identified, such as yellow (stripe) rust of wheat (*Puccinia striiformis*), yet these are able to generate sufficient genetic diversity without sex. Evidence from studies in Australia, where this fungus arrived in 1979, have shown how rapidly a pathogen can evolve without a sexual cycle, and this pathogen has subsequently evolved into more than 20 known new pathotypes. Similarly, from the 1840s, when potato blight first arrived in Europe, until 1976, only one mating type was known to exist in Europe and no sexual cycle occurred outside Mexico, yet this pathogen was readily able to generate resistance to fungicides and new races. Only in 1976 did the second mating type arrive in Europe allowing this pathogen to go through its sexual cycle.

Sexual reproduction has the capacity to generate genetic diversity much more rapidly than asexual methods, through recombination and reassortment of alleles, and many fungi use their sexual cycles to generate new races. However, whilst novel genotypes are produced more rapidly in sexual than asexual populations, selection is generally considered to work more efficiently in asexual populations because all the offspring are identical, and favourable gene combinations are not broken up through recombination events. The optimum condition for a fungus is probably to have a combination of sexual and asexual reproduction allowing both rapid generation of variability and the capacity to fix favourable mutations when they occur.

4.5 Mating-type genes

Sexual cycles in fungi depend upon mating systems in which nuclear genes prevent mating between genetically identical mycelia. Most plant-pathogenic fungi utilise a heterothallic mating system in which isolates of opposite mating type are required for completion of the sexual cycle, although there are a few fungi that are homothallic (self-fertile) and can produce male and female gametes on the same mycelium. The heterothallic mating systems vary between the ascomycete fungi and the basidiomycete fungi. In the ascomycetes, the asexual state of the fungus is haploid, whilst in basidiomycetes, the fungi are generally dikaryotic, containing two sexually compatible nuclei per cell (*Figure 4.2*). By contrast, oomycetes have completely diploid lifecycles.

In the heterothallic ascomycetes, there is a single *MAT* locus, represented by two idiomorphs, such that the presence of isolates of two opposite mating types (i.e. each of the two idiomorphs), usually referred to as MAT1 and MAT2, is required for the sexual cycle to occur (*Figure 4.3*). The function of these genes in mating and induction of gene expression as part of the sexual cycle is unclear in most cases, although pheromone and pheromone receptors analogous to the mating system in the yeast *S. cerevisiae* have been identified in the chestnut blight fungus, *Cryphonectria parasitica*.

In the heterothallic basidiomycete, *U. maydis*, a tetrapolar mating system is essential for pathogenicity of the fungus on plants. The diploid teliospores produced from infected plants germinate, undergo meiosis and produce haploid sporidia which have a yeast-like morphology and are able to grow as saprophytes in the soil and are non-pathogenic on plants. Two sporidia of opposite mating type must then conjugate, attracted to each other by a pheromone gradient, undergo fusion and produce a dikaryotic hypha. This penetrates plants, either through stomata or directly on younger plants, and ultimately forms tumours on the plant containing masses of asexual teliospores.

Mating is controlled by two loci, *a* and *b*, and there are two idiomorphs at the *a* locus and at least 25 at the *b* locus. For mating to occur, partners have to be different at both loci, i.e. a1b1 will mate with a2b20, but not with a2b1 or a1b20. The a locus encodes a lipopeptide pheromone (Mf) and a pheromone receptor (Pr), such that the receptor from a2 recognises the

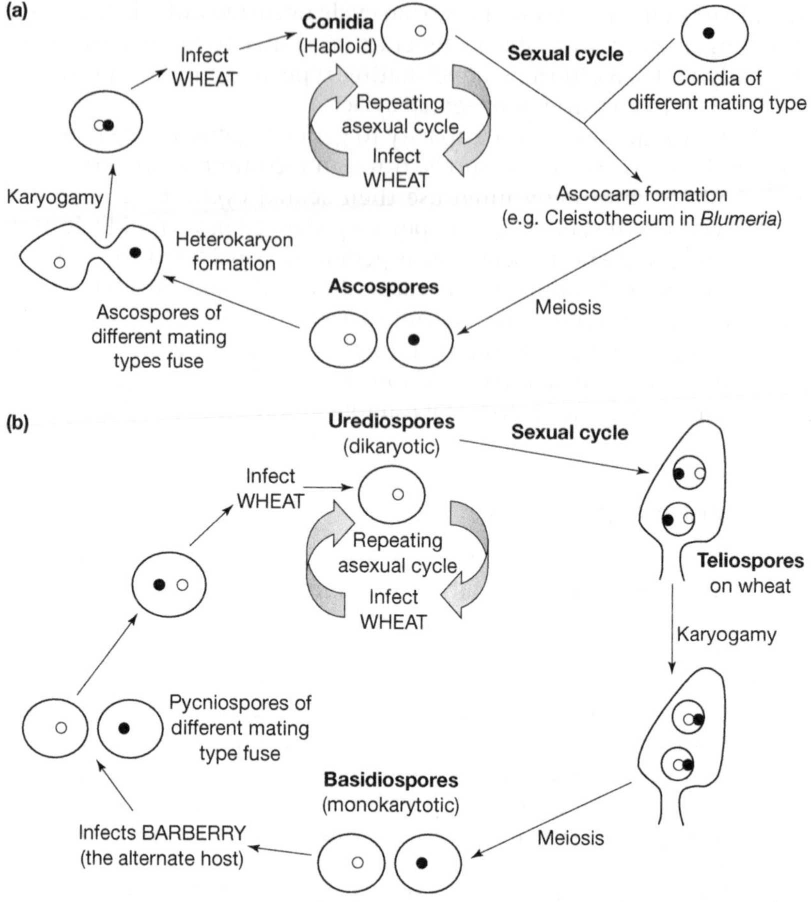

Figure 4.2

The life cycles of fungi. The life cycle of (a) a heterothallic ascomycete fungus (e.g. *Blumeria*) compared to (b) a basidiomycete (*Puccinia graminis*) to show the monokaryotic/dikaryotic nature of the different stages. Nuclei are represented by closed and open circles.

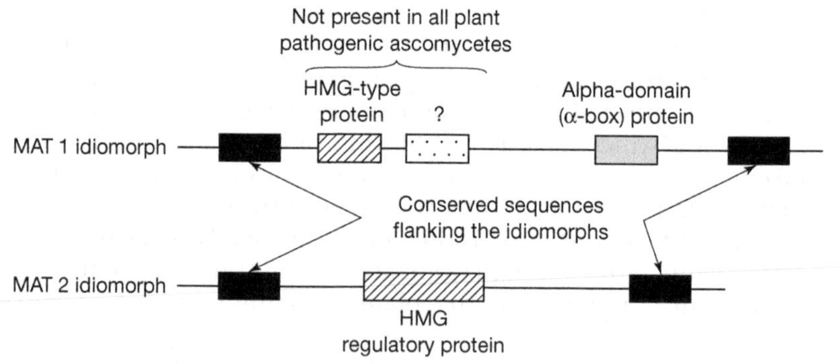

Figure 4.3

The organisation of mating-type genes in plant-pathogenic ascomycetes. For mating to occur, isolates containing the MAT 1 idiomorph have to find partners containing the MAT 2 idiomorph. (HMG = high mobility group.)

a1 pheromone and vice versa (*Figure 4.4a*). The b locus also has two open reading frames bW and bE, that produce polypeptides containing domains of more than 90 per cent identity and variable domains. The four peptides produced by the opposite mating-type b alleles form a complex that acts directly as a transcription regulator (*Figure 4.4b*). The a locus works through two signalling cascade pathways involving G proteins and cyclic AMP, and also MAP kinases (*Figure 4.4c*), and much work has been undertaken to analyse mutants deficient in the signalling pathways, and to identify homologues of signalling components in other fungi. The *Fuz7* gene is homologous to the *Ste7* MAP kinase kinase from yeast, and *Kpp2* is homologous to the *PMK1* MAP kinase from *M. grisea* (see Section 2.7.1). The kinase and cyclic AMP pathways appear to be interconnected, since *kpp2* mutants are not completely sterile and can still cause tumours, indicating that there is a great deal of cross-talk and co-ordination in the signalling. The result is activation of the Prf1 transcription activator that recognises pheromone responsive elements via its DNA-binding domain and activates gene expression resulting in cell fusion. A number of genes have been shown to be

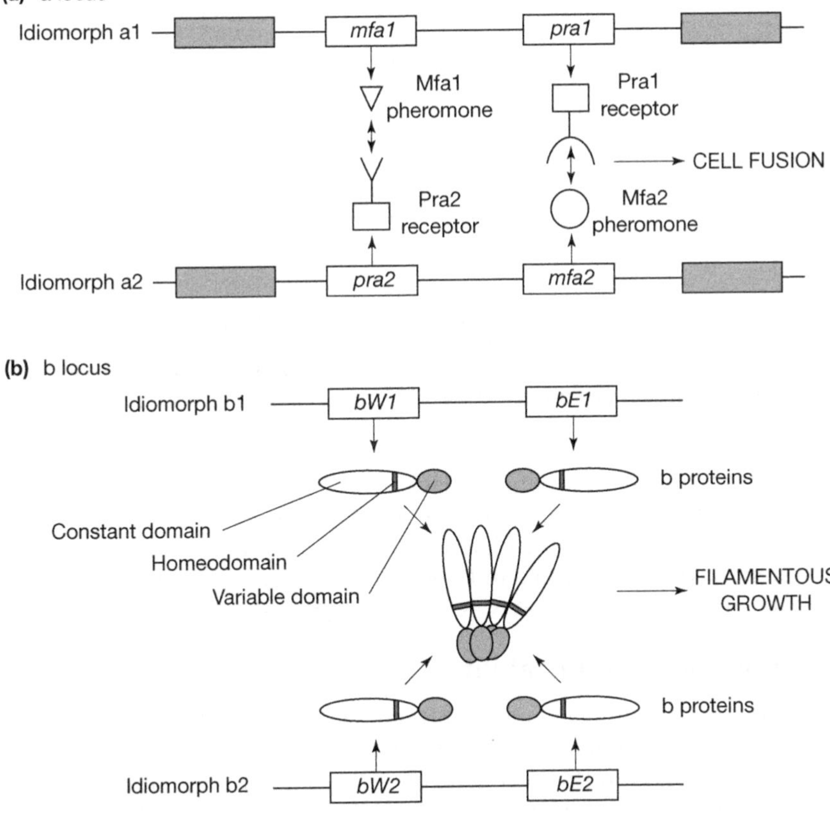

Figure 4.4a,b

The mating-type loci of *Ustilago maydis*. (a) The a locus encodes pheromones and receptors involved in inducing cell fusion. (b) The b locus encodes peptides that form a transcription regulator of filamentous growth. See text for further details.

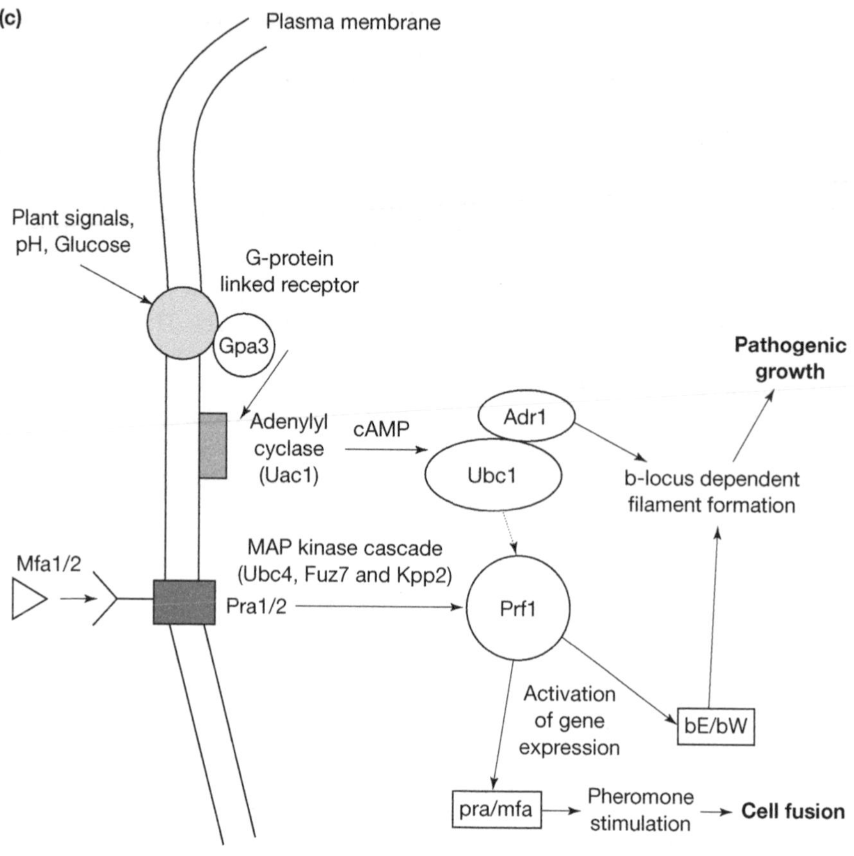

Figure 4.4c

The mating-type loci of *Ustilago maydis.* (c) The signalling cascades controlling mating in *Ustilago maydis.* The Pra1 receptor detects the Mfa2 pheromone and Pra2 detects the Mfa1 pheromone, which instigates a MAPK cascade. Other signals are detected via a G-protein linked receptor.

up-regulated, including an endonuclease, a hydrophobin and a repellent, but the role of these in establishing filamentous growth and pathogenicity of the fungus has yet to be elucidated. Interestingly, a similar tetrapolar mating system has been identified in *U. hordei* but the a and b loci are linked so that they propagate as one locus, referred to as MAT-1 and MAT-2.

4.6 Chromosome instability

There are a number of mechanisms through which genetic variation appears to be generated in fungi in the absence of sex. Point mutation within genes is one example, and in the *Avr4* gene of *C. fulvum,* point mutations within cysteine codons occur in races virulent on *Cf-4* plants so that although the gene is expressed, no active peptide is produced. Similarly, in *AVR-Pita,* point mutations in the protease consensus sequence abolish function. However, other *AVR-Pita* sequences, along with genes such as *Avr9* in *C. fulvum,* and *Avr-CO39* in *M. grisea* are lost in many virulent races through chromosomal deletion. In the case of *AVR-Pita*, this is related to the fact that the gene is only 48 bases upstream of the telomeric sequences on chromosome 1, so the gene can be lost by deletion of the tip of a chromosome. Indeed, in *M. grisea*, it has been suggested that the chromosomes can be divided into the 'A' chromosomes that are largely uniform in number and size of between 2–10 Mb, and the 'B' chromosomes, whose number and size vary considerably between isolates. In some races of *C. fulvum* that have lost the *Avr9* gene, it has been shown by Southern blots of pulse-field gel electrophoresed chromosomes of the fungus, probed with *Avr9*, that up to 500 kb of genomic DNA has been lost.

There is also evidence for chromosomal instability in other fungi. In *U. hordei*, pulse-field gel electrophoresis (electrophoretic karyotyping) of different strains has shown that chromosome numbers and sizes differ between isolates, and in some isolates of *C. heterostrophus*, the smallest chromosome (16) which is 1.4 megabases in length is absent with no obvious phenotypic changes. Furthermore, in *Nectria haematococca*, the *PDA1* gene responsible for the ability to detoxify pisatin (see Section 3.3.3), along with three other pea pathogenicity (*PEP*) genes, are linked within a 25 kb region of a dispensable (or supernumeracy) chromosome, and whilst strains that lack this chromosome grow normally in culture, they have reduced virulence on pea. In *C. gloeosporioides* there is evidence for transfer of whole chromosomes between fungi. There are two biotypes, A and B, that are clearly distinct from each other morphologically and are vegetatively incompatible. However, some naturally occurring B types have acquired a 2.0 Mb chromosome from the A type and laboratory experiments using marked chromosomes have confirmed that this can occur.

4.7 Alien genes/horizontal gene transfer

Both sexual and asexual fungi undergo chromosomal rearrangements, which are the main cause of karyotype variability within populations. So what mechanisms might be responsible for chromosomal instability in fungi? One mechanism that occurs in fungi is horizontal gene transfer of alien genes, that is the ability of fungi to pick up genes from other organisms, and the presence of 'selfish' operons is a well-known phenomenon (e.g. in bacterial genomes), in which genes for a particular characteristic are clustered together initially after transfer, and only become dissipated as the organism evolves. One line of evidence for horizontal gene transfer in fungi is the presence of secondary metabolite gene clusters. These clusters generally contain the genes for biosynthetic enzymes, the regulatory genes for these pathways and genes for autoresistance. For example, the genes for gibberellin biosynthesis in *Gibberella fujikuroi*, mycotoxin production in ergot, *Aspergillus flavus* and *Fusarium* spp., and the toxin biosynthetic genes in *Alternaria* and *Cochliobolus* spp. are in clusters, as are the melanin biosynthetic genes in *Alternaria*, although in other fungi such as *M. grisea*, *C. heterostrophus* and *Colletotrichum* species, this linkage for melanin genes does not occur. In fungi such as *A. nidulans* and *N. crassa*, the genes for nitrate, proline, quinate and ethanol utilisation are also clustered, although whether this occurs in plant pathogens is not known.

The secondary metabolite gene clusters appear to be entirely absent in non-pathogenic strains within these fungal species, and also tend to have distinctive codon usage and guanine plus cytosine (G+C) content. The

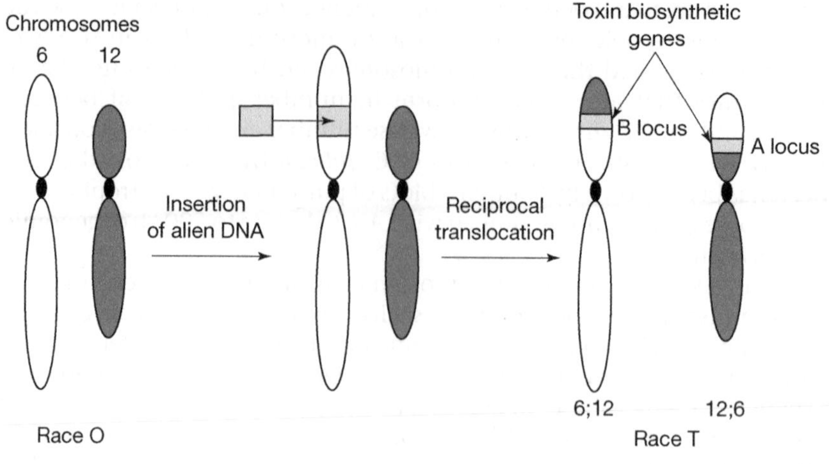

Figure 4.5

The nature of race T of *Cochliobolus heterostrophus*. The alien DNA gene cluster is believed to have inserted into chromosome 6, and a reciprocal translocation has resulted in the cluster being split between chromosomes 6 and 12 in race T.

nature of race T of *C. heterostrophus* is one such example (*Figure 4.5*). In this race, there is an additional 1.2 Mb of DNA present in race T that is not in race O of the fungus. This region contains the genes required for T-toxin biosynthesis, although because of what appears to have been a reciprocal translocation, the locus has in fact become split into two loci on separate chromosomes, 6 and 12. The Tox1B locus contains a decarboxylase gene (*DEC1*), whilst the Tox1A locus contains the *CPS1* peptide synthase genes along with at least 15 other open reading frames for genes such as a thioesterase, a laccase and a DNA-binding protein. Whilst this 1.2 Mb of DNA has fungal-like introns within it, it also has unusual codon usage for fungal DNA, so its origin remains unknown. Similarly, the *PDA1* and *PEP* gene cluster in *N. haematococca* has unusual codon usage and G+C content, as well as some regions with similarity to transposase sequences, analogous to the features of pathogenicity islands in bacteria (see Section 6.9).

The clustering of gibberellin biosynthetic pathway genes in the rice pathogen *G. fujikuroi* is also unusual, in that gibberellins are natural plant hormones required for normal plant growth and development, and are found in very few fungal genera. It has been postulated that they have been acquired in fungi by horizontal gene transfer from plants, although sequence comparisons with the corresponding genes in plants have yet to confirm this. The ability of fungi to assimilate genes from plants has also been suggested from the sequencing of genes from the bean rust fungus, *Uromyces fabae,* where the fungally encoded ATPase was found to have greater homology to bean ATPases than to those from other fungi. A second line of evidence for horizontal gene transfer comes from experiments in which *Aspergillus niger* was co-cultured in sterile soil with transgenic hygromycin B-resistant *Brassica* plants. Amongst the colonies of *Aspergillus* that were subsequently isolated were some that possessed hygromycin resistance and others than contained plasmid and/or plant DNA.

4.8 Role of transposable elements

A further mechanism for transfer of DNA between organisms, and for chromosomal deletions and abnormalities, is through the action of transposable elements. Transposable elements are widespread amongst living organisms, making up 10–15% of the genome of *Drosophila melanogaster* (the fruit fly), and more than 60% of the maize genome. They are generally subdivided into Class I and Class II elements. Class I elements transpose through reverse transcription of an RNA intermediate, and include retrotransposons with long terminal repeats (LTRs), long interspersed nuclear elements (LINEs) and short interspersed nuclear elements (SINEs). Class II elements have short inverted terminal repeats (ITR) and transpose by excision from a donor site and re-integration into another site in the genome, a process catalysed by a transposase enzyme. Transposons can cause mutations in genomes by two means. Recombination between transposons of the same family can cause deletion, inversion, duplication and translocation depending on the relative position and orientation of the elements that are recombining, whilst excisions and reinsertions of mobile elements can cause gene disruption directly if the element inserts into the coding or regulatory region of a gene.

Transposons in both classes have been found in plant-pathogenic fungi, for example the class I elements *grasshopper* and *MAGGY* in *M. grisea* and *skippy* in *F. oxysporum,* and the class II elements *Pot2, Pot3* in *M. grisea* and *Fot-1, Fot-2* and *impala* in *F. oxysporum*. Polymorphisms occur between isolates in the numbers and distribution of these elements, and there is evidence for chromosomal rearrangements caused by some of them, suggesting that they may have a significant role to play in the evolution of fungal genomes. Furthermore, in *F. oxysporum,* environmental stresses such as treatment with toxic compounds, have been shown to induce amplification of the *skippy* element and chromosomal rearrangements. It is believed that the rearrangement of genome sequences caused by transposons in response to stresses may be a general phenomenon in eukaryotic genomes because of its potential to occasionally result in increased fitness of the organism and a selective advantage.

4.9 Role of heterokaryosis

A unique feature of fungal and oomycete mycelia is that as they extend by apical growth through a substrate, they often come into contact with other mycelia, either of themselves or of other isolates of the same species or different species. In some cases these mycelia fuse and undergo a process known as anastomosis. This can only occur between vegetatively compatible groups, but it provides an opportunity for the exchange of genetic material such as nuclei, mitochondria, plasmids and viruses in the absence of sex. For successful anastomosis, strains must have the same vegetative incompatibility (*vic*) or heterokaryon (*het*) genes. If these genes are different in the strains, pre-fusion incompatibility, or more often post-fusion incompatibility occurs, resulting in a zone of cell death and an impenetrable barrier where the colonies meet.

One result of successful anastomosis is heterokaryosis, where two or more genetically different nuclei occur in the same cytoplasm, indeed the asexual stages of most basidiomycetes are heterokaryotic. If these nuclei fuse and genetic recombination occurs, novel variants of the fungus can be generated by what is referred to as the parasexual cycle. Whilst this process may be relatively rare and in most cases will not produce advantageous phenotypic or physiological alterations to the fungus, there may occasionally be spores produced that do have significant phenotypes.

4.10 Role of mitochondrial DNA

Mitochondria can also be transferred during anastomosis, and there are some examples of the effect of such transfer on fungal pathogenicity. One case is in the chestnut blight fungus, *Cryphonectria parasitica,* in which there are isolates referred to as 'hypovirulent', that have reduced growth rates, sporulation, aggression and colouration. These isolates are able to transmit this trait cytoplasmically to aggressive isolates and convert them into hypovirulent isolates. Initially, this property was found to correlate with the transfer of specific double-stranded (ds) RNA elements between isolates (see Section 4.11). However, isolates were subsequently found that lacked dsRNAs yet were able to transmit the property, and some of these isolates contain small mitochondrial plasmids (such as the 4.2 kb circular pCRY1 plasmid) associated with hypovirulence. Such plasmids are widespread in fungi, normally linear, and some have been shown to inactivate genes through integration into the mitochondrial chromosome. In another fungus, the conifer root and butt rot, *Heterobasidion annosum,* it has been shown that the origin of the mitochondria determines what type of virulence occurs. Those possessing S-type mitochondria cause disease on *Abies* and *Tsuga* species, whilst those with P-type mitochondria cause disease on *Pinus* species. Production of hybrid isolates has confirmed this dependence on mitochondrial origin.

4.11 Role of mycoviruses

The most common fungal viruses (mycoviruses) have dsRNA genomes, and are transmitted between fungi through anastomosis. In most cases these dsRNAs are encapsidated in virus-like particles, although some dsRNAs, such as those of *C. parasitica* appear to be in a form of fungal vesicle. Because mycoviruses are generally uninfectious, never leaving a cytoplasmic environment, they resemble plasmids of bacteria in their transmission characteristics. Most fungal species contain mycoviruses, and these viruses have been grouped into four taxonomic families, the *Totiviridae, Partitiviridae, Narnaviridae* and *Hypoviridae.* Often, their presence in a fungus has no obvious effect on the pathogenicity of that organism. However, the presence of such large amounts of apparently functionless genetic material (the *Puccinia* rusts for example appear to possess up to 50 kb of dsRNA in every isolate) has led others to suggest that they provide some selective advantage upon the host.

In the case of maize smut, *U. maydis,* a function has been identified. Here, as in some strains of yeast, certain dsRNAs encode the production of secreted killer toxins (*Figure 4.6*). Three killer strains, P1, P4 and P6 are known, which produce the KP1, KP4 and KP6 toxins respectively. Production of toxins is dependent on recessive nuclear genes (*p1r, p4r* and *p6r* respectively) and these also confer immunity. These genes are believed to encode toxin receptors, possibly the calcium channel proteins that these toxins block. The result is that in sensitive isolates, calcium channels and cAMP-regulated growth pathways are blocked to inhibit cell division. Sensitive isolates include isolates of

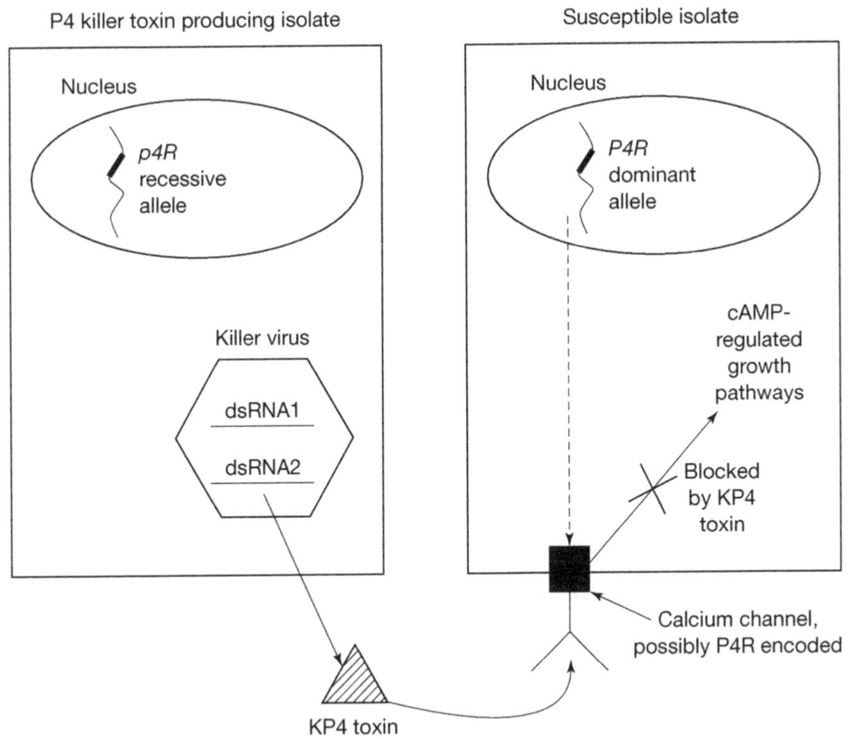

Figure 4.6

Mode of action of *Ustilago maydis* killer toxins. The KP4 toxin is encoded by the mycovirus genome. This isolate is immune to the toxin because it has the recessive allele of the *P4R* gene. Susceptible isolates have the dominant allele, which encodes the KP4 receptor.

U. maydis, and other members of the Ustilaginaceae that infect wheat, oats and barley, allowing killer strains to eliminate potential competitors from occupying the same niche. However, since only about 1% of isolates possess this capacity, it has been suggested that under other conditions, the presence of the killer system confers a selective disadvantage on the fungus.

In *C. parasitica,* the presence of dsRNAs is correlated in some isolates with hypovirulence, and there are other fungi in which the presence of dsRNAs has been associated with reduced virulence, such as *Rhizoctonia solani*. The dsRNAs in *C. parasitica* appear to interfere with signal transduction pathways, reducing the levels of the α subunit of the heterotrimeric G protein. This in turn results in the down-regulation of various pathogenicity genes such as reduced accumulation of the surface hydrophobin, cryparin, which is possibly responsible for a reduced ability to penetrate (see Section 2.8.2) and the resultant hypovirulent phenotype. A similar situation has been found in the Dutch elm disease fungus, *Ophiostoma novo-ulmi,* in which the hypovirulent fungi have drastically reduced amounts of cytochrome c oxidase in their mitochondria. It has been proposed that this is due to the presence of dsRNAs interfering with cytochrome c oxidase production, possibly through an RNA-quelling mechanism (see Section 8.8.3). It may be that in other isolates, the interference is caused alternatively by mutations within the mitochondrial cytochrome oxidase gene.

References and further reading

Bertrand, H. (2000) Role of horizontal DNA in the senescence and hypovirulence of fungi and potential for plant disease control. *Annual Review of Phytopathology* **38**: 397–422.

Casselton, L.A. (2002) Mate recognition in fungi. *Heredity* **88**: 142–147.

Gray, Y.H.M. (2000) It takes two transposons to tango: transposable-element-mediated chromosomal rearrangements. *Trends in Genetics* **16**: 461–468.

Keller, N.P. and Hohn, T.M. (1997) Metabolic pathway gene clusters in filamentous fungi. *Fungal Genetics and Biology* **21**: 17–29.

Laugé, R. and de Wit, P.J.G.M. (1998) Fungal avirulence genes: structure and possible functions. *Fungal Genetics and Biology* **24**: 285–297.

Leach, J.E., Vera Cruz, C.M., Bai, J. and Leung, H. (2001) Pathogen fitness penalty as a predictor of durability of disease resistance genes. *Annual Review of Phytopathology* **39**: 187–224.

Luderer, R. and Joosten, M.H.A.J. (2001) Avirulence proteins of plant pathogens: determinants of victory and defeat. *Molecular Plant Pathology* **2**: 355–364.

Martinez-Espinoza, A.D., Garcia-Pedrajas, M.D. and Gold, S.E. (2002) The Ustilaginales as plant pests and model systems. *Fungal Genetics and Biology* **35**: 1–20.

McCabe, P.M., Pfeiffer, P. and Van Alfen, N.K. (1999) The influence of dsRNA viruses on the biology of plant-pathogenic fungi. *Trends in Microbiology* **7**: 377–381.

McDonald, B.A. and Linde, C. (2002) Pathogen population genetics, evolutionary potentials, and durable resistance. *Annual Review of Phytopathology* **40**: 349–379.

Rosewich, L.U. and Kistler, H.C. (2000) Role of horizontal gene transfer in the evolution of fungi. *Annual Review of Phytopathology* **38**: 325–363.

Tyler, B.M. (2002) Molecular basis of recognition between *Phytophthora* pathogens and their hosts. *Annual Review of Phytopathology* **40**: 137–167.

Tyler, B.M. (2001) Genetics and genomics of the oomycete–host interface. *Trends in Genetics* **17**: 611–614.

Tudzynski, B. and Hölter, K. (1998) Gibberellin biosynthetic pathway in *Gibberella fujikuroi:* Evidence for a gene cluster. *Fungal Genetics and Biology* **25**: 157–170.

Walton, J.D. (2000) Horizontal gene transfer and the evolution of secondary metabolite gene clusters in fungi: an hypothesis. *Fungal Genetics and Biology* **30**: 167–171.

Bacterial diseases – establishing infection

5

Of the approximately 1600 known bacterial species, there are only about 100 that cause disease on plants. There are many more that can live on the leaf surface of plants or in the rhizosphere as saprophytes, and some of these may become internalised into plants, although there is no evidence that they replicate, and they cause no apparent damage to plants so are not considered as plant pathogens. However, like the mycotoxin-producing fungi, some of these are potentially harmful to consumers, and evidence has been accumulating of food-poisoning and death to humans through the consumption of food in which bacteria such as *E. coli* strain 0157 have become internalised.

Bacteria tend to enter plants either through wounds or natural openings, and then colonise the intercellular space and/or the xylem. Unlike fungi, bacteria are not able to directly penetrate the cuticle of plants and most are rod-shaped, Gram-negative, and possess flagella. *Table 1.3* in Chapter 1 lists the different symptoms and characteristics associated with the main groups of plant-pathogenic bacteria. Most molecular research into infection and pathogenicity has been performed on the more amenable *Pseudomonas, Xanthomonas, Erwinia, Pantoea, Ralstonia* and *Agrobacterium* species, and it is the molecular mechanisms governing the pathogenicity of these that form the focus for Chapters 5 and 6.

5.1 Bacterial–bacterial communication – quorum sensing

Prior to colonisation of plants, bacteria that are potential pathogens generally inhabit the leaf surface or the soil, often at low density as saprophytes. In some cases, it appears that for colonisation to occur high density and/or the presence and expression of essential virulence factors are required, and one mechanism that has been identified for controlling this is quorum sensing. Quorum sensing is a common communication mechanism used by bacteria that enables them to sense population density and respond through the regulation of expression of particular genes. In Gram-negative systems the bacteria produce autoinducers, which are diffusible signal molecules that can easily pass in and out through bacterial membranes. At high cell density, these reach a threshold level within the external environment that is detected by the bacteria and this results in the regulation of gene expression (*Figure 5.1a*). In Gram-positive bacteria, the system is different, involving modified oligopeptides secreted via ABC transporter mechanisms and detected via two-component histidine kinase signal transduction systems.

The key protein components of Gram-negative systems are the LuxI family of AHL synthases and the LuxR family of transcriptional activators.

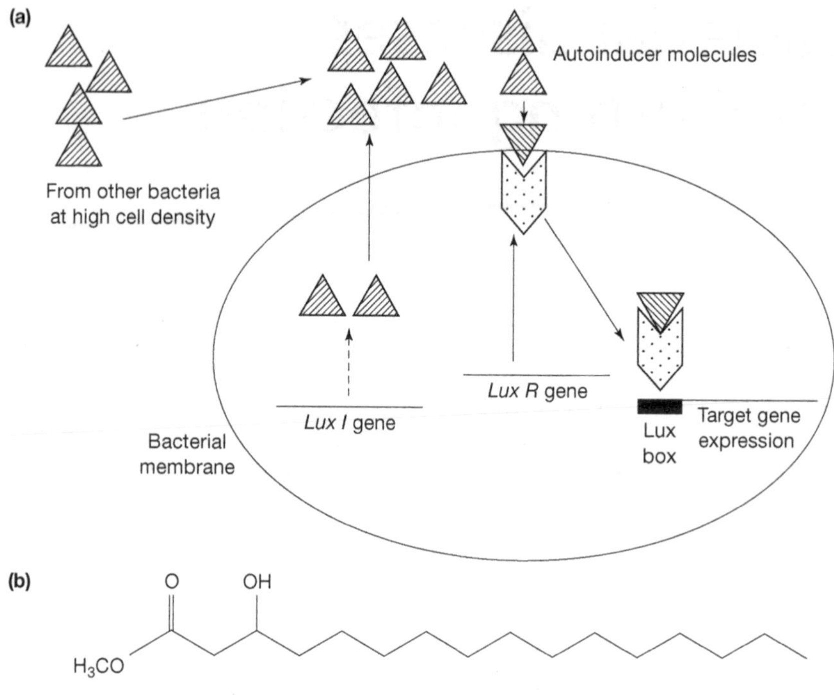

Figure 5.1

The *LuxI/LuxR* quorum sensing system of Gram-negative bacterium. (a) *LuxI* results in autoinducer production, which is detected at high cell density by the *LuxR*-encoded transcriptional activator. (b) The *Ralstonia solanacearum* signal molecule 3-OH PAME.

The AHL synthases catalyse the formation of AHL (N-acylhomoserine lactone) signal molecules from S-adenosyl methionine (*Figure 5.1b*), in which the N-acyl chains (4–14 carbons long) are provided via acyl–acyl carrier proteins or acyl-coenzyme A. The LuxR transcriptional activator proteins function as a dimer and possess an amino-terminal membrane-bound regulatory domain that binds the AHLs, and a carboxy-terminal DNA-binding domain. This domain interacts directly with a target sequence, the 'lux' box that is present in the regulatory regions of specific genes and results in activation of transcription.

In many cases, further regulatory complexities have been added to this basic LuxI/LuxR signal–response circuit, and in the case of plant-pathogenic bacteria, these often involve signals emanating from the plants. Amongst the quorum-sensing systems that have been identified in plant-pathogenic bacteria are the regulation of cell-wall-degrading enzymes (CWDEs) and exopolysaccharides (EPSs) in *Erwinia* spp., *Pantoea* spp. and *Ralstonia solanacearum* (see Sections 5.5 and 5.8), and the conjugal transfer of the Ti (tumour inducing) plasmid in *Agrobacterium tumefaciens* (see Section 5.2.2). A further feature of AHL quorum-sensing mechanisms is that some plants such as peas appear to have developed the capacity to secrete exudates that mimic AHLs. These may act to promote AHL-regulated behaviour in some cases and inhibit them in others, and this is yet more evidence of the intricate signalling systems that exist between plants and microbes.

5.2 Plant penetration

5.2.1 Foliar bacteria

Bacteria utilise a range of strategies for entry into plants. The surfaces of leaves can support large populations of many bacterial species that may have arrived in the air, been splashed on or carried by insects. Many of these can multiply on the surface to form microcolonies or larger aggregates which may passively enter and exit plants through stomata and other natural openings, and through wounds, particularly in water droplets formed on the leaf surfaces. In some cases (e.g. bacterial brown spot of beans, *P. syringae* pv. *syringae*, and halo blight of oats, *P. syringae* pv. *coronafaciens*), it has been shown that the probability of disease occurring is directly related to the external population size, indicating that a passive process of internalisation is the most likely cause of disease. However, mutants of *P. syringae* pv. *tomato* in which pilus formation is defective are more readily washed from the surface of plants than wild-type bacteria, suggesting that in other bacteria some form of adhesion to plant surfaces may be necessary.

In some *Xanthomonas* bacteria, it appears that there is only limited capacity to increase bacterial numbers on the leaf surface. Some of these (e.g. black rot of crucifers, *Xanthomonas campestris* pv. *campestris*) therefore have a more active process for invading the plant, in this case through the hydathodes, which are the structures containing water pores located at leaf margins. Under suitable weather conditions, particularly early in the morning, copious amounts of fluid are exuded through the hydathodes that collect as guttation drops around leaf margins. The bacteria move chemotactically towards these drops, which are later drawn back in through the hydathodes carrying the bacteria with them into the vascular system. Mutants in the *rpf* genes (which have reduced synthesis of extracellular enzymes and polysaccharides), show a reduced ability to enter through hydathodes, which is presumed to be because they are less well protected against plant defence responses in the hydathode.

5.2.2 Soil-borne bacteria

Soil-borne pathogens, such as *A. tumefaciens* (crown gall) and *R. solanacearum* (bacterial wilt), which tend to be broad-host-range pathogens, enter through wound sites to which they are attracted by chemical signals. For *A. tumefaciens,* the initial pre-penetration event that occurs in the rhizosphere is conjugal transfer of the Ti plasmid. The presence of the Ti plasmid is essential for *A. tumefaciens* to cause disease, and isolates that lack it are soil-inhabiting saprophytes. Conjugal transfer ensures the spread of the Ti plasmid to these isolates, increasing the number of pathogenic isolates. The LuxI/LuxR proteins, known as TraI and TraR, induce expression of genes required for mating between bacterial cells and for mobilisation of the plasmid. This circuit also responds to opine hormone signals, produced in infected plants through the action of genes encoded by the T-DNA integrated into the plant genome (see Section 5.7). These opines either activate or inhibit a repressor of the *traR* gene depending on which opines they are and which strain of *A. tumefaciens* is present, ensuring that transfer only occurs at the plant/bacterial interface and that only the plasmid encoding particular opines will be transferred into new isolates. They do this because *traR* is on an operon (the *arc* operon) in the

Ti plasmid, the expression of which is controlled by AccR, a transcriptional repressor that responds to opines. Thus, in the absence of opines, AccR represses the *arc* operon and TraR is not produced at sufficient levels to activate the transfer genes.

Both *A. tumefaciens* and *R. solanacearum* possess swimming motility, mediated by flagella, structures that consist of a long helical filament anchored in the cell envelope by a flexible hook and basal body complex. The flagellar filament is a hollow tube comprised of about 20 000 copies of a single protein known as flaggelin (FliC), polymerised into a complex helix. Rotation of the flagellum is controlled by a motor switch composed of three proteins, FliG, FliM and FliN. The precise mechanisms of chemotaxis and motility in *A. tumefaciens* is unclear. Migration towards the wide range of sugars and amino acids that accumulate in the rhizosphere around plant roots has been demonstrated, and there is evidence that some strains may also be attracted to specific exudates from wounded plants such as acetosyringone and also to opines. Non-motile mutants are fully virulent and indistinguishable from wild-type using artificial inoculation procedures, but using indirect inoculation methods through soil, these mutants are unable to cause disease, suggesting that motility and chemotaxis are important in some cases for finding host plants.

In the case of *R. solanacearum*, it has been shown that motility is important for finding plants but not for movement within plants. Non-motile mutants created through gene disruption of *fliC* (the subunit of the flagellar filament) and *fliM* (the flagellar motor switch protein) are significantly reduced in their ability to cause disease. However, if inoculated directly into plants, they have normal virulence, again indicating that the key role of flagella is in the early stages of plant invasion and colonisation. *R. solanacearum* motility is regulated through PhcA, a Lys-R type global regulator that directly and indirectly controls expression of a number of genes, both positively and negatively, including those involved in extracellular polysaccharide (EPS) production (see Section 5.8) and plant cell wall degrading polygalacturonases (PGs) (see Section 5.5). The Lys-R global regulators are a family of genes identified in almost all bacteria, that act as dual regulators, able to activate and repress different promoters. They possess a helix-turn-helix DNA binding motif in their N terminus, along with additional motifs that may be involved in metabolite recognition or in multimerisation. The expression of PhcA is in turn controlled by environmental factors through a quorum sensing mechanism. Whilst PhcA induces expression of EPS biosynthetic genes (see Section 5.8), it represses expression of PG and motility genes, by reducing the transcription of the two-component regulator of these genes, known as PehSR (*Figure 5.2*). Thus, at low bacterial density (i.e. in the rhizosphere), *phcA* is not expressed, so motility and PG genes are expressed. As bacterial populations increase in the xylem, the quorum-sensing mechanism induces expression of *phcA*, resulting in the down-regulation of motility genes and up-regulation of EPS production.

5.3 Attachment

In *A. tumefaciens,* it has been shown that not only are motility and chemotaxis important for finding a wound site, but that the bacteria then progress through a site-specific attachment process. The process appears similar

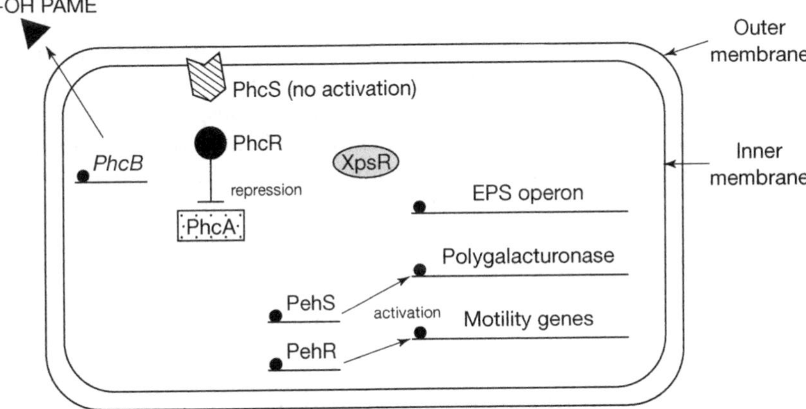

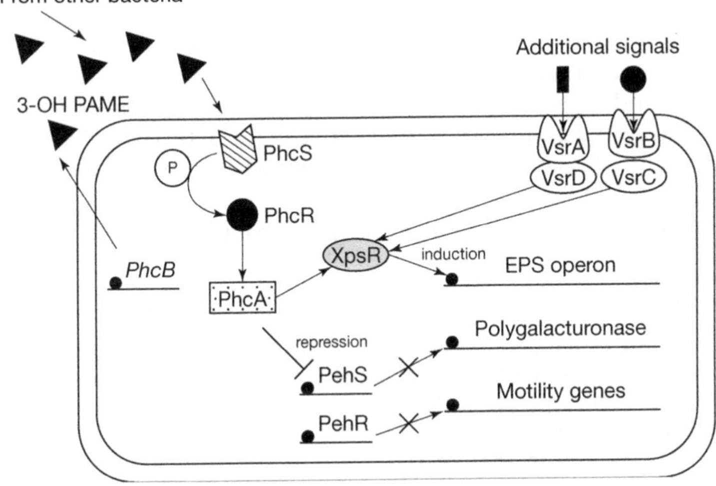

Figure 5.2

The role of the PhcA global regulator. At low cell density (a), the quorum sensing mechanism is inactive and motility and polygalacturonase genes are expressed. At high cell density (b), the quorum sensing mechanism is active and the EPS operon is induced.

to that used by mutualistic *Rhizobia* spp., although it is not clear whether site-specific attachment is required by other plant-pathogenic bacteria. In *A. tumefaciens,* attachment is a two-step process. Following an initial weak and reversible attachment, the bacteria synthesise cellulose fibrils that anchor them to the wounded plant cell surface. A number of bacterial chromosomal genes required for this process have been identified through mutant screening, notably *chvA, chvB, pscA* and *att,* and mutations in any of these genes results in a bacterium deficient in attachment. *chvA, chvB* and *pscA* appear to be involved in the synthesis and/or secretion of the cellular and extracellular polysaccharide β-1,2-D-glucan. Indirectly, this results in the production of an active version of the Ca^{2+}-dependent outer membrane protein rhicadhesin, a protein that has been shown to be important for attachment in rhizobium. *att* mutants are also deficient in a number of outer membrane proteins, but as yet there is no direct evidence that any of these proteins are involved in attachment. Plant molecules involved in the attachment process appear to be vitronectin-like molecules, as addition of anti-vitronectin antibodies or human vitronectin blocks attachment. Vitronectin is an adhesive glycoprotein component of the extracellular matrix normally involved in the cohesion of plant cells. This protein is likely to become exposed when plants are wounded, and it has been suggested that *Agrobacterium* utilises this for its attachment, which in turn stimulates the bacterium to produce cellulose fibrils that anchor it more firmly to the plant and also enable more bacteria to attach, producing a higher inoculum density.

5.4 Stimulation of gene expression in response to host factors

Having colonised a plant, bacteria need to obtain nutrients for their own growth and replication, and at the same time avoid or suppress the plant defence mechanisms. As with fungal pathogens, there are a wide variety of strategies that involve complex signalling between plants and pathogens and the stimulation and repression of various genes. Many of the mechanisms require secretion systems, and in Gram-negative plant-pathogenic bacteria, there are four main types of secretion system defined as shown in *Table 5.1*.

Table 5.1 Secretion systems of Gram-negative bacterial plant pathogens

System	Classification	Examples
Type I	Requires three or four accessory proteins to form a transmembrane channel through which the secreted proteins move No classical signal peptides	Protease secretion in *Erwinia chrysanthemi*
Type II	Sec-dependent pathway moves the proteins across the inner membrane, and further proteins, including a pilus, are involved in movement across the outer membrane Signal peptides required	Pectinase and cellulase secretion in *Erwinia*
Type III	Contact-mediated protein secretion directly from the bacterial cytoplasm. Numerous proteins involved including a pilus No signal peptide	Hrp and Avr proteins in many bacteria
Type IV	Sec-dependent pathway, in which signal peptide recognised by Sec is required for movement across inner membrane, and a complex including a pilus is required for Sec-independent secretion of DNA-protein across outer membrane	Transfer of T-DNA from *A. tumefaciens*

Two-component sensory systems are also important, and there are some examples of equivalents to the GacS/GacA two-component system which is a widespread and well-studied system in other Gram-negative bacteria, as shown in *Table 5.2*. This involves a histidine sensor kinase (*Figure 5.3*), and a conformational change in GacS caused by an environmental trigger results in transfer of a phosphate from a histidine to an aspartate residue and then to a second histidine prior to its transfer to an aspartate in the response

Table 5.2 Phenotypes controlled by GacS/GacA two-component systems in plant-pathogenic bacteria

Bacterium	GacS homologue	GacA homologue	Function
Erwinia carotovora	ExpS	ExpA	Controls expression of cell wall degrading enzymes, pectate lyases, polygalacturonases and cellulases. Mutants are less pathogenic
Pseudomonas syringae	GacS (LemA)	GacA	Controls exoprotease, tabtoxin, syringomycin, syringolin, and N-acyl homoserine lactone production. Mutants are less pathogenic

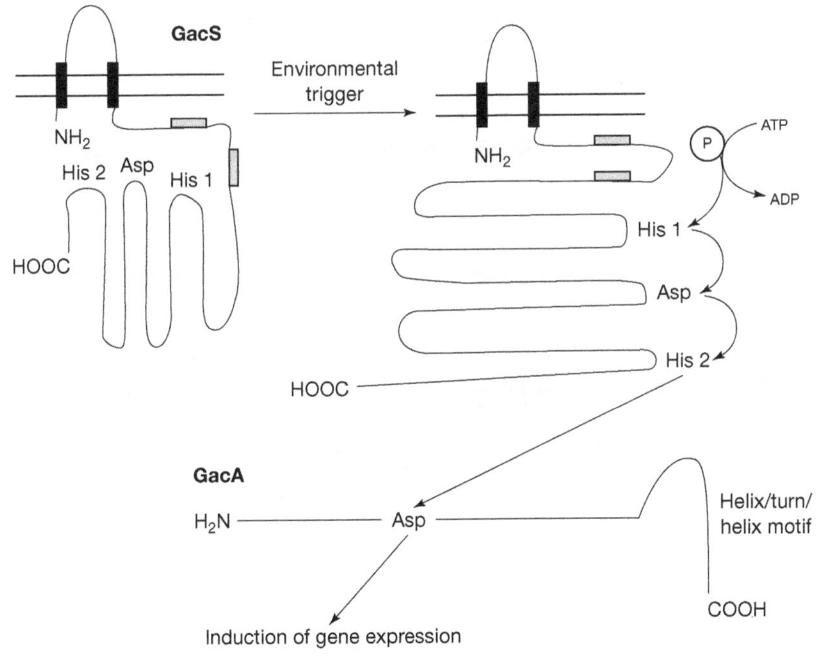

Figure 5.3

The bacterial GacS/GacA two-component histidine kinase signal transduction pathway.

regulator GacA. GacA contains a helix-turn-helix DNA-binding motif that regulates gene expression of as yet unknown genes.

Having noted some of the general mechanisms used by bacteria for signalling and secretion, how are these involved in their wide range of infection strategies, which include the production of cell-wall-degrading enzymes, toxins, exopolysaccharides or changes to the levels of plant hormones in the infected plant?

5.5 The role of cell-wall-degrading enzymes (CWDEs)

Necrotrophic bacteria such as the soft-rots *Erwinia carotovora* and *E. chrysanthemi* cause damage to plants by use of CWDEs to soften and macerate the plant tissue. However, whilst both pathogens use the same strategy, they do not use the same cocktail of enzymes. In *E. chrysanthemi*, pectinases (pectin methylesterases, pectate lyases, a pectin lyase and polygalacturonases), cellulases, four proteases and a phospholipase are produced. The pectate lyases (Pels) appear to be particularly important for soft rot and there are at least seven of these, of which five are major isozymes (PelA–PelE). Mutations in single *pel* genes do not eliminate the symptoms completely, and the evidence from these mutagenesis experiments suggests that each Pel makes a contribution to infection and that these relative contributions vary depending upon the host species, accounting for the wide host range of the pathogen.

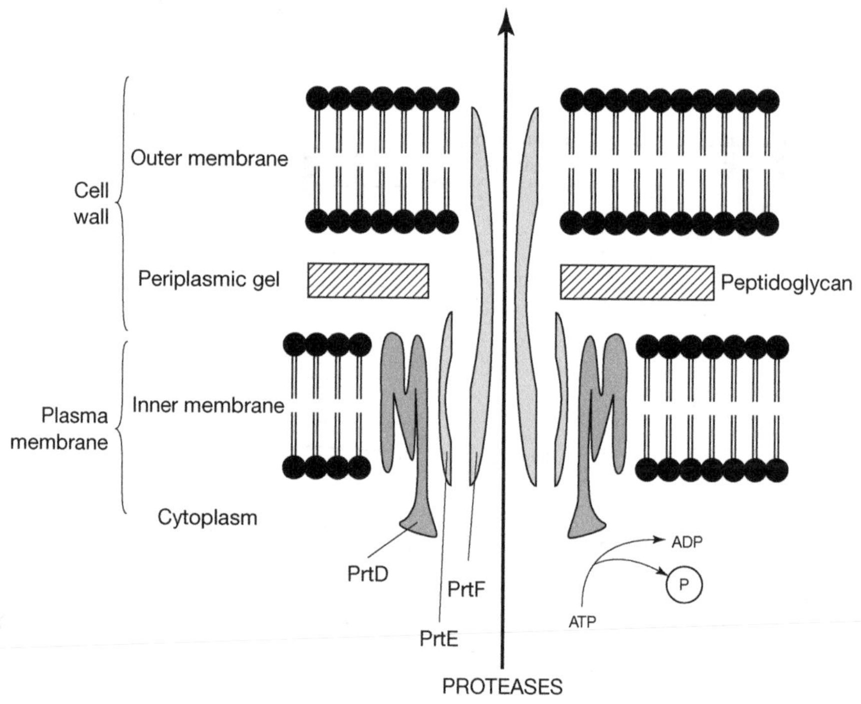

Figure 5.4

The Type I secretion mechanism for *Erwinia chrysanthemi* proteases. The PrtD, PrtE and PrtF proteins form a channel through which the proteases are secreted, driven by ATP hydrolysis.

For example, PelA is most important on peas, whilst all except PelE are important on chicory.

The CWDEs have to be secreted from the bacterium in order to reach the plant cell, and for the proteases, which contain C-terminal secretion signals, this is by a type I mechanism in which the products of the *prtD, prtE* and *prtF* genes form a channel through which the proteins pass, driven by ATP hydrolysis (*Figure 5.4*). By contrast, most of the pectinases and cellulases are secreted by a type II Out secretion mechanism (*Figure 5.5*). The *out* locus consists of 15 genes organised in five transcription units (*outS, outB, outT, outCDEFGHIJKLM* and *outO*), and the products of these genes have homology to gene products with similar functions from *E. coli*. This secretion requires the proteins to have a cleavable N-terminal signal peptide. The regulation of CWDE gene expression is complex, involving pectin, plant cell wall breakdown products, other plant factors, catabolite repression, environmental factors such as temperature, anaerobiosis, iron limitation, osmolarity and nitrogen starvation. The response to pectin breakdown products is particularly important, and experiments using fusions of *pel* gene promoters to the reporter gene *uidA* (GUS), have shown that the product of *pelE*, which has a high basal level of expression, starts the initial degradation of pectin, and that this subsequently induces expression of the other *pel* genes (*Figure 5.5*).

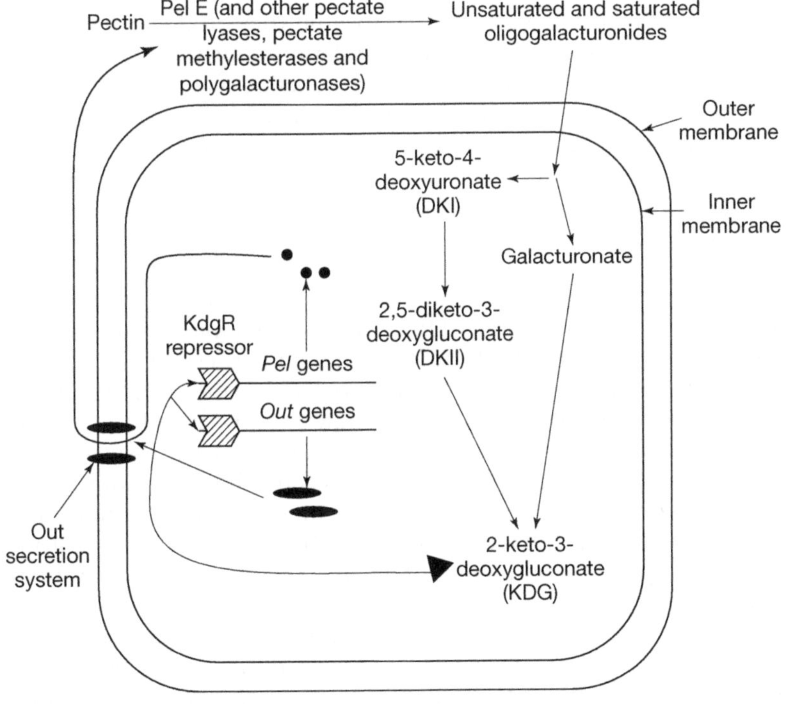

Figure 5.5

The role of 2-keto-3-deoxygluconate (KDG) in removal of the KdgR repression of pectinase genes, and the Out secretion mechanism. KdgR is normally bound to the conserved KdgR-box in the promoter region of some *out* and *pel* genes. Upon DKI, DKII or KDG binding, this repression is released, and the *out* and *pel* genes are expressed.

Many other signalling pathways are also involved in regulating CWDE expression, along with extensive cross-talk between them. For example, there are 'controller' genes that are required in addition to transcriptional regulators. One such gene is *dsbA,* which encodes a protein involved in controlling the proper folding of secreted and periplasmic proteins through its regulation of disulphide bond formation. Mutations in this gene result in defects in pectate lyase secretion and therefore virulence of *Erwinia* spp. Catabolite repression is believed to function through cAMP and a catabolite activator protein (CRP), and a quorum-sensing mechanism and LysR-type global regulators have also been identified in *E. chrysanthemi.*

5.6 The role of toxins

Toxins are a particularly important weapon used by many *P. syringae* pathovars, and whilst these are generally non-host-specific toxins that are not essential for pathogenicity, they induce chlorosis in plants and increase the severity of disease. The best studied bacterial toxins in plant pathology have been coronatine, tabtoxin, phaseolotoxin, syringomycin and syringopeptin (*Figure 5.6*).

Coronatine is produced by a number of *P. syringae* pathovars such as *glycinea, maculicola, atropurpurea, morsprunorum* and *tomato,* which infect soybean, crucifers, ryegrass, *Prunus* spp. and tomato, respectively. *Tn5* insertion mutants of various pathovars that are coronatine-deficient have been produced, and whilst these are still pathogenic, they produce smaller necrotic lesions and reach smaller bacterial population densities. Coronatine consists of two structural components, the polyketide coronafacic acid and the amino acid derivative coronamic acid, produced as shown in *Figure 5.6a.* In *P. syringae* pv. *glycinea,* the biosynthetic genes are clustered in a 32 kb region of a 90 kb plasmid. This cluster can be separated into two regions. The CMA region consists of four genes for CMA biosynthesis, and the CFA region is a single transcriptional unit containing 11 open reading frames. Between these two domains is a regulatory region, and expression is regulated by temperature, with maximal expression occurring at 18°C. Regulation is through a two component histidine protein kinase regulatory system, and there is also evidence that plant factors stimulate coronatine production through as yet unknown mechanisms. Coronatine has structural and functional homologies with methyl jasmonate (MeJA), which is a signalling molecule produced by higher plants (see Section 11.7). However, the cell wall thickening, changes in chloroplast structure, and chlorosis induced by coronatine are more severe than the affects of MeJA, indicating that it is more than merely a mimic. It has been suggested that one function of coronatine may be to suppress the induction of defence-related genes by plants, allowing greater pathogen ingress and multiplication.

Tabtoxin can also be synthesised by a wide range of *P. syringae* pathovars, such as *tabaci* (wild fire of tobacco) and *coronafaciens* (halo blight of oats). It is a monocyclic β-lactam (*Figure 5.6b*) of which the tabtoxine-β-lactan (Tβl) moiety generated through aminopeptidase cleavage is the active part. Genes for toxin biosynthesis are clustered in a 31 kb region of the genome, which appears to be quite unstable, excising at high frequency and resulting in spontaneous *tox*-minus strains that produce milder symptoms on plants. The toxin appears to be synthesised from the lysine biosynthetic pathway, regulated through a histidine protein kinase by environmental factors, and

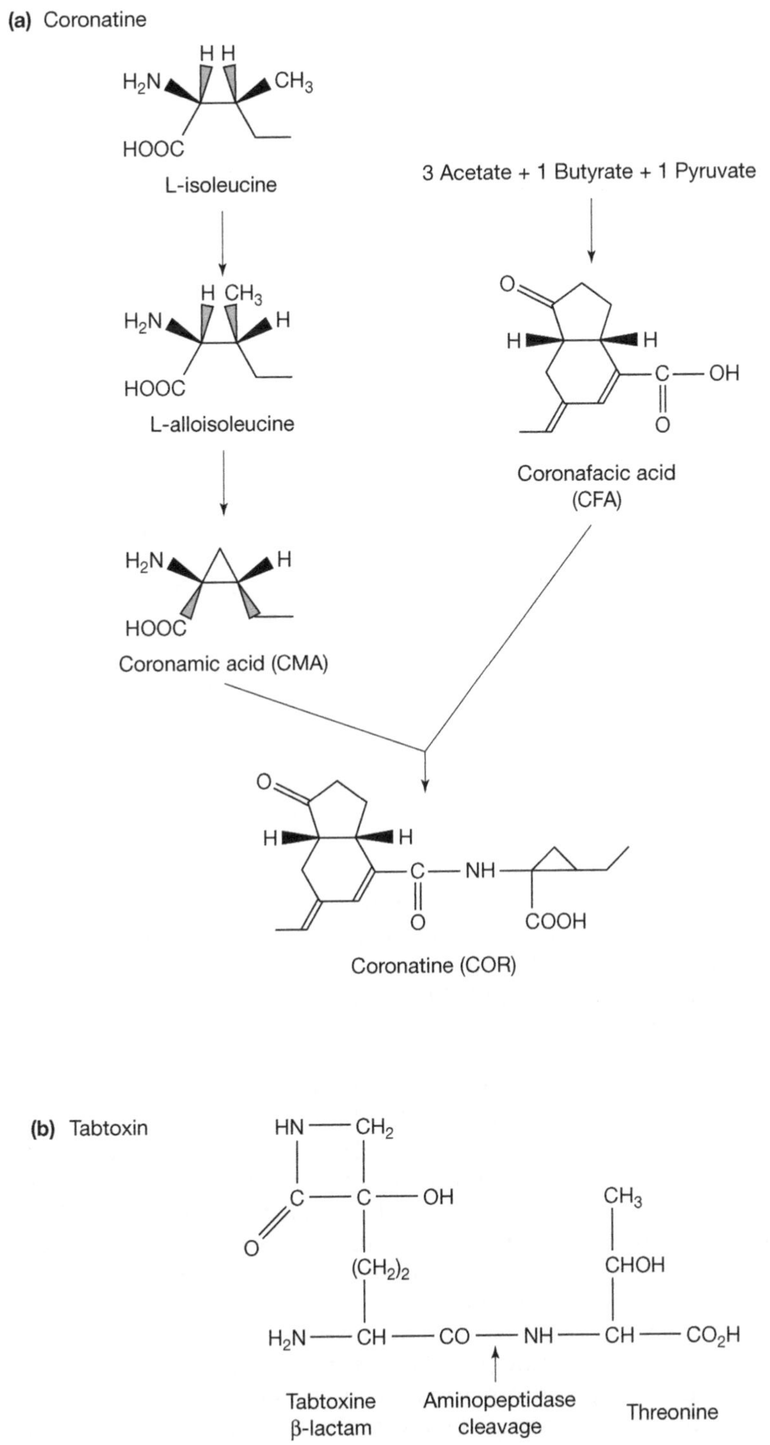

(a) Coronatine

H_2N HH CH_3 HOOC

L-isoleucine

3 Acetate + 1 Butyrate + 1 Pyruvate

H_2N H CH_3 H HOOC

L-alloisoleucine

Coronafacic acid (CFA)

H_2N H HOOC

Coronamic acid (CMA)

Coronatine (COR)

(b) Tabtoxin

HN — CH_2 C — C — OH O $(CH_2)_2$ H_2N — CH — CO — NH — CH — CO_2H CHOH CH_3

Tabtoxine β-lactam

Aminopeptidase cleavage

Threonine

Figure 5.6

The chemical structures of some bacterial non-specific toxins. (a) The formation of coronatine by *P. syringae*. (b) Tabtoxin, (c) Phaseolotoxin and (d) Syringomycin.

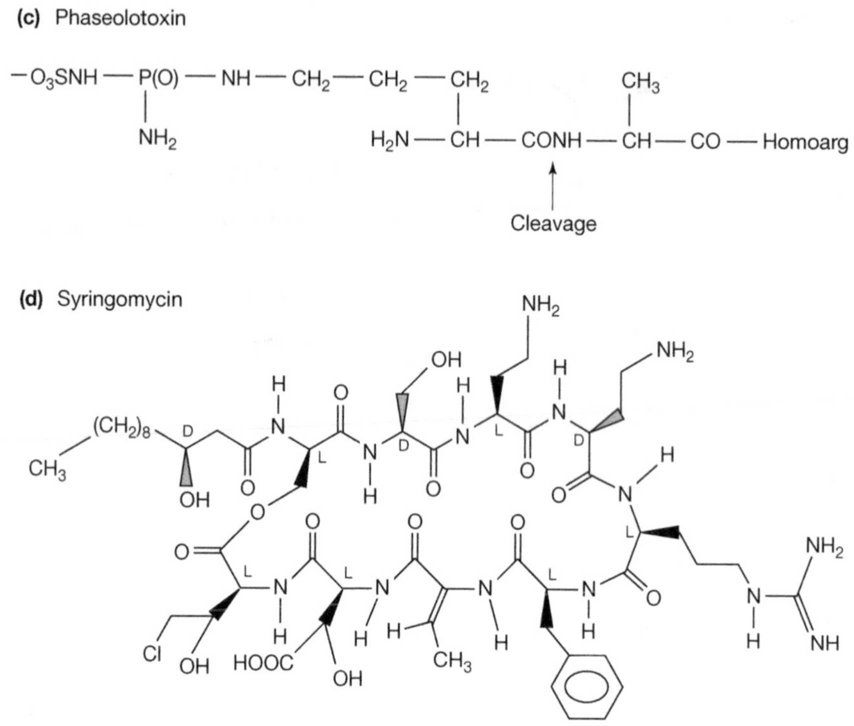

(c) Phaseolotoxin

(d) Syringomycin

Figure 5.6

(Continued)

acts by irreversibly inhibiting glutamine synthase. This enzyme is required in plants for detoxification of ammonia, and the accumulation of ammonia causes disruption of thylakoid membranes in the chloroplasts, uncoupling of photophosphorylation and chlorosis.

Phaseolotoxin is produced by two pathovars of *P. syringae*, pv. *phaseolicola* (halo blight on legumes) and pv. *actinidiae* (bacterial canker on kiwifruit) (*Figure 5.6c*). Synthesis is temperature regulated occurring more readily at lower temperatures, and the toxin inhibits ornithine carbomyltransferase (OCTase) in plants, resulting in accumulation of ornithine and a deficiency in arginine, that causes the chlorosis. *P. syringae* pv. *phaseolicola* copes with the toxic affects of OCTase inhibition by producing two isozymes, ROCTase (resistant to toxin) and SOCTase (sensitive to toxin). When the bacteria is actively synthesising the toxin, it produces the ROCTase isozyme to ensure that it does not succumb to the toxin.

Syringomycin (*Figure 5.6d*) and syringopeptin, produced by *P. syringae* pv. *syringae* are lipopeptide phytotoxins that cause necrotic symptoms on plants. They form pores in plasma membranes resulting in ion fluxes and lysis of cellular membranes. The biosynthetic genes for the two classes of toxins (which include large peptide synthetase genes) are clustered adjacent to each other in the bacterial genome in a 135 kb genomic pathogenicity island. Regulation of expression appears to be in response to plant phenolic glycosides such as arbutin, enhanced by sugars, and involves a complex regulatory cascade that has not been fully characterised.

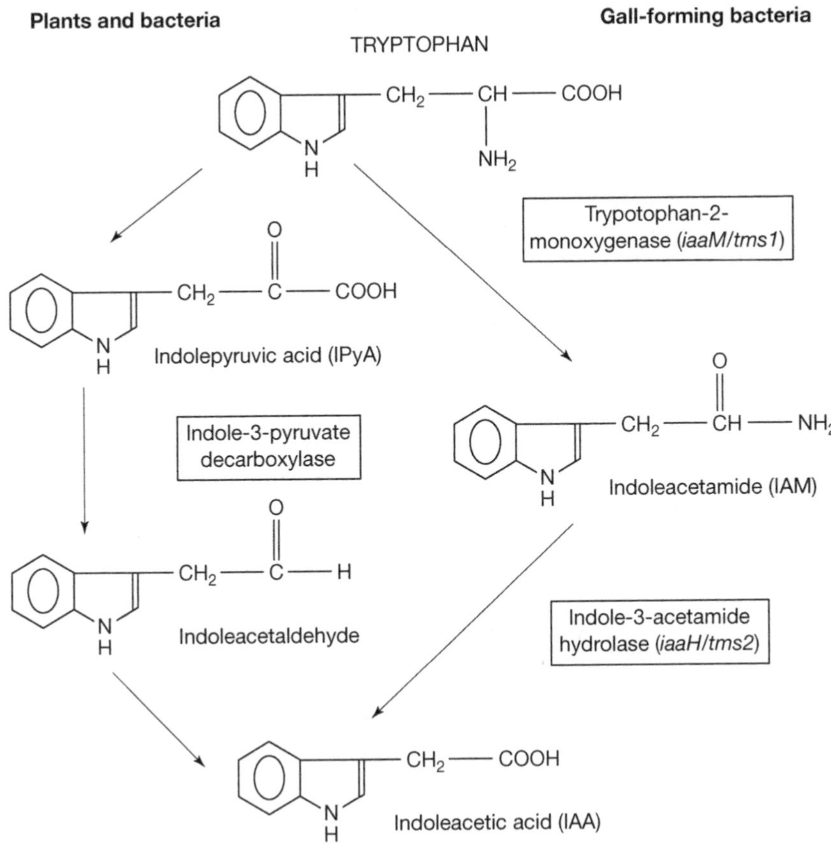

Figure 5.7

The two pathways for production of auxin (indoleacetic acid) from tryptophan. The enzymes involved are indicated in boxes, with *iaaM* and *iaaH* being the genes that encode these in *P. syringae,* and *tms1* and *tms2* the equivalent genes in *A. tumefaciens.*

5.7 The role of hormones

Alterations in levels of plant hormones are particularly important for those bacteria that cause uncontrolled proliferation of plant tissue, resulting in galls and knots, such as *P. syringae* pv. *savastanoi* (olive and oleander knot), *A. tumefaciens* (crown gall), *Pantoea herbicola* pv. *gypsophilae* (galls on table beet and *Gypsophila paniculata,* a perennial flower), and the Gram-positive nocardiform bacterium *Rhodococcus fascians* (leafy galls on many plants).

In these, *Pseudomonas, Agrobacterium and Pantoea* species, it is production of the auxin indole-3-acetic acid (IAA) that is important for pathogenesis. IAA is synthesised from tryptophan by two main routes in bacteria, via indoleacetamide (IAM) or via indolepyruvic acid (IpyA) (*Figure 5.7*). In many non-gall-forming isolates of *P. syringae* and *P. herbicola,* low levels of IAA are produced via the IpyA pathway, which is generally assumed to be the common pathway for IAA biosynthesis in plants. The IAM pathway however, appears to be quite rare in plants and this is the pathway used by the gall-forming *Pseudomonas* spp., *Pantoea* spp. and *Agrobacteria*. It has been suggested that this is because plants do not have the regulatory mechanisms to control the IAM pathway, allowing the pathway to proceed in an uncontrolled way in infected plants, resulting in the proliferation of plant tissue.

In *P. syringae* pv. *savastanoi* (olive strains), the *iaaM* and *iaaH* genes required for this pathway are chromosomally located, whilst in oleander strains and in *P. herbicola* pv. *gypsophilae,* the genes are on plasmids. These bacteria infect the hosts through wounds and it is believed that gall production depends on expression of *iaaM* and *iaaH* by bacteria *in planta* in combination with an *hrp* type III secretion mechanism (see Section 6.3), and that this is stimulated by factors from wounded plants, since the bacteria are able to live as epiphytes on unwounded plants without inducing symptoms. In A. *tumefaciens,* the equivalent genes (*tms1* and *tms2)* are present on the Ti (tumour inducing) plasmid, and the approach used for expression of these genes in plants (and also by the related species *A. rhizogenes* and *A. vitis*) involves the transfer of the section of DNA on which these genes reside, known as the T-DNA, from the genome of the plasmid into that of the plant (*Figure 5.8*). Because of the value of *A. tumefaciens* in biotechnology and

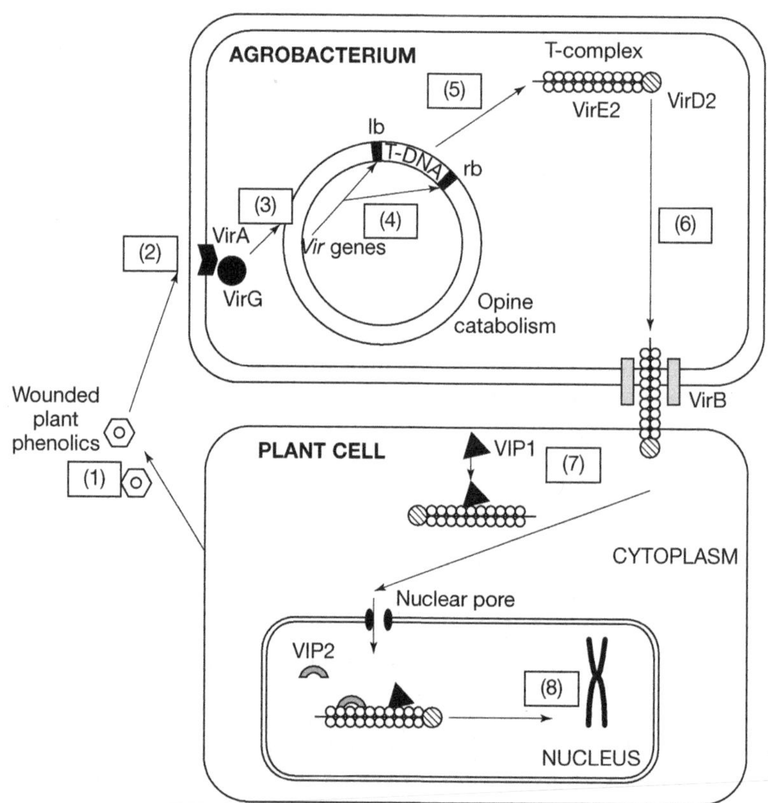

Figure 5.8

The *Agrobacterium tumefaciens* infection process. Phenolics (1) are detected by VirA/VirG (2). This induces *vir* gene expression (3). The T-DNA is excised from the plasmid (4), and the T-complex (5) is then transported out of the bacterium (6). Once in the plant cytoplasm, the complex is targeted to the nucleus (7), and integrates into DNA (8).

genetic manipulation of plants, this process has been extensively studied, and will be described here.

Following attachment of *A. tumefaciens* to wounded plant cells (see Section 5.3), it is genes on the Ti plasmid that are essential for mobilisation and transfer of the T-DNA into the genome of the plant. The *Vir* region (which occupies between 28 and 44 kb of this 180 kb plasmid, depending on the strain) contains between six and eight operons, *virA–virH* (nopaline strains lack *virF* and *virH*). The *virA* and *virG* operons are responsible for a two-component histidine kinase regulatory system, in which VirA is a membrane sensor protein, and VirG a cytoplasmic response regulator. VirA detects the presence of wounded plant exudates, including lignin and flavonoid precursors such as acetosyringone, either directly or through two chromosomally encoded phenolic binding proteins, P10 and P21. Monosaccharides, such as glucose and arabinose lower the concentrations of these phenolics required for activation of *vir* gene expression, through the activity of ChvE, a chromosomally encoded sugar-binding protein that interacts with the periplasmic domain of VirA. The signal is transmitted through the transmembrane domain of VirA to the kinase domain, which is normally autophosphorylated at histidine residue 474. The signal results in this phosphate transferring to an aspartate residue on VirG, and the activated VirG protein then specifically interacts with 12 bp DNA sequences (the *vir* box) located in the promoter regions of the *vir* genes, including itself, resulting in amplification of the signal. A further regulator of *virG* gene expression that has recently been identified is the chromosomal *chvD* gene. This appears to act as an ABC (ATP-binding cassette) transporter, though its role in regulation has yet to be elucidated.

Expression of the *vir* genes leads to production of a single-strand copy of the T-DNA (the T-strand) and transfer of this into the plant genome. The sizes of T-DNA elements vary from a single 22 kb T-DNA in nopaline strains, to three independently transferred T-DNAs of 13, 1.5 and 7.8 kb in octopine strains. These T-DNAs are all flanked by conserved 25 bp sequences known as the right and left borders, with the right border essential for transfer. To release the T-strand, a complex of VirD2 and VirD1 proteins nicks the DNA of the opposite strands between the third and fourth base of the right and left borders, and relaxes the supercoiled plasmid. VirD2 then covalently attaches to the exposed 5′ ends and the protein bound at the right border removes the T-strand, whilst that attached at the left border promotes the bacterial DNA machinery to repair the gap in the plasmid. As the T-strand is removed, it becomes surrounded by molecules of the single-strand DNA binding protein VirE2 to form a semi-rigid, hollow, cylindrical filament (the T-complex). This complex, resembling a single-strand DNA bacteriophage, is then exported by a type IV secretion mechanism.

For export of the T-DNA, the T-transporter is assembled from 11 proteins encoded by the *virB* operon and VirD4 (*Figure 5.9*). The resultant pilus is believed to act as a conduit that opens in response to plant signals (possibly contact with putative receptors on the plant cell surface) to allow transfer of the T-complex into the cytoplasm of the plant. Once in the cytoplasm, the T-complex has to pass through the nuclear envelope into the nucleus. Import of proteins into the nucleus of plants is a tightly regulated process in which large proteins normally require a nuclear localisation signal. In the T-complex, these are present in both VirD2 and VirE2, enabling the complex to pass into the nucleus through the nuclear pores. VirD2 is believed to

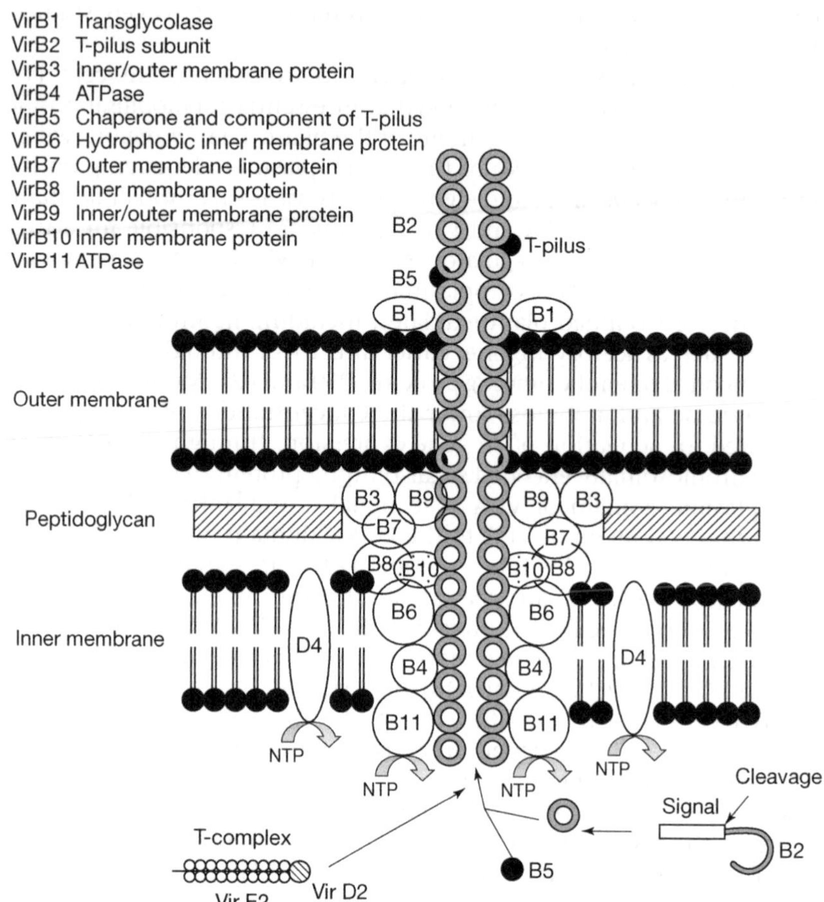

Figure 5.9

The formation of the T-pilus type IV transporter in *A. tumefaciens.* The role of the 11 VirB proteins is indicated. VirB1 initiates the assembly. VirB2 is the major structural protein in the pilus and is cyclised following removal of the signal sequence. The other VirB proteins are involved in forming the transport complex that promotes pilus assembly.

target the T-complex to the nuclear pore. VirE2 then helps mediate passage through the pore, and has been shown to be particularly important for the transfer of large ssDNA substrates.

The T-DNA then becomes integrated into the genome of the plant essentially at random. Because the T-DNA complex does not encode any of the enzymatic activities necessary for this to occur, it is presumed that this is controlled by host factors, and mutants of *Arabidopsis* that are resistant to *A. tumefaciens* transformation at different stages are currently being characterised to determine what genes and processes are required for this process. From this, a VIP1 basic zipper transcription factor has been identified that appears to be involved in facilitating entry of the T-complex into the nucleus and its association with chromatin, whilst a second protein, VIP2 may be involved in targeting this to transcriptionally active DNA. Once integrated into the plant genome, the auxin (*tms1* and *tms2*) and cytokinin (*ipt*) biosynthetic genes present on the T-DNA are expressed within the plant, resulting in uncontrolled proliferation. Opine biosynthetic genes on the T-DNA are also expressed, and these opines are used as a source of nutrients by the bacterium as it proliferates in the gall.

An homologous isopentenyltransferase (*ipt*) gene has also been identified in *Rhodococcus fascians* as being important for cytokinin biosynthesis. This pathogen causes fasciation, a disease characterised by loss of apical dominance and the development of multiple shoots giving rise to leafy galls. The *ipt* gene is present on a 200 kb linear plasmid, and its expression is induced by plant extracts, but the mechanisms through which this occurs have yet to be identified. Furthermore, levels of cytokinins do not appear to correlate with *ipt* expression, suggesting that other factors and perhaps other cytokinins are produced by the bacterium.

5.8 The role of extracellular polysaccharides (EPSs)

Extracellular polysaccharides (EPSs) are large polymers that are important for many phytopathogenic bacteria, such as *R. solanacearum, X. campestris, P. stewartii* and *E. amylovora* as both capsules around the bacteria and as fluidal slime released by the bacteria. They usually consist of a mixture of sugars precisely arranged in repeating subunits. For example, EPS I of *R. solanacearum* consists of a 10^6 Da polymer with a trimeric repeat unit of N-acetyl galactosamine, 2-N-acetyl-2-deoxy-L-galacturonic acid and 2-N-acetyl-4-N-(3-hydroxybutanoyl)-2-4-6-trideoxy-D-glucose. The role of EPS in bacteria appears to be two-fold, and in some bacteria, mutants that are deficient in them have reduced virulence. They provide a barrier against desiccation for bacteria on the surface of leaves or in the rhizosphere, and secondly they provide a defence against toxic plant compounds and induced host defences. The consequence of EPS accumulation in the plant is the formation of large water-soaked lesions and/or blockage of the xylem causing wilt, although how this benefits the bacteria is not clear. There are many genes required to encode the biosynthetic machinery for EPS production and for the transport of these to the cell surface, and there are well-characterised environmentally sensitive regulatory systems to ensure that this energy-intensive process is utilised at the correct time of plant invasion.

In *R. solanacearum*, the bacteria enter the xylem as described in Section 5.2.2, and then rapidly spread upward into the stem. During this process, and when bacterial cell densities are low, the cells are motile and highly pectolytic. However, as the density of bacteria increases in the xylem, the global regulator PhcA is expressed, inducing EPS production and reducing motility (see Section 5.2.2). Genes for the biosynthesis of EPS I (the principal EPS) are encoded in a 16 kb gene cluster, and mutations in these genes have shown the importance of their role in pathogenicity. Control of the PhcA transcriptional regulator is through the quorum sensing system detailed in *Figure 5.2*.

In *P. stewartii*, the causal agent of Stewart's wilt and leaf blight of sweet corn, the bacterium colonises the xylem after entry through leaves, and produces the EPS stewartan. The *cps* gene cluster that encodes the biosynthetic pathway is similar to that in *E. coli* for colanic acid biosynthesis, and also that of amylovoran biosynthesis in *E. amylovora*, the causal agent of fireblight, a highly destructive disease of pears and apples. These gene clusters contain genes encoding sugar transferase activities, a lipid-carrier transferase and components of a postulated ABC transporter system believed to be responsible for secretion. Regulation of these clusters was believed to be through two-component membrane-bound sensor/histidine kinase transmitter systems, in which the environmental sensor RcsC phosphorylates RcsB, which then binds the enhancer protein RcsA, and the resulting heterodimer binds to the promoter regions and stimulates transcription. However, evidence in *P. stewartii* indicates that *cps* gene expression and stewartin production are also controlled by a quorum-sensing system at high cell densities, with LuxI and LuxR homologues designated EsaI and EsaR.

Xanthomonas campestris pv. *campestris,* the causal agent of black rot in crucifers, also produces an EPS known as xanthan gum, which is used commercially in the food industry. The *gum* locus for biosynthesis and export of xanthan contains 12 open reading frames (*gumB* to *gumM*), and *gum* mutants have reduced pathogenicity, producing smaller and slower-developing lesions. Regulation of these genes also appears to involve a number of mechanisms including a two-component system and a quorum-sensing autoregulation system.

It is clear that all of these bacteria have mechanisms for regulating EPS production in response to cell density, and it is believed that, as in *R. solanacearum,* this may be a mechanism to ensure that EPS is not produced in large quantities when the bacteria are outside the host plant, so that they are motile and able to swim towards plants, and that they are only produced once the bacteria are in the plant, as a protection against plant defences. Additionally, activation only when bacteria reach a high density may help to ensure that the defence responses of the plant are not triggered too early in pathogen invasion, and that the bacteria have time to prepare themselves for a 'mob' approach to overcome these defences.

Lipopolysaccharides (LPSs) have also been found to be important in *R. solanacearum* pathogenicity, and mutants with altered LPS composition may be unable to survive in plants making them effectively non-pathogenic. They are also important in *X. campestris,* where genes for their production appear to be clustered in a 35 kb region of the *Xanthomonas* genome, which has all the hallmarks of a pathogenicity island (see Section 6.9).

References and further reading

Basler, B.L. (2002) Small talk: Cell-to-cell communication in bacteria. *Cell* **109**: 421–424.

Bender, C.L. (1999) Chlorosis-inducing phytotoxins produced by *Pseudomonas syringae. European Journal of Plant Pathology* **105**: 1–12.

Cao, H., Baldini, R.L. and Rahme, L.G. (2001) Common mechanisms for pathogens of plants and animals. *Annual Review of Phytopathology* **39**: 259–284.

Denny, T.P. (1999) Autoregulator-dependent control of extracellular polysaccharide production in phytopathogenic bacteria. *European Journal of Plant Pathology* **105**: 417–430.

Heeb, S. and Haas, D. (2001) Regulator roles of the GacS/GacA two-component system in plant-associated and other Gram-negative bacteria. *Molecular Plant Microbe Interactions* **14**: 1351–1363.

Hentschel, U., Steinert, M. and Hacker, J. (2000) Common molecular mechanisms of symbiosis and pathogenesis. *Trends in Microbiology* **8**: 226–231.

Hugouvieux-Cotte-Pattat, N., Condemine, G., Nasser, W. and Reverchon, S. (1996) Regulation of pectinolysis in *Erwinia chrysanthemi. Annual Review of Microbiology* **50**: 213–257.

Lai, E-M. and Kado, C.I. (2000) The T-pilus of *Agrobacterium tumefaciens*. *Trends in Microbiology* **8**: 361–369.

Loh, J., Pierson, E.A., Pierson, L.S., Stacey, G. and Chatterjee, A. (2002) Quorum sensing in plant-associated bacteria. *Current Opinion in Plant Biology* **5**: 285–290.

Schell, M.A. (2000) Control of virulence and pathogenicity genes of *Ralstonia solanacearum* by an elaborate sensory network. *Annual Review of Phytopathology* **38**: 263–292.

Ward, D.V., Zupan, J.R. and Zambryski, P.C. (2002) *Agrobacterium* VirE2 gets the VIP1 treatment in plant nuclear import. *Trends in Plant Science* **7**: 1–3.

Withers, H., Swift, S. and Williams, P. (2001) Quorum sensing as an integral component of gene regulatory networks in Gram-negative bacteria. *Current Opinion in Microbiology* **4**: 186–193.

Zupan, J., Muth, T.R., Draper, O. and Zambryski, P. (2000) The transfer of DNA from *Agrobacterium tumefaciens* into plants: a feast of fundamental insights. *The Plant Journal* **23**: 11–28.

Bacterial diseases – determinants of host specificity

6

Having established the basic requirements for bacterial entry into plants, and some of the factors that determine pathogenicity, we can now consider host-specific determinants, and address why plants are able to defend themselves against some bacteria but not others. The same basic gene-for-gene concept applies to many bacterial pathogens as for fungal pathogens (see Section 4.1). Bacteria, particularly those that use stealth as their means of attack, possess avirulence genes (*avr*), expression of which culminates in the production of elicitors, and plants possess resistance genes, the products of which are able to recognise these elicitors and respond. In bacteria, it has been found that *avr* gene function is generally dependent on the function of a further group of genes, the *hrp* genes (*h*ypersensitive *r*esponse and *p*athogenicity) that are responsible for their export by a type III secretion mechanism. It is the nature of the avirulence genes and these secretion mechanisms, along with the genetics of bacterial plant pathogens that we focus on in this chapter.

6.1 The cloning of avirulence genes

Prior to the identification of type III secretion mechanisms and *hrp* genes, the first bacterial avirulence gene, *avr6* was cloned in 1984 from *Pseudomonas syringae* pv. *glycinea*. Since then, more than 50 *avr* genes have been cloned from a range of pathovars of *P. syringae* and *X. campestris*, the main bacteria that exhibit gene-for-gene resistance systems.

6.2 The products of avirulence genes

Based on initial sequence analysis of *avr* genes and gene products from a range of bacteria, it soon became apparent that there was no clear motif that defined these proteins and that there was often no significant homology to genes of known function held in databases. *Table 6.1* lists some of the putative functions of these gene products based on sequence homologies. Of these, AvrBs2 is unique in having a defined predicted enzymatic role, being similar to agrocinipine synthase from *A. tumefaciens*. Members of the AvrRxv type have sequence similarity to the YopJ and YopP proteins of the mammalian pathogens *Yersinia pseudotuberculosis* (tuberculosis) and *Y. pestis* (bubonic plague), that have been shown to inhibit MAP kinase kinases, and these may be involved in suppression of plant defence responses that operate via these kinase cascades (see Section 11.2). AvrD of *P. syringae* also appear to have a defined role, in the synthesis of low-molecular-weight syringolide elicitors.

Table 6.1 Features of cloned bacterial avirulence genes

Gene	Resistance gene	Structural information
X. campestris		
pv. *vesicatoria*		
avrBs1	*Bs1* (pepper)	Two open-reading frames (ORFs) required
avrBs2	*Bs2* (pepper)	Two ORFs required. Related to agrocinipine synthase and glycerol-phosphodiesterase
avrBs3	*Bs3* (pepper)	Nuclear localisation signal/*hrp*-dependent
avrRxv	*Rxv* (pepper)	MAP-kinase binding/*hrp*-dependent
pv. *malvacearum*		
avrb6	*b6* (cotton)	*avrBs3*-type
avrB4	*B4* (cotton)	*avrBs3*-type
avrB5	*B5* (cotton)	*avrBs3*-type
avrB101	*B101* (cotton)	*avrBs3*-type
avrBln	*Bln* (cotton)	*avrBs3*-type
avrB102	*B102* (cotton)	*avrBs3*-type
avrB7	*b7* (cotton)	*avrBs3*-type
X. citri		
pthA	? (citrus)	*avrBs3*-type
X. oryzae		
avrXa7	*Xa7* (rice)	*avrBs3*-type
avrxa5	*xa5* (rice)	*avrBs3*-type
avrXa10	*Xa10* (rice)	*avrBs3*-type
P. syringae		
pv. *maculicola*		
avrRpm1	*Rpm1* (*Arabidopsis*)	Myristolation motif, kinase binding, *hrp*-dependent
pv. *tomato*		
avrPto	*Pto* (tomato)	Myristolation motif, kinase binding, *hrp*-dependent
avrRpt2	*Rps2* (tomato)	Hydrophobic protein, *hrp*-dependent
avrE	? (soybean)	Two ORFs required, *hrp*-dependent
avrA	*Rpg2* (soybean)	*hrp*-dependent
avrD	*Rpg4* (soybean)	Encodes enzyme for syringolide biosynthesis, *hrp*-dependent
pv. *phaseolicola*		
avrPphF	*R1* (bean)	Two ORFs required, *hrp*-dependent
E. amylovora		
dspE/F	? (various hosts)	Virulence factors/*hrp*-dependent

Perhaps more significant is the fact that if the proteins encoded by these *avr* genes are applied externally to plants carrying the corresponding resistance genes, they do not elicit a hypersensitive response. However, if they are expressed transgenically within plant cells, they are active. Furthermore, the products of two groups of *avr* genes have signals that suggest that they are specifically targeted to host cell organelles. The AvrPto group have a myristoylation motif that would target them to the cytoplasmic surfaces of membranes, whilst the AvrBs3 type (restricted to *Xanthomonas* pathogens) have nuclear localisation signals in their carboxyl coding regions, and have been shown to interact with the importin α chaperone to enter the nucleus. These proteins also have a repeated sequence of 34 amino acids of unknown function that varies in copy number between *avr* genes, along with a C-terminal transcriptional activation domain, suggesting a possible function in activation of gene expression in plants.

The evidence clearly indicates that avirulence gene products are both secreted from bacteria and targeted into plant cells, and this has led to the realisation that they probably represent important components of bacterial virulence, and that their 'function' as avirulence gene products is because the plant defence response has learnt to recognise them. This is supported by the fact that both enhancement of virulence and initiation of the hypersensitive defence responses in plants require the presence of a bacterially encoded type III secretion system. In addition, it has recently been shown in *E. amylovora*, that the *DspE/F* (disease-specific) locus (also called *DspA/B*) that is required for virulence on pears and apples acts as an avirulence determinant when expressed in *P. syringae* pv. *glycinea* race 4.

6.3 Type III secretion mechanisms

Studies of factors required for pathogenicity of Gram-negative mammalian pathogens have identified gene clusters that are closely related between species such as *Yersinia*, *Salmonella*, *E. coli*, *Pseudomonas* and *Shigella*. Several of these loci encode a type III secretion mechanism for the translocation of polypeptides across the bacterial double membrane envelope and injection of virulence factors into the cytosol of host cells. Formation of this secretion mechanism and secretion itself is triggered by contact between the pathogen and the host cell and is referred to as 'contact-dependent' secretion. The mechanism itself is highly conserved with that used for biogenesis of flagella in bacteria, such that of the 11 conserved proteins that make up the virulence-associated system, ten are conserved in the flagella export apparatus (see *Table 6.2*).

Studies in *Yersinia* spp. have identified the key proteins (Ysc) that make up the export system (*Figure 6.1*). The YscL, YscQ and YscN components are cytoplasmic or peripheral membrane proteins, and YscN appears to be an ATPase and probably plays an important role in energisation of the export system. The YscD, YscJ, YscR, YscS, YscT, YscU and YscV components are integral inner membrane proteins, with YscJ acting as a bridge across the periplasmic space. YscC is an outer membrane protein that forms as multimers in a ring-shaped structure with an external diameter of about 200 Å and central pore of about 50 Å. These structures require the YscW outer membrane lipoprotein to form properly and are believed to form channels through which proteins are secreted across the outer membrane.

6.4 Type III secretion in plant pathogens

In plant-pathogenic bacteria, a random mutagenesis approach led to the identification of a particular group of mutants that were: (1) unable to cause disease on genetically susceptible hosts, at the same time as being (2) unable

Table 6.2 Similarity of *P. syringae hrp/hrc*-proteins to type III secretion proteins from *Yersinia pestis* and flagellar biosynthesis proteins

P. syringae	*Y. pestis*	Flagellar	Location
HrcC	YscC		Outer membrane secretin
HrpQ	YscD	FliG	Inner membrane
HrcJ	YscJ	FliF	Inner and outer membrane
HrpE	YscL	FliH	Cytoplasmic
HrcN	YscN	FliI	Cytoplasmic ATPase
HrcQ	YscQ	FliN	Inner membrane
HrcR	YscR	FliP	Inner membrane
HrcS	YscS	FliQ	Inner membrane
HrcT	YscT	FliR	Inner membrane
HrcU	YscU	FlhB	Inner membrane
HrcV	YscV	FlhA	Inner membrane
HrpA	YscF		Secreted/pilus subunit
HrpB	YscI		Secreted
HrpD	YscK		Cytoplasmic
HrpF			Cytoplasmic
HrpG			Cytoplasmic
HrpO	YscO		Secreted
HrpP	YscP		Secreted
HrpT			Outer membrane

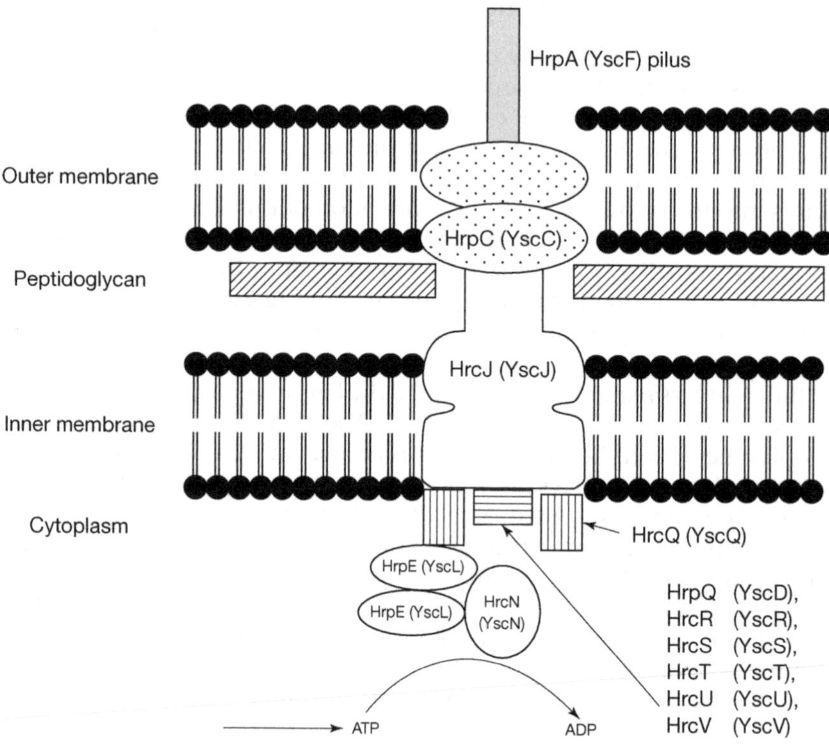

Figure 6.1

The type III secretion system of *P. syringae* (Hrp and Hrc proteins), based on analogies to *Yersinia pestis* proteins (Ysc proteins). Secretion of specific proteins is driven by ATP hydrolysis through a mechanism composed of *hrp* and *hrc* gene products as shown.

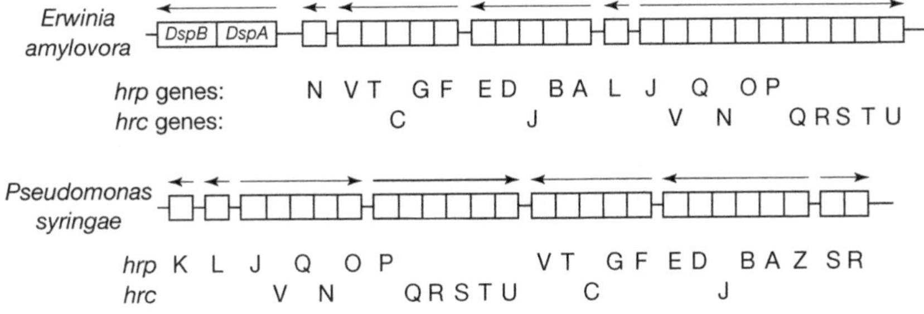

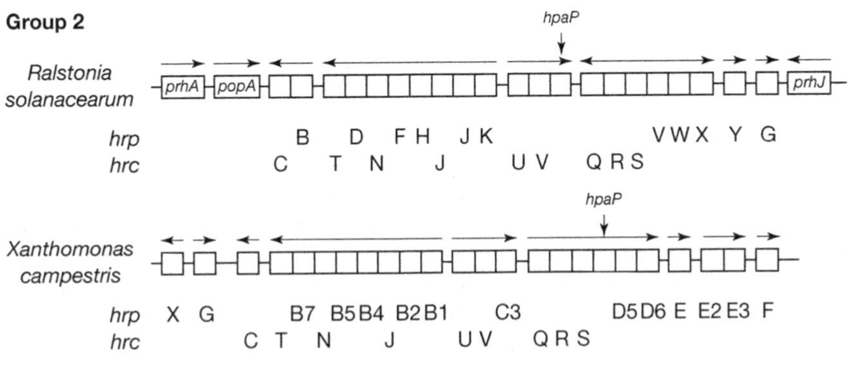

Figure 6.2

Organisation of the *hrp* gene clusters in plant-pathogenic bacteria. The arrows show the direction of transcription of the various operons, and the letters show the locations in these clusters of the various *hrp* and *hrc* genes.

to elicit a hypersensitive response on genetically resistant and non-host plants. These mutants were designated *hrp* (*h*ypersensitive *r*esponse and *p*athogenicity) mutants, and the genes involved the *hrp* genes. Subsequently, it has been found that some of these genes are highly conserved with those from other type III secretion systems (*Table 6.2*), and these genes have been renamed *hrc* (*h*ypersensitivity *r*esponse, pathogenicity and *c*onserved) genes. Based on sequence similarities, operon structures and regulatory systems, the *hrp* genes of plant-pathogenic bacteria have been grouped into Group 1 (*P. syringae* pvs., *Erwinia* species and *P. stewartii*) and Group 2 (*R. solanacearum* and *X. campestris* pvs.). There are around 20 *hrp* genes in each case, organised into a number of transcription units and clustered in approximately 25 kb regions of the genome or on plasmids. *Figure 6.2* shows the organisation of these gene clusters in *E. amylovora*, *P. syringae* pv. *syringae*, *X. campestris* and *R. solanacearum*.

6.5 Hrp-pili

A significant feature of plant-pathogenic bacterial type III secretion systems is the formation of structures called hrp-pili. These are long (at least 2 μm) whip-like proteinaceous appendages (unlike the shorter needle complexes produced by mammalian pathogens), encoded by *hrpA* in *P. syringae* and *hrpY* in *R. solanacearum*. Mutants that lack these pili are still able to attach to plant cells, showing that they are not attachment structures. However, these mutants are deficient in the ability to secrete proteins and are non-pathogenic. Given that plant-pathogenic bacteria, unlike mammalian ones, have the additional barrier of a 200 nm-thick plant cell wall through which their virulence factors must pass, it has been suggested that these hrp-pili, which appear to be unique to plant pathogens and essential for pathogenicity, may be the 'needles' through which the pathogenicity factors are translocated.

6.6 Regulation of *hrp* genes

In mammalian pathogens, the formation of type III secretory systems is controlled by contact between the pathogen and the host cell. In plants, it has also been shown that expression of *hrp/hrc* genes is tightly regulated by environmental factors and by host plant factors. *hrp* genes are not expressed when bacteria are grown in rich artificial media, but are expressed in minimal media that mimic the contents of plant apoplastic fluids.

In *R. solanacearum*, a protein PrhA has been identified as a key component in the regulation of *hrp/hrc* gene expression. This gene has homology to outer membrane hydroxamine siderophore receptors such as those in *E. coli* and *Pseudomonas putida*. Siderophores are low-molecular-weight compounds that scavenge iron, and the receptors are involved in the transcriptional regulation of iron transport genes in the bacterium. By using *hrp-gfp* (green fluorescent protein) reporter gene fusions in *R. solanacearum*, it has been shown that *hrp* gene expression does not occur in *prhA* mutants, but is induced within 90 minutes in wild-type strains, following contact with plant cell walls. Induction is not host-specific and occurs in a wide range of plant species, such as *Arabidopsis*, tomato, tobacco and *Medicago truncatula* (the latter two being non-hosts for *R. solanacearum*). This indicates that the biochemical or physical component of the cell walls required for elicitation is highly conserved amongst plant species. Importantly, this also implies that host range and host specificity is not a consequence of the inability of bacteria to form type III secretion systems *in planta*, but is presumably due to the nature of the pathogenicity factors that they secrete.

The *E. coli* outer membrane siderophore receptor (FecA) that has homology to PrhA binds to ferric citrate and operates through two membrane proteins, FecR and FecI to transmit the surface signal. Similar proteins (PrhR and PrhI) have been identified in *R. solanacearum*, and PrhI then directs transcription of *prhJ* (*Figure 6.3*). PrhJ is in turn a transcriptional activator of the OmpR-type response regulator HrpG, which is in the same group of RO_{II} regulators as VirG of *A. tumefaciens*. This in turn activates *hrpB*, which activates transcription of the remaining *hrp* operons through interaction with the upstream PIP-boxes (TTCGC-N_{15}-TTCGC).

In *Erwinia* and *P. syringae*, *hrp* regulation appears to go through a different mechanism involving responses to low levels of carbon and nitrogen, low pH (5.5) and low temperature (18°C). In *E. amylovora*, mutations in four genes, *hrpS*, *hrpL*, *hrpX* and *hrpY* abolish secretion, and these are believed to be the key components of the regulatory systems. *hrpS* encodes a protein similar to σ^{54}-dependent transcriptional activators, whilst HrpL is homologous to

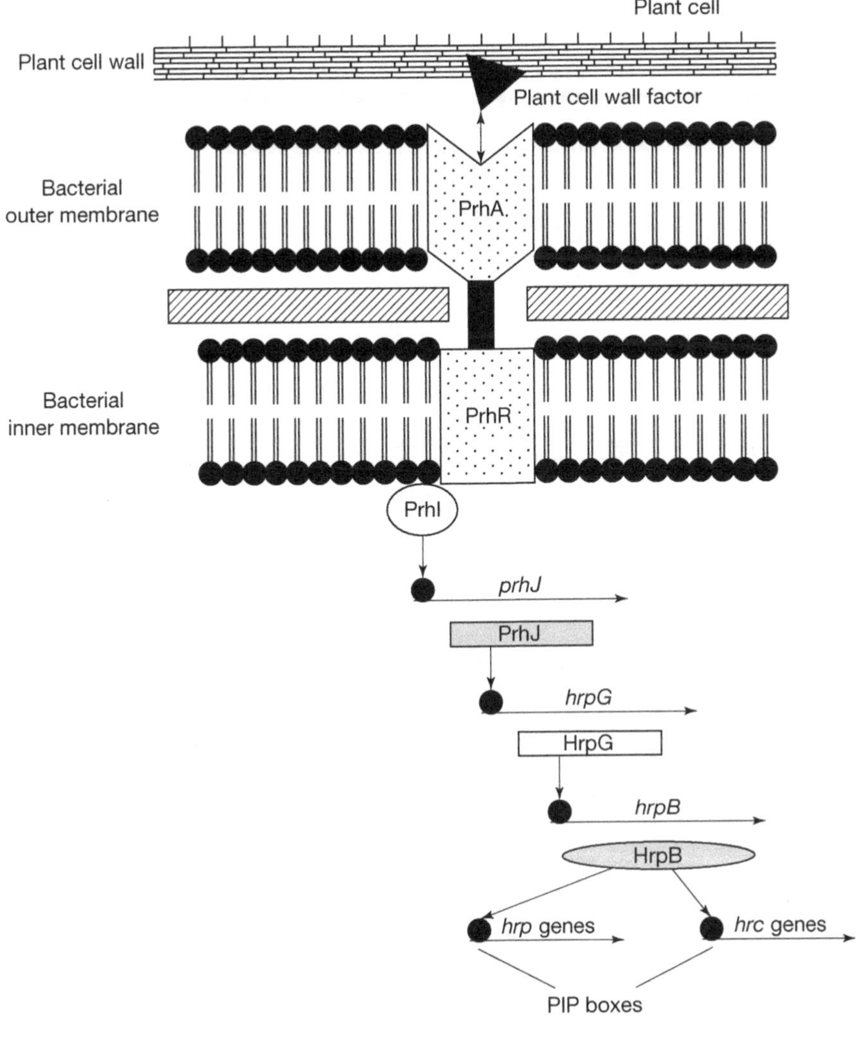

Figure 6.3

Model for the role of PrhA in induction of *hrp/hrc* gene expression in *R. solanacearum*. Factors from plant cell walls bind to the bacterial outer membrane receptor, PrhA (see text for details).

the ECF subfamily of sigma factors and is believed to direct transcription through *hrp* promoter consensus boxes (GGAACC-N_{15}-CCACTAT) upstream of the *hrp* secretion operons and certain pathogenicity genes that are co-regulated (*Figure 6.4*). *hrpX* and *hrpY* encode a two-component regulatory system that operates with HrpS to activate HrpL. HrpX is the sensor protein, with an N-terminal PAS domain and a C-terminal histidine kinase domain. PAS domains of other bacteria are highly conserved regions believed to be involved in environmental sensing. HrpY is typical of the RO_{III} type response regulator, unlike the *hrp* regulator in *R. solanacearum* (HrpG), which is of the RO_{II} type. The means through which HrpS and HrpX/Y interact are not entirely clear, but there is good evidence that the two sensor systems work

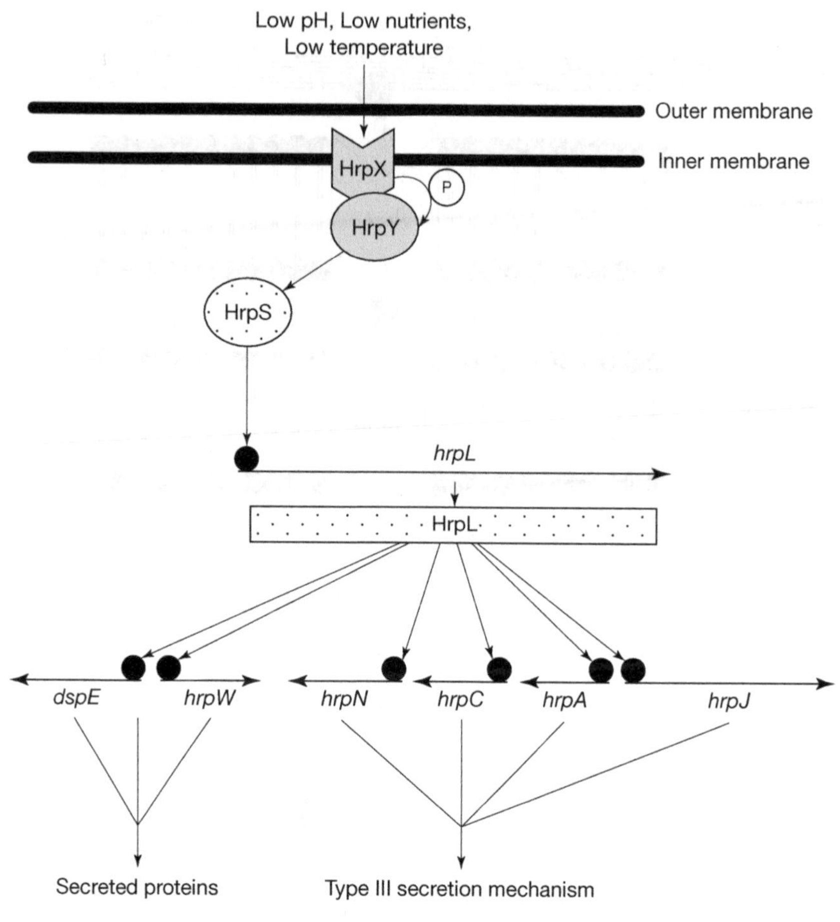

Figure 6.4

Model for *hrp* gene regulation in *Erwinia* and *P. syringae.*

independently to detect environmental signals before converging to regulate *hrpL* expression.

6.7 Secreted proteins

In mammalian pathogens, numerous proteins are secreted through type III systems, including essential virulence determinants and pathogenicity factors. In plant pathogens, it has become clear that avirulence gene products are also secreted by these mechanisms. Indeed, immunogold-labelling of AvrPto has shown a clear physical association between this protein and the hrp pilus in *P. syringae*. Amongst the proteins that have been shown to be secreted by type III mechanisms in addition to avirulence gene products, are the harpins of *E. amylovora* and *P. syringae* pv. *syringae*, PopA, PopB and PopC in *R. solanacearum*, HrpA in *P. syringae* pv. *tomato* and HrpW, DspA/E and DspB/F of *E. amylovora*. Of these, the harpins appear to be non-specific, glycine-rich, cysteine-lacking, heat-stable elicitors of the hypersensitive response (HR) in plants (see Section 9.3), causing extracellular alkalinisation and potassium efflux, and inducing programmed cell death. However, their precise role in pathogenicity remains unclear. PopA is a 33 kDa protein similar to the harpins, but not essential for pathogenicity, since PopA-deficient mutants have normal virulence. PopB has nuclear localisation signals like some of the avirulence genes, whilst PopC has 22 leucine-rich repeats analogous to those in plant resistance genes (see Section 10.3.1), and it has been suggested that through these, PopC may interfere with plant defences. HrpA appears to be involved in pilus formation, whilst the function of Dsp A/E is unclear, except that it appears to require the chaperone protein Dsp B/F for its secretion and is essential for virulence on pears and apples.

6.8 Secretion signals

In mammalian pathogens, there are two known pathways that signal proteins for secretion by type III mechanisms, a secretion signal in the first 15 codons of the protein, or the presence of chaperones. Through heterogeneous expression of secreted proteins derived from plant pathogens in animal systems, it has been shown that these same basic signals are conserved between bacteria. The *Yersinia* system can secrete *P. syringae* AvrPto and AvrB, whilst the *E. chrysanthemi hrp* system can secrete YopE and YopQ derived from *Yersinia*. However, the efficiency with which proteins are secreted in the different systems is very variable, which may reflect some form of protein sorting by the different mechanisms.

Some of the proteins in plant-pathogenic bacteria appear to use chaperones, such as the reliance of Dsp A/E on Dsp B/F. Others (e.g. AvrPto) appear to use signals in the first 15 codons. It has been shown that frameshift mutation in the first 15 codons of AvrPto do not alter secretion, suggesting that it is not the protein composition but something about the mRNA of secretion substrates that is important for function. Computer comparisons of *avrPto* and *avrB* genes has revealed secondary structures similar to those of *Yersinia yop* genes which place the Shine-Dalgarno 16S ribosomal RNA binding site and the AUG translation start sequences in a hyphenated stem-loop structure, and it is believed that this may be the signal that is important for determining secretion. However, there are still many functions of *hrp* secretion mechanisms and *avr* gene function that have not been elucidated.

6.9 Pathogenicity islands

The organisation of avirulence and regulatory genes in plant-pathogenic bacteria exhibits some unusual features. These genes are often in blocks in the bacterial genomes, referred to as pathogenicity islands (PAIs). Some of these PAIs are found on large plasmids, such as the 2100 kb megaplasmid of *R. solanacearum* strain GMI1000 or the 154 kb plasmid of *P. syringae* pv. *phaseolicola*, and removal of these plasmids has been shown to result in loss of pathogenicity. These PAIs generally have different guanine plus cytosine (G+C) contents to the rest of the bacterial genomes, and the sequences are flanked by direct repeats or insertion sequences, suggesting that they may have been acquired through horizontal gene transfer. For example, avirulence genes of *P. syringae* possess 40–52.5 mol % G+C, whilst the rest of the genome averages 59–61 mol %. The PAIs often possess tRNA genes, which may act as the integration sites into the bacterial genome.

Table 6.3 Important traits carried on plasmids in some pathovars of plant pathogenic bacteria

Species	Genes
P. syringae	Avirulence genes, e.g. *avrD, avrPphF, hrp* genes
	Coronatine biosynthetic genes
	Copper resistance
P. savastanoi	IAA and cytokinin biosynthetic genes
X. campestris	Avirulence genes, e.g. *avrBs1, avrBs3*
R. solanacearum	*hrp* genes
A. tumefaciens	T-DNA and *vir* genes

6.10 The role of plasmids

Plasmids are extrachromosomal DNA elements that are usually stably inherited within bacterial cell lines and can be transferred between strains, species or genera. They have been found to have a significant role in pathogenicity of many bacteria that infect plants, harbouring a number of avirulence genes, a range of toxin and hormone biosynthetic genes, pathogenicity genes, resistance to chemical control strategies such as copper and antibiotics, and insertion elements as shown in *Table 6.3*, and this has led to the belief that transposon-based mechanisms and loss/gain of plasmids may be the means through which many bacteria acquire or lose pathogenicity factors. These plasmids can range in size from a few kb to 2100 kb megaplasmids. In some cases, such as *A. tumefaciens*, it is the presence of these plasmids that determines whether the isolate is pathogenic on plants or not. In many bacteria, cryptic plasmids with unknown functions have also been found, and it remains to be seen whether there are genes on these that are also crucial for pathogenicity.

References and further reading

Francis, M.S., Wolf-Watz, H. and Forsberg, A. (2002) Regulation of Type III secretion systems. *Current Opinion in Microbiology* **5**: 166–172.

Genin, S. and Boucher, C. (2002) *Ralstonia solanacearum*: secrets of a major pathogen unveiled by analysis of its genome. *Molecular Plant Pathology* **3**: 111–118.

Hacker, J. and Kaper, J.B. (2000) Pathogenicity islands and the evolution of microbes. *Annual Review of Microbiology* **54**: 641–679.

Innes, R.W. (2001) Targeting the targets of Type III effector proteins secreted by phytopathogenic bacteria. *Molecular Plant Pathology* **2**: 109–115.

Kjemtrup, S., Nimchuk, Z. and Dangl, J.L. (2000) Effector proteins of phytopathogenic bacteria: bifunctional signals in virulence and host recognition. *Current Opinion in Microbiology* **3**: 73–78.

Koebnik, R. (2001) The role of bacterial pili in protein and DNA translocation. *Trends in Microbiology* **9**: 586–590.

Lahaye, T. and Bonas, U. (2001) Molecular secrets of bacterial type III effector proteins. *Trends in Plant Science* **6**: 479–485.

Plano, G.V., Day, J.B. and Ferracci, F. (2001) Type III export: new uses for an old pathway. *Molecular Microbiology* **40**: 284–293.

Staskawicz, B.J., Mudgett, M.B., Dangl, J.L. and Galan, J.E. (2001) Common and contrasting themes of plant and animal diseases. *Science* **292**: 2285–2289.

Vivian, A., Murillo, J. and Jackson, R.W. (2001) The roles of plasmids in phytopathogenic bacteria: mobile arsenals. *Microbiology* **147**: 763–780.

White, F.F., Yang, B. and Johnson, L.B. (2000) Prospects for understanding avirulence gene function. *Current Opinion in Plant Biology* **3**: 291–298.

Plant viruses – structure and replication

7

Pioneering work in many aspects of molecular plant pathology such as protein purification and sequencing, X-ray crystallographic determination of structures, and nucleic acid sequencing have all been developed using plant viruses such as *Tobacco mosaic virus* (TMV), because of the relative simplicity of these infectious agents. Their small nucleic acid genomes (normally encoding between four and ten polypeptides) combined with their simple protein capsids (coats) has resulted in a comprehensive understanding of many of their structural properties. However, this 'simplicity' means that they have an absolute requirement on host cell components and metabolism for their infection cycle, and understanding the nature of this intricate signalling between viruses and plants has become the major focus of modern molecular plant virology. In this chapter, we shall examine the structure and replication of plant viruses and other sub-microscopic infectious agents such as viroids, before discussing the ways in which they interact with host plants and vectors in Chapter 8.

7.1 The structure of plant viruses

Plant viruses have been organised into 73 genera, and 49 of these genera have been classified amongst 15 families, with the remaining 24 unassigned. These groupings are based on characteristics such as size, shape, nucleic acid composition, sequence comparisons, transmission characteristics, host range and serological relationships (see *Table 1.4* in Chapter 1). Almost 50% of known plant viruses are rod-shaped (flexuous or rigid), and most of the remainder are isometric (*Figure 7.1*). In addition, there are three genera of *Geminiviridae* that possess geminate (twinned) particles, and a small number of cylindrical bacillus-like rods. In most cases, the capsids consist of multiple copies of one or two protein subunits, but a few genera, notably the *Tospoviruses* and the bacillus-like viruses, have more complex protein structures surrounded by outer lipoprotein membranes. The genomes of most plant viruses (approximately 90%) are single-stranded RNA, mainly in the plus (+) or sense polarity. Others such as the phytoreoviruses have dsRNA genomes, some (the *Geminiviridae*) – ssDNA genomes, and the caulimoviruses and badnaviruses have dsDNA genomes. In many cases, the genome is a single segment of nucleic acid, but some genera possess split genomes in either a single capsid, or multicomponent viruses (e.g. the tobraviruses and *Comoviridae*) in which the split genomes are encapsidated in particles of different sizes, all of which must be present in the plant for the virus to replicate and spread.

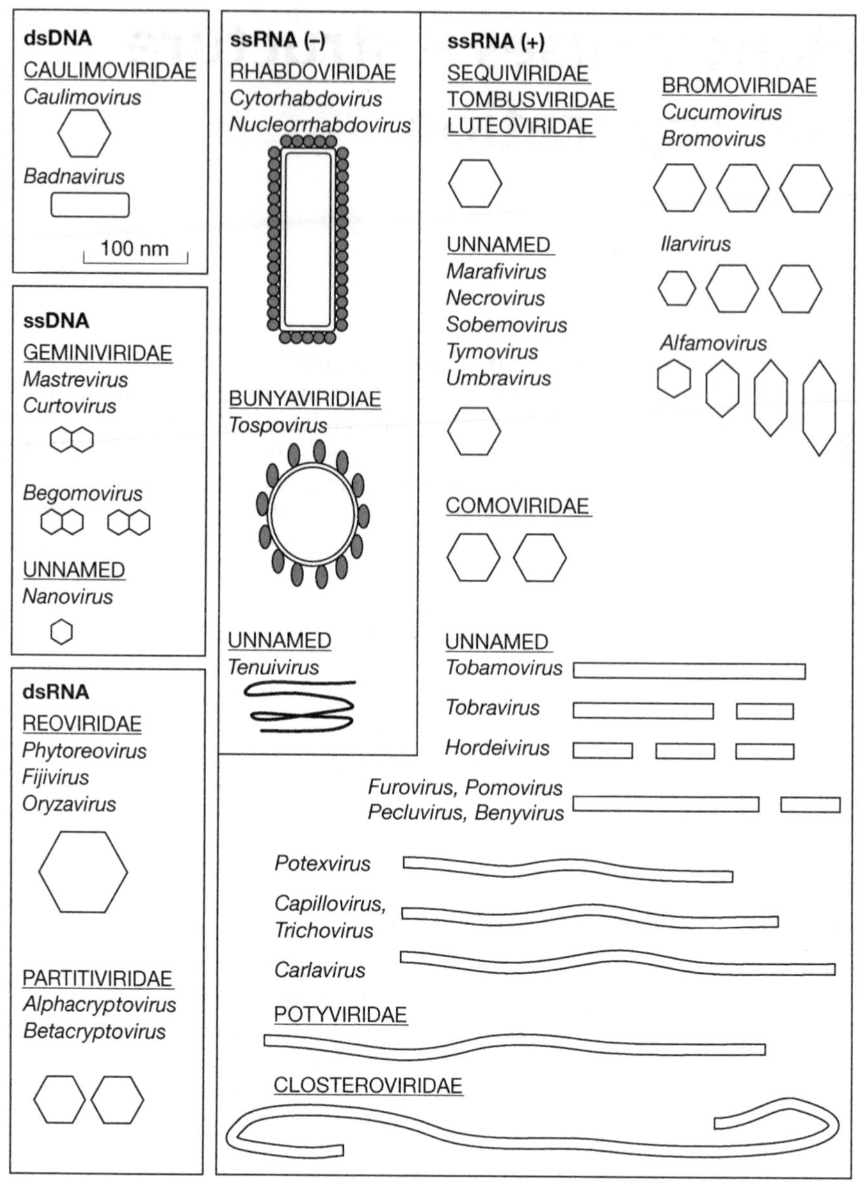

Figure 7.1

Shape and relative size of plant virus families. Famlies (underlined) and genera within these families (italics).

7.2 Virus infection of plants

Most plant viruses are transmitted by vectors (in particular insects, nematodes and fungi), or through seed and pollen (see Sections 8.1–8.5). The basic infection process involves the virus entering the plant cell through wounds made mechanically or by the vector (*Figure 7.2*). In the case of some of the RNA viruses, it has been shown that the capsid is then removed in a

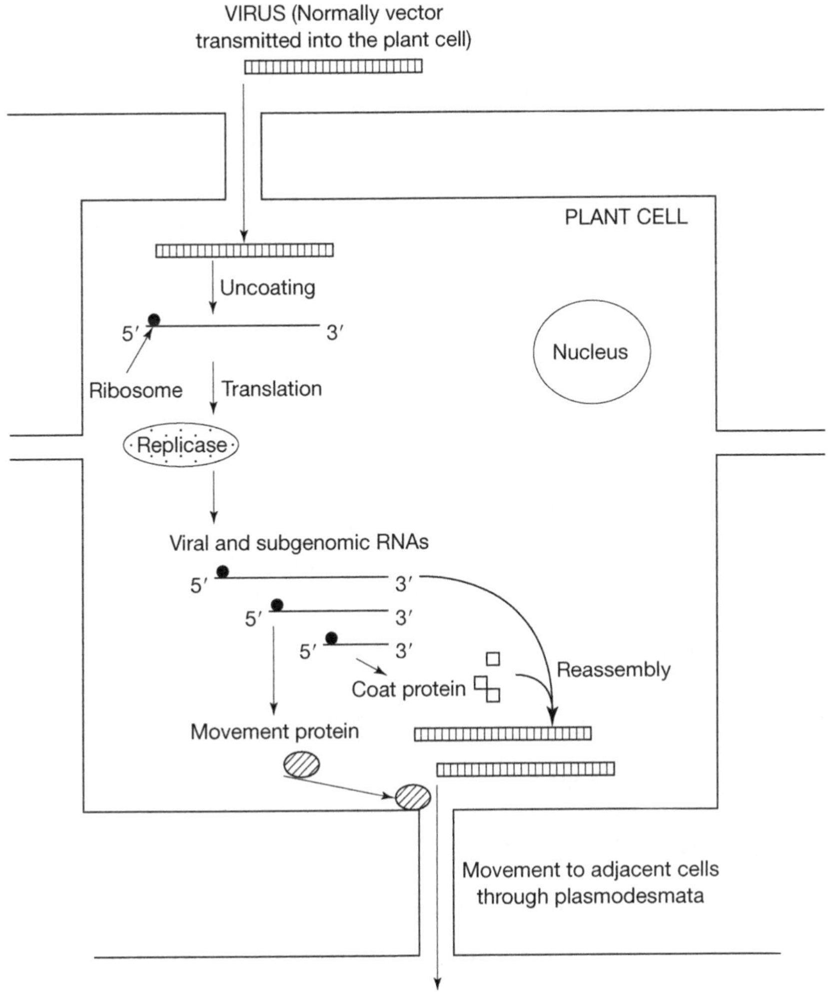

Figure 7.2

General scheme of virus spread within plants. The virus enters the plant cell, and following replication and production of more coat protein, the virus reassembles and moves between adjacent cells through the plasmodesmata, facilitated by the movement protein.

process known as co-translational disassembly, in which the RNA is surrounded by ribosomes to replace the coat protein and protect the RNA from nuclease digestion. The genes encoded by the nucleic acid are then expressed, resulting in genome replication, and production of the viral proteins required for pathogenicity. The resultant genomes are then packaged or form into ribonucleoprotein complexes, and spread through the plant.

7.3 Translation and replication of positive-strand RNA viruses

Plant RNA virus genomes are generally very compact, with the genes separated by very short intergenic regions or overlapping with each other, such that there appears to be very little, if any, redundant sequence. This compact nature, and the fact that they encode only a small number of polypeptides means that they rely on many host cell components for their translation, replication and movement. The majority of plant viruses possess single positive-strand RNA genomes, and replication and translation of these viruses invariably take place in the cytoplasm of the plant, often associated with membrane-bound cytoplasmic inclusions (viroplasms) containing host endoplasmic reticulum. However, a fundamental problem faced by viruses when using the eukaryotic translation machinery is that it is generally only the 5′-most cistron in a polycistronic message that will be translated, and translation will cease at the first in-frame stop codon, such that all the subsequent cistrons will remain untranslated. To circumvent this, plant viruses have developed a range of strategies to ensure that all their genes are translated.

7.3.1 The production of sub-genomic RNAs and virus replication

One strategy used by many RNA viruses is to produce sub-genomic RNAs through the process outlined in *Figure 7.3*. In this process, the first cistron

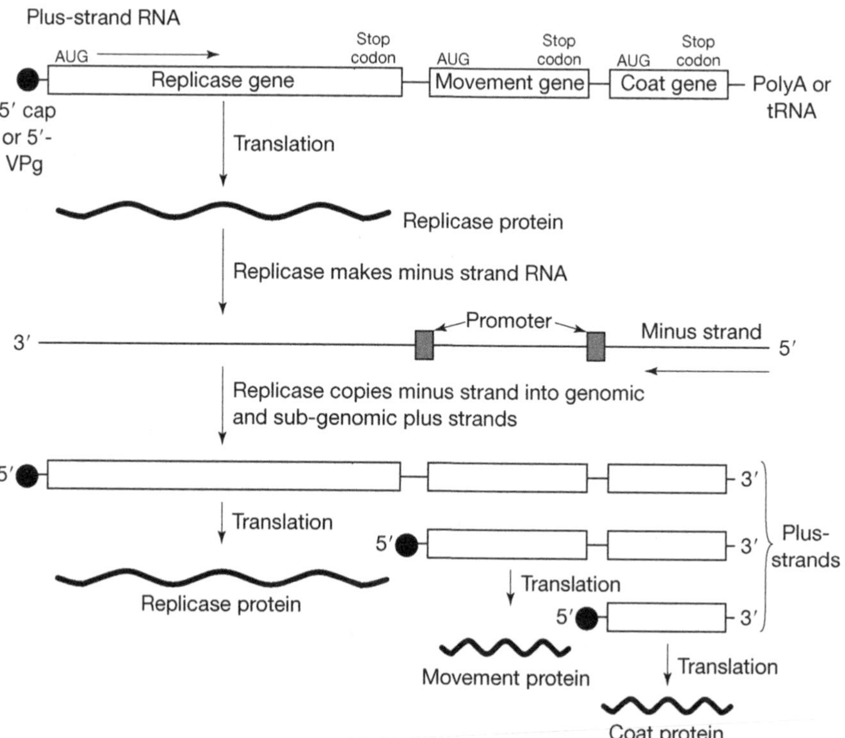

Figure 7.3

General scheme for plus-strand RNA virus replication through sub-genomic RNAs. The first cistron translated is the replicase, which produces a full-length minus strand that is then replicated into more full-length genomic RNAs and subgenomic plus-strands. These are then translated into the different functional proteins.

that is translated from the plus-strand RNA is the replicase. The replicase then acts on the positive-strand template to produce the complementary minus strand, and this in turn is replicated into more full-length genomic plus-strands and also sub-genomic plus-strands. These sub-genomics are positive-strand RNAs for each of the downstream genes, and because each of these sub-genomic RNAs has its own promoter and terminator sequence, it will in turn be translated into the functional proteins, such as the movement and coat proteins.

The fact that the 5′-most cistron encodes the replicase function has been confirmed by structural and functional studies in some viruses and inferred through sequence motif comparisons and mutational analysis for many others. In the majority of plant viruses, the replicase function has an RNA-dependent RNA polymerase (RdRp) component and a helicase component. There are however some viruses, in particular the *Tombusviridae* and the *Luteoviridae* for which no obvious helicase component has been identified although it is possible that one of the open reading frames encodes a helix destabilising protein in these viruses. In some viruses (e.g. *Potato virus X*), the helicase and RdRp functions are encoded in a single polypeptide and it is likely that the protein acts as a dimer in order to perform both functions. In others (e.g. TMV), the helicase activity is present in the 126 kDa protein, whilst the RdRp activity is in the read-through portion of the 183 kDa protein that is produced from the same initiation point. The role of the helicase is believed to be to unwind the duplex RNA that forms during replication and also to remove any secondary structures in the ssRNA. The RdRp is responsible for the catalytic activity and synthesis of RNA from NTP substrates.

In some cases host proteins are required in the replication complexes, in particular to initiate RNA synthesis, and amongst the host factors that have been identified are the eukaryotic translation initiation factor eIF-3, which associates with BMV and TMV RdRps. However, these two viruses utilise different subunits of eIF-3, suggesting different roles in BMV and TMV. The RdRp of TMV interacts with the plant equivalent of the yeast 55 kDa GCD10 subunit, which has RNA-binding activity, and it may be that this GCD10-related protein has similar RNA-binding properties and a role in initiation of RNA synthesis in plants. The host protein synthesis elongation factor EF1α also associates with the TMV RdRp (as it does with *Turnip yellow mosaic virus*, TYMV). However, the role of this factor in replication complexes is unclear, and it may reflect a role in associated protein synthesis rather than in the replication. In other cases, such as *Cowpea mosaic virus* and *Cucumber mosaic virus*, it has been shown that the virus-encoded RdRp alone is enough for synthesis of viral RNA without the need for host factors.

Most eukaryotic mRNAs possess 5′ cap structures, which protect the RNA and act as signals for translation. Many plant viral RNAs, including sub-genomic RNAs also possess 5′ caps consisting of 7MeGpppN, where N is usually adenine (A) or guanine (G). The capping of most cellular RNAs and those of DNA viruses and retroviruses occurs in the nucleus. Capping of plus-strand viral RNAs however occurs in the cytoplasm, and for this to occur, these viruses possess methyltransferase activity within their replicase function. In the case of TMV, the methyltransferase forms the N-terminal part of the 126 kDa protein, with the helicase domain at the C-terminal end, and this organisation in which the methyltransferase and helicase domain are within the same protein is common in plant RNA viruses.

Untranslated (UTR) regions of plant virus genomes are also important for replication. Sequences present in the 3′-UTR regions can form pseudoknots and terminal tRNA-like structures in many cases, which are believed to be involved in initiation of RNA synthesis as viruses in which these regions are mutated or deleted have reduced replication efficiency. In some viruses, the 3′-UTR has an additional role in translation initiation. 5′-UTRs also contain domains that are important for replication including translation enhancer sequences. However, the mechanisms through which these enhancer regions, translation factors and other structures control transcription and replication have yet to be elucidated. What is clear is that the replicase complexes bind to cytoplasmic inclusions (viroplasms), which enlarge during infection to form 'X-bodies'. This compartmentalisation to membranes is likely to help with recruitment of necessary factors to a local area resulting in a more efficient replication process. It may also be that fluidity of these membranes is important for replication, since some of the host proteins with which viral replicases associate appear to be involved in regulating membrane fluidity.

7.3.2 Segmented genomes

A second approach to ensure replication and translation of all the viral RNA is to have a segmented (multipartite) genome. Members of the *Bromoviridae* family (e.g. *Brome mosaic virus* and *Cucumber mosaic virus*) have three genomic RNAs, packaged into individual isometric particles composed of the same protein subunits, and between them these RNAs encode four proteins (*Figure 7.4*). In these cases, RNA 1 encodes the 1a protein that has the methyltransferase and helicase motifs. RNA 2 encodes the RdRp function, and these proteins are independently localised to the plant endoplasmic reticulum, where replication occurs. In the genera *Cucumovirus* and *Ilarvirus* within this family, this RNA also contains an overlapping second open reading frame at the 3′ end, which is expressed from a low abundance sub-genomic RNA and may have a role in suppressing the host post-transcriptional gene silencing (PTGS) resistance mechanism (see Section 8.8.3). The third RNA in these viruses encodes the movement and coat proteins, the latter being produced from a sub-genomic RNA, although in the cucumoviruses, this sub-genomic RNA is packaged into isometric particles alongside RNA 3. As with some other plus-strand viruses (such as the bromo- and alfamoviruses), the RNAs have tRNA-like sequences at their 3′ ends that contain the replicase binding site and initiate synthesis of the negative strands, starting at the penultimate cytosine residue in the plus-strand. Once synthesised, the resultant negative strands also have a penultimate cytosine residue at their 3′ ends, which are in turn used to initiate synthesis of more genomic-plus strands. The sub-genomics are also initiated from cytosine residues, and the replicase protein interacts with specific promoter sequences directly upstream of these.

7.3.3 Polyprotein processing

The third strategy adopted by viruses such as the *Potyviridae* and the *Tymovirus* genera, is to produce a single polyprotein from the plus-strand RNA and then to cleave this at the post-translational level into the different polypeptides (*Figure 7.5*). Cleavage is triggered by virally encoded

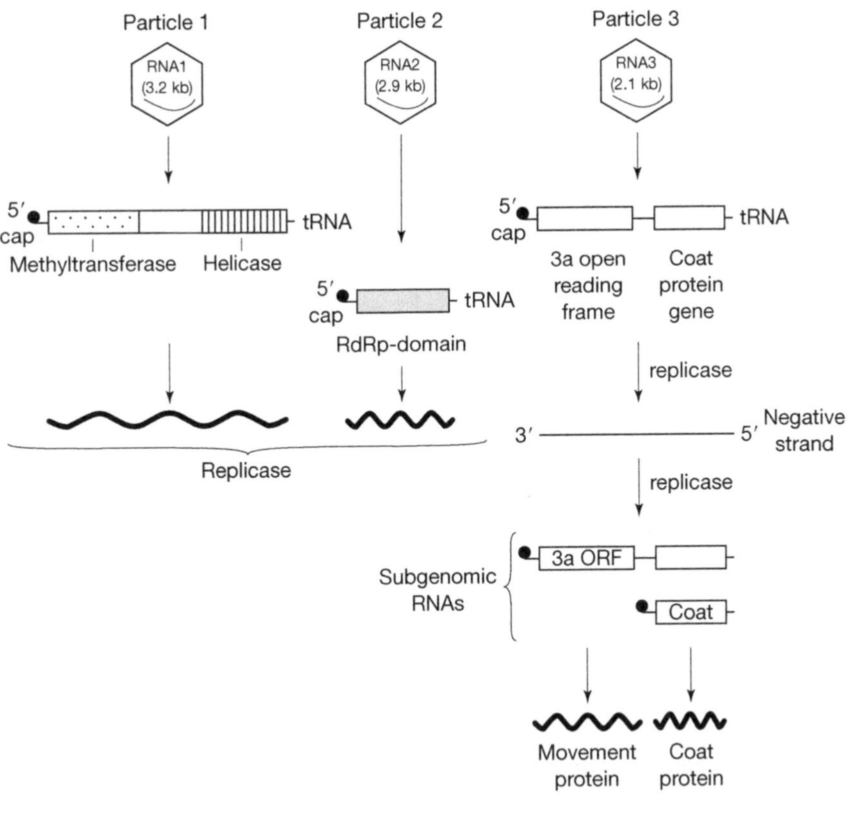

Figure 7.4

Organisation of the genome in *Brome mosaic virus,* a typical multipartite virus. The three particles required for infection each contain a different RNA.

proteinases. In the case of the tymoviruses, the papain-like cysteine proteinase activity is located between the methyltransferase and helicase domains encoded by open reading frame (ORF) 1, and this enzyme cleaves the C-terminal RdRp domain away to promote replication. In the *Potyviridae* there are three proteinase activities within the polyprotein, and the replicase activity is towards the C-terminal end of this polyprotein. The NIa chymotrypsin-like serine proteinase is responsible for cleavage of the replicase functions from this polyprotein. In the *Comoviridae,* a similar gene order exists to that in the *Potyviridae* (helicase, VPg, proteinase, RdRp), but these functions are on a separate RNA (RNA B) to the movement and coat proteins (RNA M). The role of the VPg, which is linked to the 5′ end of the viral RNAs is believed to be analogous to that of the 5′-cap in other viruses. Yeast two-hybrid studies on the *Turnip mosaic virus/Arabidopsis* interaction support this role where it has been shown that the VPg binds to a eukaryotic translation initiation factor eIF-3.

7.3.4 Readthrough and frameshifting

Viruses in the *Tombusviridae* and the *Dianthovirus, Luteovirus* and *Necrovirus* genera possess a short open reading frame at the 5′ end encoding a

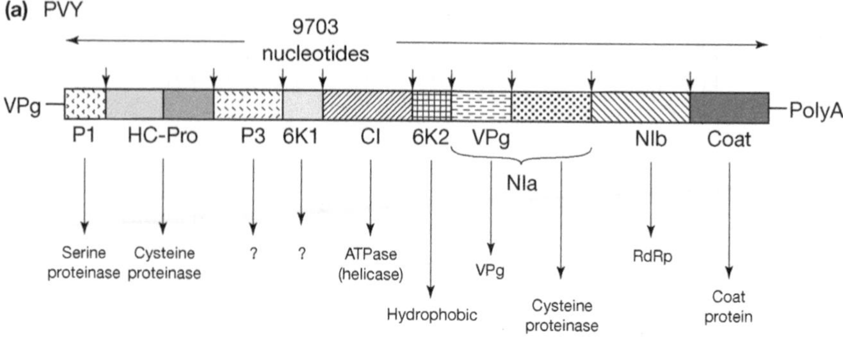

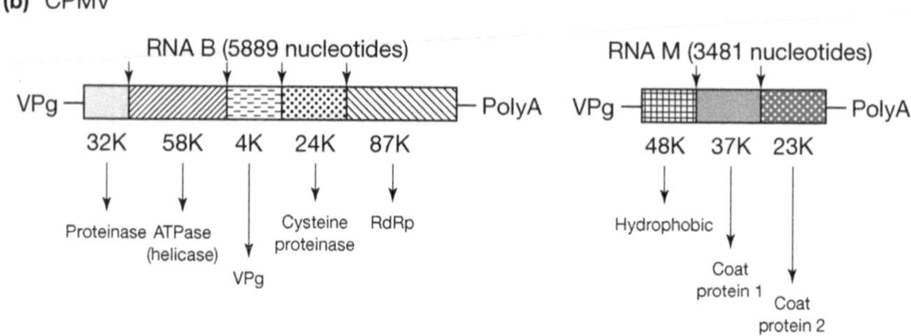

Figure 7.5

Organisation of genomes in examples of viruses that go through polyprotein processing: (a) *Potato virus Y* and (b) *Cowpea mosaic virus.* One large polypeptide is produced, which is then cleaved by the proteinases within it into the different peptides.

22–33 kDa protein. In some cases, the stop codon at the end of this ORF is read through, whilst in others, a ribosomal frameshift occurs prior to the stop codon to result in a fusion protein of 82–99 kDa. The carboxy-terminal region of this larger protein appears to possess the RdRp domain, whilst the smaller protein appears to substitute for the absence of a helicase domain in helix-destabilisation. The presence of readthroughs, frameshifts, overlapping ORFs and shunt mechanisms that allow scanning ribosomes to 'ignore' certain regions of RNAs are all mechanisms used by plant viruses to maximise the amount of information encoded by them. It is intriguing to speculate that mechanisms which essentially allow ORFs to encode more than one functional polypeptide may exist in genomes of other organisms, but that they have not yet been revealed because of the lack of sufficient proteomic and genomic sequence information.

7.4 Negative-strand RNA viruses

Monopartite negative or ambisense-strand viruses are classified as *Rhabdoviridae*, whilst the multipartite viruses such as *Tomato spotted wilt virus* (TSWV) are classified as *Bunyaviridae*. The genomes of all these viruses consist of essentially three parts (gene blocks) (*Figure 7.6*). Block 1 encodes the nucleocapsid and phosphoprotein, block 2 the envelope membrane

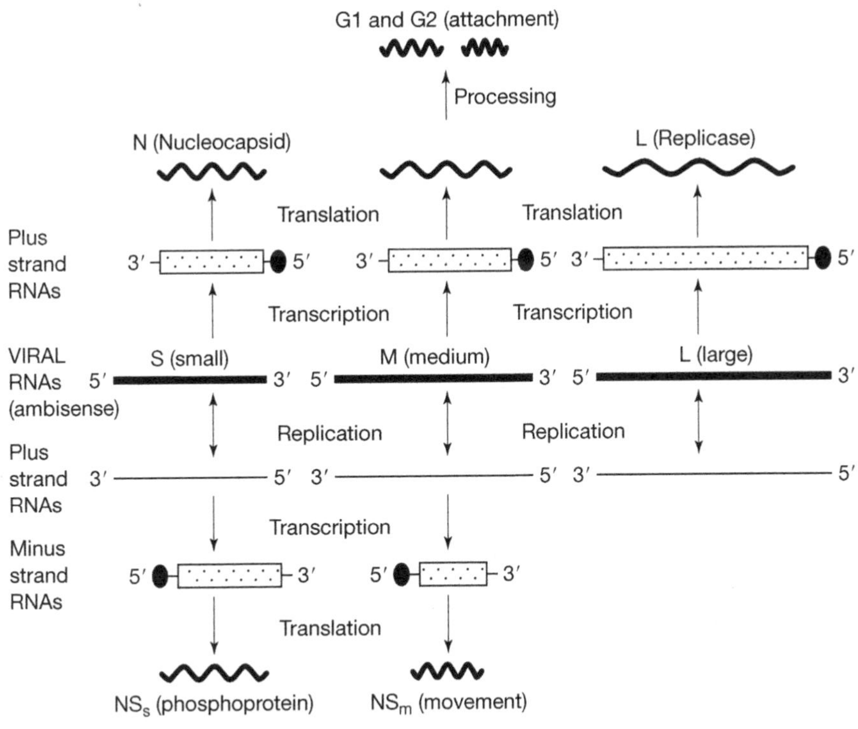

Figure 7.6

Organisation of *Bunyaviridae* genomes. The virus encodes some functions in its 'negative' strand that are translated as shown in the bottom half of the diagram, and some in the positive sense (top half of the diagram).

proteins and movement protein, and block 3 the replicase. The initial replication of the RNA is promoted by the presence of an estimated 10–20 copies of the viral RdRp protein packaged along with the viral RNA in the capsids. This results in the production of plus-strand RNA for translation, which in turn leads to the production of sub-genomic RNAs and full-length negative strands for packaging. In the monopartite viruses, the polymerase also catalyses polyadenylation and capping of mRNAs, whilst in the segmented viruses, the mRNAs do not possess poly-A tails and obtain their caps by a mechanism known as cap-snatching.

7.5 Double-strand RNA viruses

The double-stranded RNA viruses are grouped into two families; the *Reoviridae* which have genomes of 26–28 kb in 10–12 segments, and the *Partitiviridae* that have 3–4 kb genomes in 2–3 segments. Furthermore, the *Reoviridae* utilise a conservative mechanism of replication whilst the *Partitiviridae* replicate semi-conservatively. Like the negative-strand RNA viruses, these viruses contain RdRp protein within their protein coats and the main difference between the two mechanisms is that in semi-conservative replication, the strand is displaced by the replication fork as it proceeds, whilst in conservative replication, the duplex RNA is only unwound transiently at the growing end of the nascent strands.

7.6 Single-strand DNA viruses

The majority of single-strand DNA viruses belong to the *Geminiviridae*, although there is a recently discovered unrelated genus called the *Nanovirus*. The *Geminiviridae* possess either one or two small circular ssDNA molecules (2.6–3.0 kb) packaged in twinned particles (*Figure 7.7*). Replication of these viruses occurs in the nucleus of the infected plant, and the first

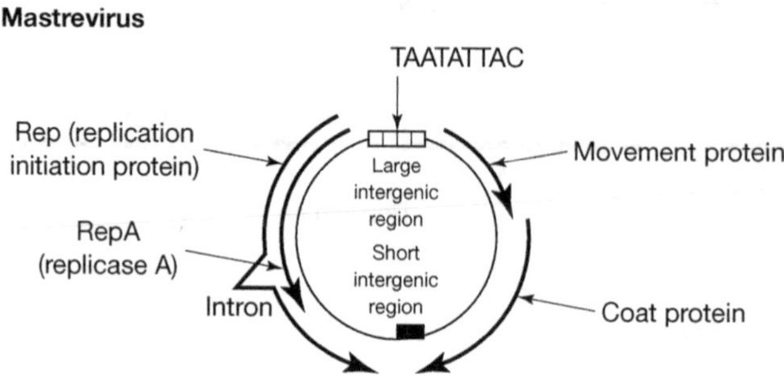

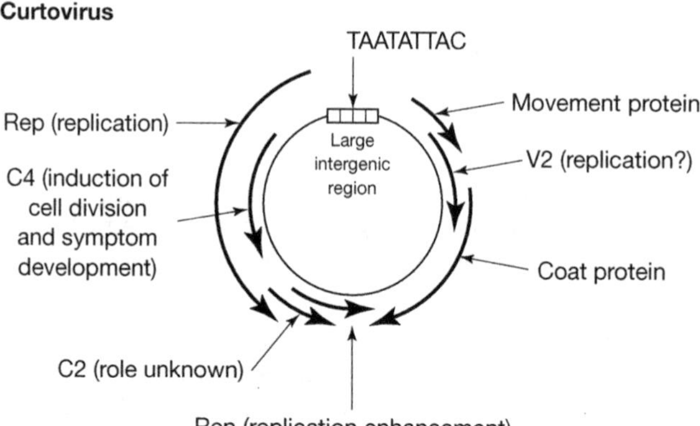

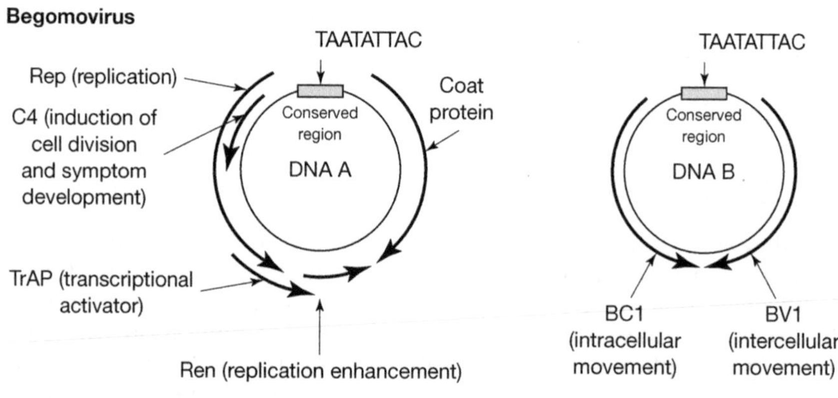

Figure 7.7

Geminivirus genome organisation. Conserved regions containing the invariant TAATATTAC initiation of replication domain are shown. The transcripts are shown as large black arrows.

stage of replication involves conversion of the ssDNA into dsDNA using plant cellular enzymes, although how and which enzymes are recruited for this is unclear. The dsDNA is then the template for bi-directional transcription from divergent promoters, which results in production of the viral replicase (Rep) protein. This protein is required for the second phase of replication, the rolling circle phase, which results in the production of new dsDNA and nascent ssDNA viral forms. This process is initiated between the seventh and eighth bases (TA) of an invariant TAATATTAC locus present in all geminiviruses, and also involves a number of cellular DNA replication factors.

A notable feature of geminivirus replication is that it does not occur in the plant meristem, where plant cell replication and division is occurring and DNA replication is active. Instead, the virus has to induce a permissive state for DNA replication in other non-proliferating, already differentiated plant cells. The evidence suggests that this occurs through an interaction between the viral Rep protein and a plant retinoblastoma-related (RBR) protein. In animal cells, these retinoblastoma proteins are involved in regulating the cell cycle from G_1 through G_1–S to the S-phase, and animal oncoviruses are able to bypass this pathway to promote cell proliferation. It is conceivable that geminiviruses have developed similar mechanisms for bypassing the normal cellular controls of replication in differentiated cells, and that by shifting mature cells to S phase they produce the permissive state for replication.

The ssDNA *Nanoviruses* such as *Banana bunchy top virus* (BBTV) and *Subterranean clover stunt virus* (SCSV) are 18–22 nm diameter isometric particles that contain at least six circular ssDNA molecules of between 1000–1200 nucleotides. Each of these DNAs contains an identical stem-loop structure, believed to be involved in initiation of replication, along with at least one ORF and poly-A addition signals, but there is little known about the replication of these viruses nor about their molecular interactions with host cells.

7.7 Double-strand DNA viruses

The double-stranded DNA *Caulimoviridae* are plant pararetroviruses that replicate via reverse transcription. Their genomes (approximately 8.0 kb) are composed of an α-strand (the template for transcription) that has a single break (discontinuity) in it and a gap of 1–2 nucleotides (*Figure 7.8*). The opposite β-strand possesses 1–3 breaks at which the strands overlap each other. Upon entering a host cell, the virus moves to the nucleus, the overlapping nucleotides are removed, the breaks closed, and the DNA forms into a supercoiled mini-chromosome. From this DNA, two RNA transcripts are produced. The 19S RNA (so-called because of its sedimentation properties in sucrose gradients) moves to the cytoplasm, where it is translated into the 58 kDa gene *VI* TAV (translation transactivator) protein. The second transcript, the 35S RNA is a full-length transcript that also includes a 180-nucleotide duplication. This transcript is processed to a form that encodes six known proteins (as shown in *Figure 7.8*), with a number of other short ORFs in the 5′ leader sequence. Unlike in many RNA plant viruses, the replicase is not the first function to be translated. Instead, two mechanisms appear to operate to ensure that all the ORFs on this poly-cistronic 35S

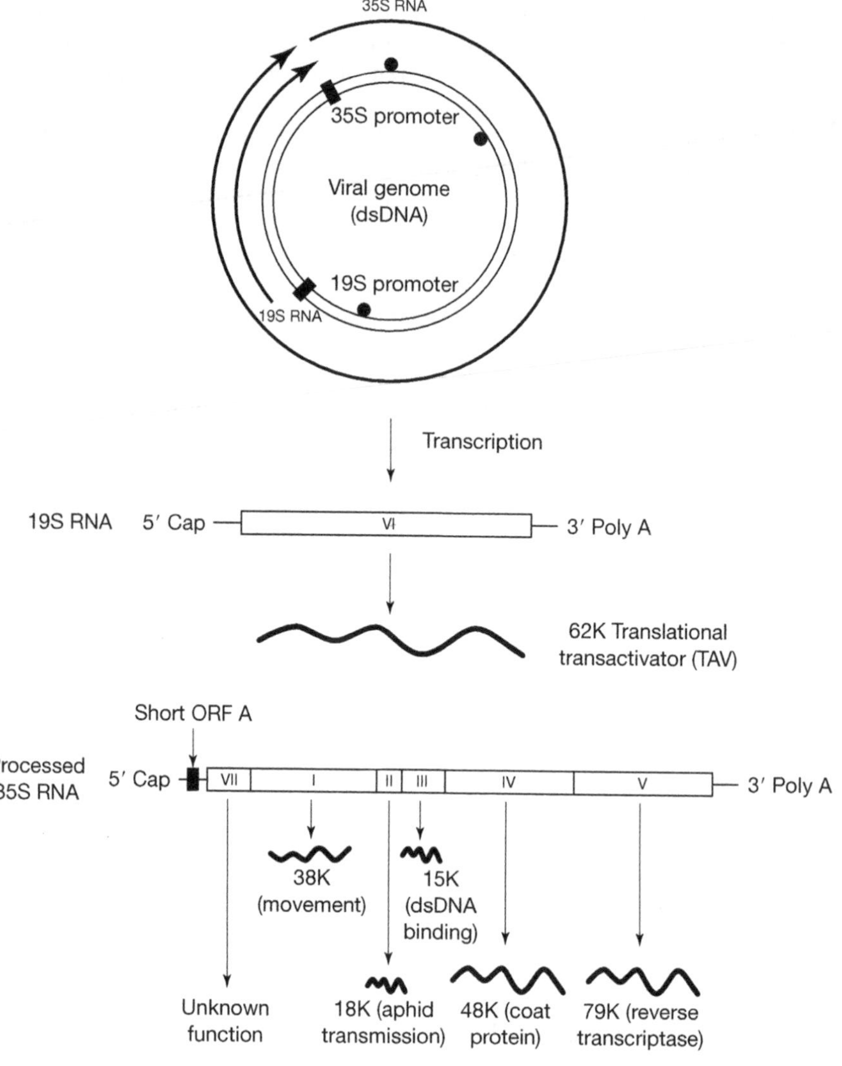

Figure 7.8

The genome of *Cauliflower mosaic virus.* The dsDNA genome is closed at the 3 discontinuities (marked as black dots) and two transcripts, the 19S and 35S RNAs are produced.

message are translated. The first of these is 'ribosome shunting'. In this, the scanning ribosome translates the first short ORF (sORF A) in the 35S leader, and then bypasses the AUG start codon of all the other short ORFs in the leader and is transferred to the AUG of gene *VII*. The secondary structure of the leader sequence is important for this to happen, along with the TAV protein encoded by the 19S RNA. This mechanism also allows ribosomes that have translated one ORF to remain competent and translate downstream ORFs in the polycistronic message. The second mechanism involves splicing of the 35S RNA, with spliced products having been detected in infected plants, and the spliced RNAs are used for translation.

The 35S RNA is also the template for replication in these DNA viruses, and the TAV/ribosome shunting mechanism is involved in regulating the use of this RNA for translation versus replication. Replication is via reverse transcription of the 35S RNA, but unlike retroviruses of vertebrates and bacteria, the infection cycle does not appear to involve integration into the host DNA, although viral sequences have been found in some plant genomes. However, integration of two badnaviruses, *Banana streak virus* (BSV) and *Tobacco vein-clearing virus* (TVCV) sequences into plant genomes is strongly correlated with virus infection. Here, it appears that environmental stresses may trigger activation of these integrated sequences (analogous to stress-induced activation of some transposable elements as discussed in Section 4.8) and these sequences may then recombine to generate novel viruses. How widespread this phenomenon is in other plant viruses is unclear, but the potential for generation of new viruses from episomal virus-like sequences could be important for generation of new plant diseases.

7.8 Viroids

Viroids are autonomously replicating, mechanically transmitted circular RNA molecules of 246–399 nucleotides. Despite this small size and the fact that they encode no proteins, they are responsible for many severe diseases. *Potato spindle tuber viroid* is a serious disease of potatoes in the USA, Russia and South Africa that results in elongated, tapered tubers and yield reduction of at least 25%. *Coconut cadang-cadang viroid* (dying disease) has killed more than 30 million coconut trees in the Philippines since its discovery in the 1930s and more than 1 million trees succumb to the viroid each year as it spreads due to the lack of any known control. How these RNA molecules exert their pathogenic effects is unclear although evidence suggests that they may interfere with protein kinases and other signal transduction mechanisms in plants causing hormonal imbalances and alterations in metabolic processes.

Viroids have been classified into two families, the *Pospiviroidae* and the *Avsunviroidae* (*Table 7.1*). The *Pospiviroidae,* which are the main group, appear to replicate in the nucleus of the plant and these viroids do not self-cleave, whilst the *Avsunviroidae* appear to replicate in chloroplasts and have ribozyme structures involved in self-cleavage during replication. The *Pospiviroidae* share a model of five structural-functional domains, the central conserved region (CCR), pathogenic region, variable region and terminal left and right regions. Whilst viroids are able to adopt a rod-like secondary structure *in vitro* the structure that they adopt *in vivo* is unknown, as is the function of the different genome regions.

Replication is believed to occur via a rolling-circle model, involving RNA intermediates. In the case of the *Pospiviroidae* this occurs via an asymmetric pathway and the α-amanitin-sensitive nuclear RNA polymerase II (*Figure 7.9*). In the *Avsunviroidae* the symmetric pathway is probably used along with an as yet unidentified α-amanitin-resistant RNA polymerase. The hairpin I structure present within the CCR region is believed to be the origin of replication, as it resembles the origin of replication in geminiviruses, and in the case of the *Pospiviroidae,* the RNAse and RNA ligase activities required for generation of unit length RNAs and their circularisation are probably encoded by enzymes from the host. In the *Avsunviroidae* however, the

Table 7.1 The classification of viroids

Viroid	Abbreviation	Size	Group	Sub-group
Avocado sunblotch viroid	ASBVd	246–250	*Avsunviroidae*	*Avsunviroid*
Peach latent mosaic viroid	PLMVd	336–339	"	*Pelamoviroid*
Chrysanthemum latent mosaic viroid	ChMVd	398–399	"	"
Potato spindle tuber viroid	PSTVd	356–360	*Pospiviroidae*	*Pospiviroid*
Tomato planta macho viroid	TPMVd	360	"	"
Citrus exocortis viroid	CEVd	370–375	"	"
Tomato apical stunt viroid	TASVd	360–363	"	"
Chrysanthemum stunt viroid	CSVd	354–356	"	"
Iresine viroid	IrVd	370	"	"
Columnea latent viroid	CLVd	370–373	"	"
Hop stunt viroid	HSVd	297–303	"	*Hostuviroid*
Coconut cadang-cadang viroid	CCCVd	246–247	"	*Cocadviroid*
Coconut tinangaja viroid	CTiVd	254	"	"
Hop latent viroid	HLVd	256	"	"
Citrus IV viroid	CVd-IV	284	"	"
Apple scar skin viroid	ASSVd	329–330	"	*Apscaviroid*
Citrus III viroid	CVd-III	294–297	"	"
Apple dimple fruit viroid	ADFVd	306	"	"
Grapevine yellow speckle 1 viroid	GYSVd-1	366–368	"	"
Grapevine yellow speckle 2 viroid	GYSVd-2	363	"	"
Citrus bent leaf viroid	CBLVd	318	"	"
Pear blister canker viroid	PBCVd	315–316	"	"
Australian grapevine viroid	AGVd	369	"	"
Coleus blumei 1 viroid	CbVd-1	248–251	"	*Coleviroid*
Coleus blumei 2 viroid	CbVd-2	301	"	"
Coleus blumei 3 viroid	CbVd-3	361–364	"	"

RNAse activity comes from the hammerhead ribozyme structures that self-cleave the RNA in both the production of the negative strand and in the production of the positive strand made from this template.

7.9 Other sub-viral entities

Along with viroids, a number of other sub-viral entities have been identified associated with plant diseases. These include satellite viruses, which are mRNAs that encode their own coat protein, but require another virus in order to replicate; satellite RNAs, which are encapsidated in virus particles along with the virus genomes; virusoids, which are viroid-like satellite RNAs; and satelloids, which are satellite RNAs in satellite viruses. None of these entities have autonomous replication, and in many cases they have no obvious effects on viral pathogenicity. However, there are some satellite RNAs and satellite viruses that do appear to attenuate virus symptoms, which has led to their use as biological control agents and also to their development in transgenic approaches for controlling plant virus diseases (see Section 14.1.4). Others, such as the *Cucumber mosaic virus* D satellite, cause systemic necrosis in host tomato plants by inducing programmed cell death.

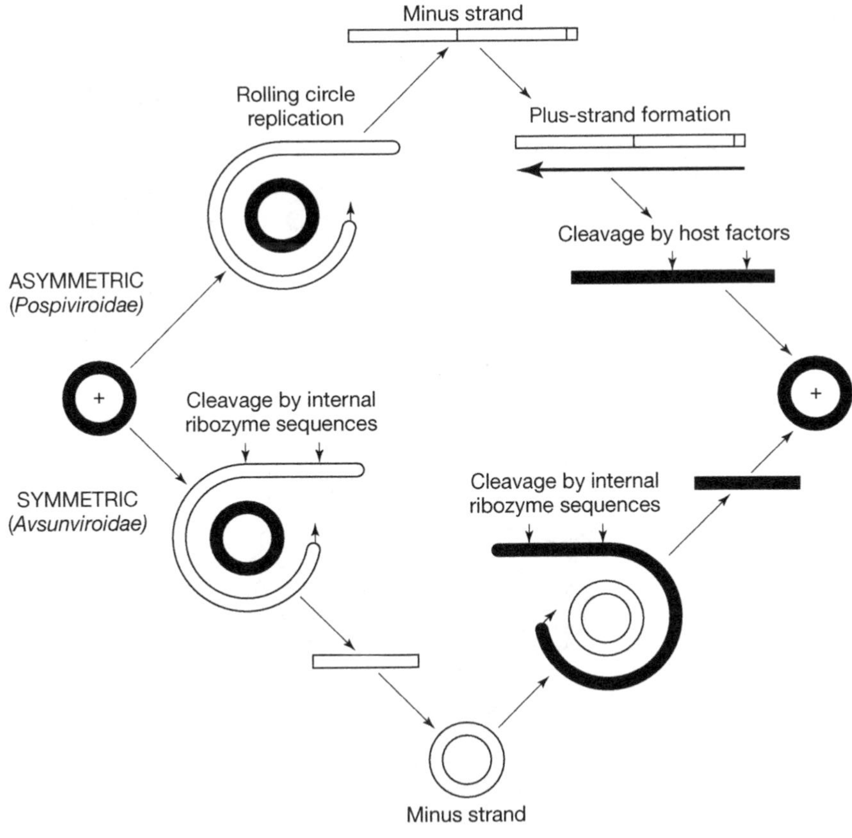

Figure 7.9

Replication of viroids. The asymmetric rolling-circle replication mechanism of *Pospiviroidae* (top), and the symmetric mechanism of *Avsunviroidae* (bottom).

7.10 Viral assembly

7.10.1 Assembly of rod-shaped viruses

Probably the best-studied example of virus reassembly stems from the studies of Butler and Klug on TMV. The virus consists of an RNA of 6400 nucleotides within a rod-shaped particle consisting of 2130 copies of the 158 amino acid, 17.6 kDa coat protein sub-unit. *In vitro* studies show that in the absence of RNA, the protein subunits assemble into discs containing 34 units in two rings of 17 (*Figure 7.10a*). In the presence of viral RNA (a condition that can be mimicked *in vitro* by lowering the pH from 7.0 to 6.5) the discs dislocate into helices of 16.33 subunits/turn. The viral RNA also possesses a specific protein-binding motif at which the packaging process is initiated. This motif, which forms into a hairpin loop structure in the RNA is not at the ends of the viral RNA, but approximately 1000 nucleotides form the 3′ end, in the movement protein ORF (in other rod-shaped viruses, similar initiation structures exist though not always in the same position). Having bound to the first 34 subunit disc and caused dislocation, a further 34 subunit discs are then recruited and dislocated as the RNA is drawn up

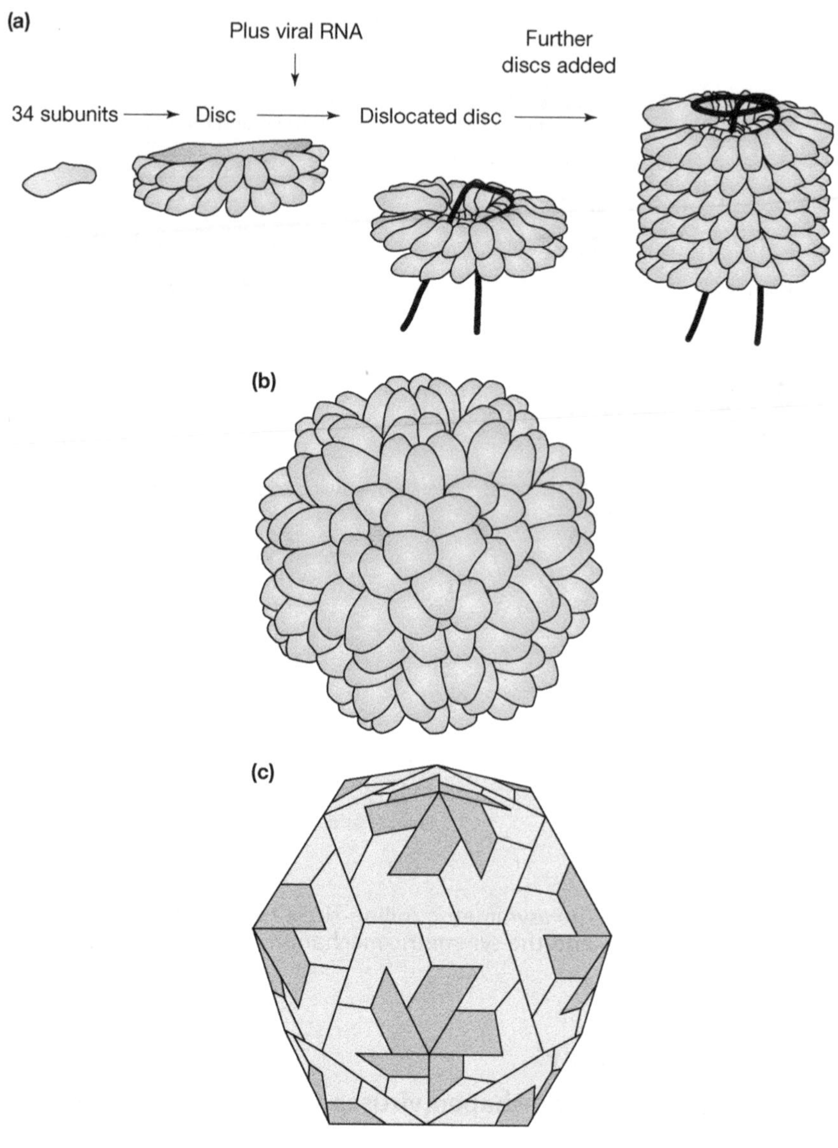

Figure 7.10

Virus assembly. (a) Assembly of *Tobacco mosaic virus.* (b) Structure of a T = 3 isometric particle, e.g. *Cowpea chlorotic mottle virus.* (c) The same symmetry for two coat proteins, e.g. *Cowpea mosaic virus.*

into a groove inside the helix, until the entire RNA is wrapped inside the helical protein coat.

7.10.2 Assembly of isometric particles

Assembly of isometric particles has also been shown in many cases to be a spontaneous process that can occur *in vitro* and in the absence of viral RNA under appropriate physiological conditions. The basic structure consists of 60T subunits, where T is an integer. For example, in *Cowpea chlorotic mottle virus* (CCMV) (a member of the *Bromoviridae*), T = 3, and the 180 identical homo-dimer subunits are arranged in 20 hexamers, and 12 pentamers at each end of the six 5-fold symmetry axes (*Figure 7.10b*). In other cases, where the coat is made of two coat proteins (e.g. *Cowpea mosaic virus*), the 12 pentamers are effectively composed of five subunits of the 23 kDa coat protein surrounded by five subunits of the 37 kDa coat protein, but the basic symmetry is the same (*Figure 7.10c*). Assembly does not occur as a single spontaneous event, but requires a number of lower order steps. In studies on CCMV, the coat proteins initially form into pentamers of dimers in solution, and these are the nucleation points for subsequent assembly. The viral RNA also has protein-binding motifs, which interact with the protein and are essential for formation of viable virus particles.

7.10.3 Assembly of membrane-bound particles

In the *Bunyaviridae* such as *Tomato spotted wilt virus*, the genomic RNA is encapsidated by multiple copies of the viral nucleocapsid protein (N), which are in turn encapsidated along with the RdRp replicase protein in a host-derived membrane bilayer. The assembly appears to be initiated through multiple RNA binding domains on the N protein and occurs at the Golgi in infected plants and also in insect vectors. Once the nucleocapsid has formed, this buds through a cellular membrane to acquire the lipid envelope, although how this process is initiated and controlled is unclear.

References and further reading

Adkins, S. (2000) Tomato spotted wilt virus – positive steps towards negative success. *Molecular Plant Pathology* **1**: 151–158.

Buck, K.W. (1996) Comparisons of the replication of positive-stranded RNA viruses of plants and animals. *Advances in Virus Research* **47**: 159–251.

Buck, K.W. (1999) Replication of tobacco mosaic virus RNA. *Philosophical Transactions of the Royal Society, London B* **354**: 613–627.

Callaway, A., Giesman-Cookmeyer, D., Gillock, E.T., Sit, T.L. and Lommel, S.A. (2001) The multifunctional capsid proteins of plant RNA viruses. *Annual Review of Phytopathology* **39**: 419–460.

Culver, J.N. (2002) Tobacco mosaic virus assembly and disassembly: determinants in pathogenicity and resistance. *Annual Review of Phytopathology* **40**: 287–308.

Drugeon, G., Urcuqui-Inchima, S., Milner, M., *et al.* (1999) The strategies of plant virus gene expression: models of economy. *Plant Science* **148**: 77–88.

Flores, R., Di Serio, F. and Hernández, C. (1997) Viroids: the noncoding genomes. *Seminars in Virology* **8**: 65–73.

Flores, R., Daròs, J.-A. and Hernández, C. (2000) *Avsunviroidae* family: viroids containing hammerhead ribozymes. *Advances in Virus Research* **55**: 271–323.

Gutierrez, C. (2000) DNA replication and cell cycle in plants: learning from geminiviruses. *The EMBO Journal* **19**: 792–799.

Harper, G., Hull, R., Lockhart, B. and Olszewski, N. (2002) Viral sequences integrated into plant genomes. *Annual Review of Phytopathology* **40**: 119–136.

Kao, C.C. and Sivakumarn, K. (2000) Brome mosaic virus, good for an RNA virologist's basic needs. *Molecular Plant Pathology* **1**: 91–98.

Lai, M.M.C. (1998) Cellular factors in the transcription and replication of viral RNA genomes: a parallel to DNA-dependent RNA transcription. *Virology* **244**: 1–12.

Mandahar, C.L. (ed.) (1999) *Molecular Biology of Plant Viruses*. Kluwer Academic Press, Dordrecht.

Mayo, M.A. and Brunt, A.A. (2001) The current state of plant virus taxonomy. *Molecular Plant Pathology* **2**: 97–100.

Palmer, K.E. and Rybicki, E.P. (1997) The use of geminiviruses in biotechnology and plant molecular biology, with particular focus on Mastreviruses. *Plant Science* **129**: 115–130.

Zaitlin, M. and Palukaitis, P. (2000) Advances in understanding plant viruses and virus diseases. *Annual Review of Phytopathology* **38**: 117–143.

Plant viruses – movement and interactions with plants

8

Having established the role of virally encoded proteins in replication and assembly of plant viruses, we need to consider the function of other viral genes and proteins in infection. These roles include facilitating the spread of the virus from plant to plant, spread of viruses within plants, and regulation of processes to overcome defence responses within the plant and to ensure a suitable environment for replication and spread.

8.1 Transmission of viruses

Unlike animal viruses, which can often rely on movement of the host animal to facilitate movement to new hosts, most plant viruses have to recruit vectors for transmission, or are carried in seed and pollen. Of the vector-transmitted viruses, 70% are transmitted by arthropods, mainly phloem-feeding homoptera, the remainder being carried by nematodes and zoosporic soil-borne fungal-like organisms. Some viruses, such as TMV may also be carried by larger animals including humans, brushing against and abrading the surface of infected plants and then transmitting the virus as they subsequently abrade healthy plants.

All normal vector transmission requires a degree of specificity in the virus/vector relationship, and it has become apparent that the amount of genetic variation that can be tolerated in a viral genome is often determined by maintaining this need to be transmitted, and by this specificity. Indeed, the host range of viruses is to a large extent determined not by which plant species the virus can replicate in, but by the host range of its vector. In cases where this has been increased through introduction of vectors into new habitats (e.g. the Old-World B-biotype of the whitefly *Bemesia tabaci* into the Americas), the virus has readily adapted to new plant species. Furthermore, there are no individual virus species that are capable of being transmitted by insects from more than one family.

8.2 Transmission by insect vectors

Insect transmission of plant viruses has traditionally been categorised according to the nature of the association with the vector (*Table 8.1*). Stylet-borne viruses that are acquired and carried on the mouthparts are termed non-persistent because they are lost from the insect once it feeds (*Figure 8.1*).

Table 8.1 Mode of transmission of plant viruses

Stylet-borne Non-persistent Non-circulative	Foregut-borne Semi-persistent Non-circulative	Persistent Circulative	Persistent Propagative
Alfamovirus	*Badnavirus*	*Begomovirus*	*Tospovirus*
Carlavirus	*Caulimovirus*	*Curtovirus*	*Marafivirus*
Cucumovirus	*Closterovirus*	*Mastrevirus*	*Phytoreovirus*
Fabavirus	*Sequivirus*	*Enamovirus*	*Fijivirus*
Machlomovirus	*Trichovirus*	*Luteovirus*	*Oryzavirus*
Potexvirus	*Waikavirus*	*Polerovirus*	*Cytorhabdovirus*
Potyvirus		*Nanovirus*	*Nucleorhabdovirus*
		Umbravirus	*Tenuivirus*
		Bromovirus	
		Carmovirus	
		Comovirus	
		Sobemovirus	
		Tymovirus	

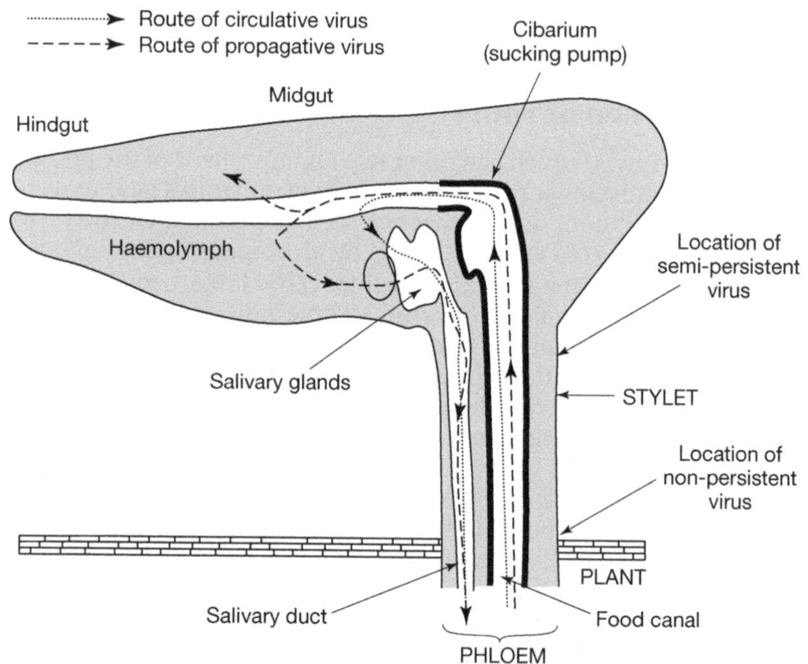

Figure 8.1

Stylised diagram of an insect's feeding system showing the extent of virus uptake. Non-persistent viruses are carried on the stylet, semi-persistent viruses on the foregut and persistent viruses are taken up into the haemolymph and salivary glands and can be divided into the circulative viruses and the propagative viruses.

Foregut-borne viruses are known as semi-persistent, whilst those that pass through the gut into the haemolymph and then to the salivary glands are known as persistent, because they can often be transmitted for as long as the insect lives. Persistent viruses in turn are sub-divided into the propagative viruses, which are able to replicate in both plants and the insect vector, and circulative, which can only replicate in plants.

Mutagenic analysis on plant viruses has shown that they encode polypeptides with domains that are essential for insect transmission. In most non-circulative viruses, these domains are in the coat protein and in a second protein known as the helper component (HC). In the *Potyvirus* genus, for example, the HC (termed HC-Pro) non-structural protein of 53–58 kDa is believed to form into a heterodimer and act as a bridge between the virus and the vector, having one domain that binds to the viral coat protein, and a second to as yet uncharacterised receptors in the vector mouthparts or saliva. The binding domain in the coat protein has been identified by mutagenesis and deletion analysis as a 7 amino acid domain encompassing the amino acid triplet Aspartic acid-Alanine-Glyine (DAG), which is exposed on the surface of the viral capsid. This interacts with a KITC sequence in the N-terminal conserved cysteine-rich domain of the HC. The HC-coat protein complex then binds to the cuticle lining the mouthparts. However, it has yet to be determined whether it is the HC protein, or the viral coat protein modified through its association with HC, that binds to the putative insect receptors. A DAG motif is also present in the coat protein of the aphid-transmitted *Potato aucuba mosaic virus* potexvirus, but not in the non-aphid transmitted close relative PVX, suggesting that this domain may be used by other viruses for aphid transmission.

The same bridging hypothesis has been invoked as the mechanism for viral attachment in semi-persistent viruses such as the *Caulimovirus* genus. Here, the gene II protein of *Cauliflower mosaic virus* has been shown to possess two functional domains, one of which interacts with viral particles and the other with specific sites in the aphid stylet. Furthermore, to increase the chances of a probing insect acquiring the virus, the gene II protein organises the virus particles into large inclusion bodies in the cell cytoplasm.

In the persistently transmitted geminiviruses, the coat protein has been shown to be the key component in vector attachment and specificity. Exchanging the coat protein from the whitefly-transmitted *Begomovirus, African cassava mosaic virus* (ACMV) with that of the leafhopper-transmitted *Curtovirus, Beet curly top virus* (BCTV) resulted in a change in the vector from whitefly to leafhopper. However, the specific binding domains in these virus coat proteins have yet to be identified. In the *Luteovirus* genus of persistent circulative viruses, the binding factors have been identified. Here, it has been shown that the readthrough domain of the surface-exposed minor capsid protein binds to the GroEL protein of the endosymbiotic bacterium *Buchnera* sp. that resides in the haemolymph of the aphid.

Persistent propagative viruses, such as *Tomato spotted wilt virus* (TSWV) often have lipid-containing protein coats and other characteristics that distinguish them from the major groups of plant viruses, and there has been a speculation that these viruses originated from insect viruses that developed the capacity to replicate in plants. TSWV has its RNA contained within a nucleocapsid, which is in turn surrounded with a host-derived lipid membrane into which two virally encoded glycoproteins, G1 and G2 are embedded. The presence of these glycoproteins is essential for acquisition, which can only occur in larval thrips. However, once acquired, the virus can be transmitted by both larvae and adults. An RGD (arginine-glycine-aspartic acid) motif present in G2 has been shown to be involved in attachment, and a 50 kDa protein(s) to which this attaches has been identified in larval midguts. It has been suggested that following binding, the virus enters the insect cells through receptor-mediated endocytosis involving clathrin-coated vesicles, similarly to many animal viruses.

8.3 Transmission by nematodes

Soil-inhabiting nematodes of the genera *Longidorus, Paralongidorus* and *Xiphinema* are able to transmit the nepoviruses and dianthoviruses, whilst nematodes of the genera *Trichodorus* and *Paratrichodorus* transmit members of the *Tobravirus* genera. They acquire the virus by feeding on infected roots, and the viruses have been shown to attach to the surface of the food canal, although there is no evidence that they replicate in the nematodes. The viral coat proteins are required for transmission and it is possible but not proven that virally encoded helper components are also required, and that attachment is similar to the HC-Pro bridge formation in aphid-transmitted potyviruses.

8.4 Transmission by zoosporic 'fungi'

A number of viruses such as the *Furovirus* genus and some genera of the *Potyviridae* are transmitted by soil-borne zoosporic protozoa *Polymyxa* and *Spongospora* species (see *Table 1.4*). These organisms possess many fungal-like characteristics (see Section 1.2), but only the necroviruses are transmitted by true fungi of the *Olpidium* species. These fungal and fungal-like organisms transmit plant viruses either non-persistently, when the virus is absorbed to the surface of the motile zoospores, or persistently, in which case the virus can become internalised into both zoospores and the resting spore stage. These 'fungi' are not themselves pathogens of plants, but when they become associated with roots, the virus is transmitted into the plant. In some viruses, such as *Beet necrotic yellow vein virus,* it has been shown that deletion of the four amino acid motif KTER in the readthrough domain of the capsid protein is important for transmission, but the details of this mechanism have not been determined. Furthermore, this domain is not present in other fungally transmitted viruses, suggesting that protein secondary structures rather than specific amino acid motifs may be important. The mechanisms of fungal transmission are not yet understood.

8.5 Seed and pollen transmission

Around 20% of all plant viruses can be transmitted through seed of one or more plant species, and are mostly contained within the embryo. They can invade the developing embryo in one of two ways, either indirectly by moving into the gametes prior to fertilisation, or directly by infecting the embryo after fertilisation. This capacity for seed transmission is determined by both host and viral genes. Genetic studies looking at the potyvirus *Pea seed-borne mosaic virus* (PSbMV) infection of pea cultivars has indicated that some as yet uncharacterised host genes expressed in maternal tissues are important for seed transmission. It has been shown that a single recessive gene in barley conditions resistance to *Barley stripe mosaic virus* (BSMV) seed transmission. Seed-transmitted viruses of *Arabidopsis* are currently being sought and characterised in an attempt to elucidate the host genes involved in this model system.

Viral components required for seed transmission have been localised to the HC-Pro and coat protein regions of the PSbMV genome, genes that are known to be important in viral replication, movement and insect transmission. Since it is likely that the virus exploits the embryonic suspensor to gain access to the embryo, it is important that the virus is transported to this region, and the role of HC-Pro and the coat protein may therefore be in ensuring effective long-distance movement of the virus by the mechanism that will be discussed in Section 8.7.

8.6 Short-distance movement of viruses in plants

To spread within plants, viruses must move from the initial infection site systemically to other parts of the plant. Although there are a few xylem-inhabiting viruses, the majority are normally introduced initially into epidermal and/or mesophyll cells in leaves, and must then find their way into the phloem (*Figure 8.2*). Once in the phloem, the virus spreads long distances along with photoassimilates from source tissues to sink tissues such as young

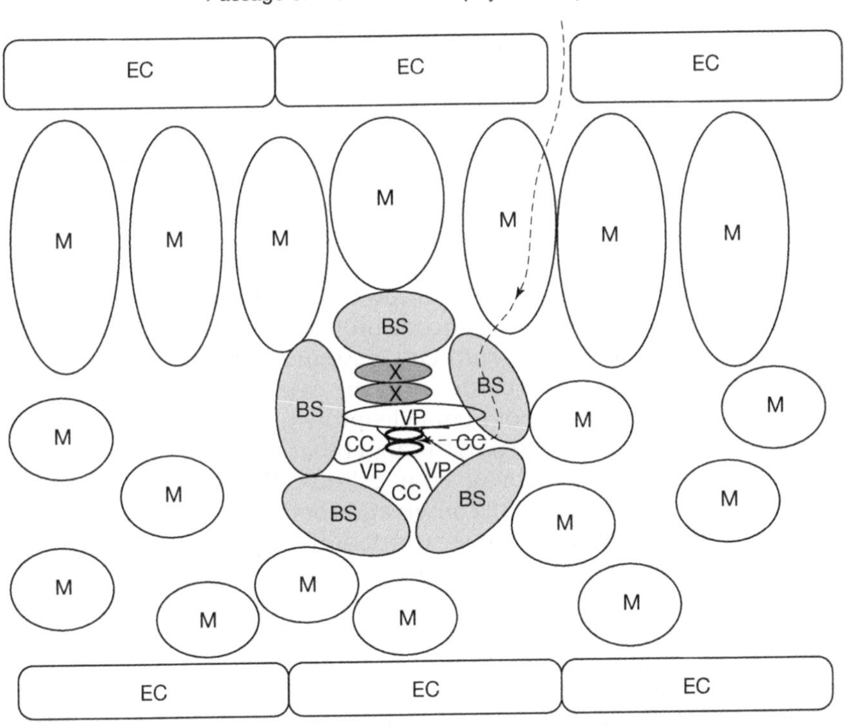

Figure 8.2

Cross section through a plant leaf showing the route of viral movement into the phloem. EC = epidermal cells, M = mesophyll cells, BS = bundle sheath cells, X = xylem, VP = vascular parenchyma cells, CC = companion cells, ⬭= sieve element (phloem).

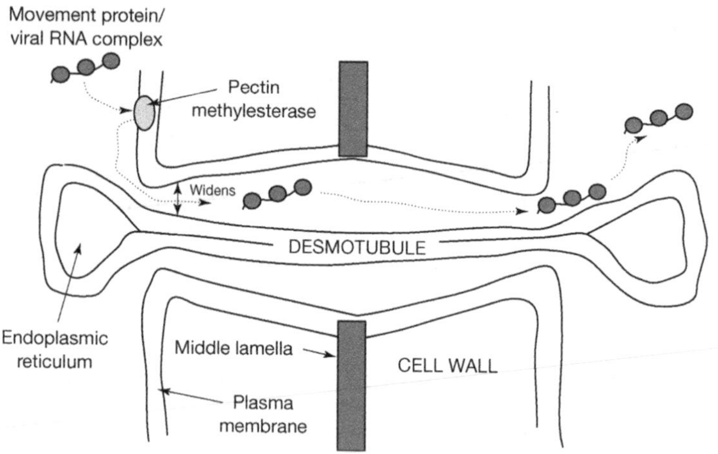

Figure 8.3

Structure of a plasmodesmata. Movement protein complexes appear to interact with proteins (e.g. pectin methylesterases) within the cell wall and then move to the plasmodesmata, which widen to allow the movement protein/viral RNA complexes through.

leaves, where it is unloaded back into the mesophyll cells. The process of movement from cell to cell into the phloem is referred to as short-distance movement, and movement in the phloem as long-distance movement.

Short-distance movement involves transfer of the virus between epidermis and mesophyll cells, into bundle sheath cells, vascular parenchyma cells and companion cells before entry into the phloem. All these cells possess symplastic connections known as plasmodesmata, which are small channels through the cell wall containing cylinders of endoplasmic reticulum called desmotubules. For example, in tobacco, the normal size exclusion of plasmodesmata is around 1 kDa, allowing solutes of less than 1.5 nm radius to pass through (*Figure 8.3*). However, this is too narrow for passage of intact viruses, which therefore encode proteins referred to as movement proteins (MPs). The role of these proteins is essentially to widen the plasmodesmata so that either intact particles (in the case of some viruses such as *Cowpea mosaic virus* (CPMV)) or ribonucleoprotein complexes (for example in the case of TMV) are able to pass through them.

In many viruses, the MPs involved in cell-to-cell movement have been identified, and there is significant conservation between them such that they have been grouped into one large superfamily of proteins. In TMV and many other viruses, there is a single movement protein encoded by a single gene (the 30 kDa protein in TMV), whilst in others such as the potexviruses, hordeiviruses, benyviruses and pecluviruses, there are three proteins encoded by overlapping reading frames, referred to as the 'triple gene block'. In PVX for example, the block encodes a 25 kDa, 12 kDa and 8 kDa protein. The 25 kDa protein has RNA-binding domains and is believed to envelope the viral RNA to move it from cell to cell, whilst the other two proteins facilitate movement through interactions with cellular components. Geminiviruses contain two separate movement proteins, and it is believed that one of these is involved in cell-to-cell movement and the other is required for transport of the ssDNA from the nucleus to the cytoplasm.

The main features of MPs are that they are non-structural, transiently expressed nucleic-acid-binding proteins that can widen plasmodesmata and move through them. It has also been found that these proteins are often not virus specific and can facilitate the movement of other unrelated viruses. In some viruses (e.g. the potyviruses) no obvious movement protein has been identified. Here, movement is associated with domains in the coat protein, the CI helicase protein and possibly the VPg and HC-Pro helper component protein. For cell-to-cell movement, the CI and core of the coat protein are important, whilst for loading into the phloem and systemic movement, mutational analysis has shown that the N- and C-terminal regions of the coat protein, the VPg and the HC-Pro are required.

The mechanisms by which MPs work have been most extensively studied for the TMV 30 kDa protein which is synthesised in the cytoplasm of infected cells and has at least five defined functional features. The first of these is to act as a nucleic-acid-binding protein to bind to TMV RNA, and studies have shown that this protein binds single-strand nucleic acids with greater efficiency than double-stranded molecules. Studies using GFP-tagged movement proteins have shown that this complex then interacts with microtubules and actin microfilaments in the host cell cytoskeleton, which may help to target the nucleoprotein complex towards the plasmodesmata. Further binding studies and work with yeast two-hybrid systems have shown that the MP then interacts with pectin methylesterase (PME) within the cell wall, which may be a specific target close to the plasmodesmata, although this has not been confirmed. The MP then increases the size exclusion of the plasmodesmata and acts as a chaperone protein as it transports the viral nucleic acid and itself through the plasmodesmata and into the adjacent cell. How this is achieved is unclear but the process appears to be controlled through phosphorylation of the 30 kDa protein of TMV by host protein kinases at key residues such as Ser 258, Thr 261 and Ser 265. Following movement, the MP accumulates within the plasmodesmata blocking further viral movement, and this may have a role in preventing spread of host factors involved in antiviral surveillance. Eventually, the plasmodesmata return to their normal size exclusion limits so that uncontrolled leakage of large solutes does not occur. MPs from other viruses have been shown to function in similar ways with phosphorylation and association with host proteins (for example *Turnip crinkle virus* MP associates with a potential membrane protein from *Arabidopsis*) being important parts of the process. A further feature of some MPs, both those involved in moving nucleoprotein complexes and intact virus particles, is that they form tubule structures at the plasmodesmata through which the viruses move. These tubules, composed of the movement protein, appear to pass through the plasmodesmata and replace the endoplasmic reticulum desmotubules to form 40–80 nm wide pores through which the virus then passes as a single row. How the viruses are moved is unclear but it has been suggested that these movement proteins may possess ATPase activity, or alternatively that actin filaments in the plant walls push the virus through a 'treadmilling' mechanism.

8.7 Long-distance movement in plants

Once they have entered the sieve elements, viruses have to move from source to sink tissues, and whilst this is partly a consequence of moving along with plant photoassimilates, there are also viral proteins involved in regulation of the process. Because the connections between adjacent sieve elements are much wider than plasmodesmata, passage of viruses is not physically impeded. However, in many cases movement proteins and also coat proteins have been shown to be essential for long-distance transport. Part of this role is protection of the viral RNA from nuclease degradation. However, some viruses, for example *Groundnut rosette virus* and *Barley stripe mosaic virus* are able to move without coat proteins, and here it is believed that other proteins are recruited by the RNA to protect it in place of the coat proteins. Viroids have also been shown to associate with specific phloem proteins into ribonucleoprotein complexes presumably to aid movement and for protection of the RNA. What is less clear is the mechanism by which viruses are unloaded from the phloem back into non-vascular tissue. In tobacco, the heavy metal cadmium blocks the exit of TMV from the vascular tissue, but has no apparent effect on its loading into the system, indicating that the mechanisms for exit and entry are different. What factors cadmium inhibits in plants to block this transport have not been determined.

8.8 Viral affects on plants

Viral infections of plants result in major alterations to host gene activities and metabolic processes. These include the activation of a vast array of defence responses in the plant along with the accumulation of starch and soluble sugars, increased respiration, decreased photosynthesis and a rise in levels of amino acids and organic acids in infected areas. As the virus infection front spreads and moves on, so the tissues in which these changes are occurring alters. At the same time, the older-infected tissues change to a state in which they are able to survive and perform normal metabolic functions whilst the virus persists as a non-replicative entity. The virus must therefore be able to induce a permissive state for its own replication in some cells and at the same time avoid or suppress defence responses. In turn, and as with bacteria and fungi, plants have developed surveillance mechanisms to allow them to detect the presence of viral components and elicit defence responses.

8.8.1 Alterations in host gene expression

Virus infection results in the diversion of resources away from normal cellular processes as they are recruited for virus replication. One process used to achieve this diversion is host gene 'shutoff' in which the expression of particular cellular genes is down-regulated through virus infection. This shutoff mechanism can be non-specific, in that it is the presence of viruses generally that results in activation of the mechanism, or specific, in which a particular virus activates a particular response. Non-specific mechanisms include the activation of double-stranded RNA-dependent protein kinase (PKR), an enzyme that phosphorylates the α-subunit of the eIF-2 initiation factor, resulting in inhibition of translation of genes. There may also be mechanisms in which host RNAs are degraded due to ribonuclease activities in the viral proteins, as occurs for some animal viruses such as Herpes Simplex Virus-1.

To ensure that the infecting virus is able to replicate, not all genes are down-regulated by these shutoff mechanisms and some escape the control.

For example, the genes encoding the heat shock chaperone protein HSP70 and an NADP$^+$-dependent malic acid enzyme are up-regulated during *Cucumber mosaic virus* infection, and during *Pea seed-borne mosaic virus* (PSbMV) infection mRNAs for HSP70, polyubiquitin and glutathione reductase 2 were shown to accumulate when other messenger RNAs were declining, and gene expression for actin and β-tubulin remained unchanged. The latter proteins are known to be important for cell-to-cell movement of viruses (see Section 8.6), which may explain why it is important for viruses to maintain levels of these. The role of heat shock proteins is less clear, but it has been suggested that the process of virus infection is very similar to abiotic stresses such as heat shock in which changes in gene expression are transient, but plant cell death is not desirable. This is an important difference between viral infection of animals compared to plants, since in animals infection generally results in necrosis and lysis of cells, whilst cells infected with plant viruses generally show recovery. Indeed, for PSbMV, it has been shown that shut-off occurs in a zone of just 6–8 cells behind the infection front in peas, viral replication occurs in a zone of 10–12 cells, and that transcript levels in cells behind this zone return to normal and at the same time, viral replication ceases.

8.8.2 Alterations in host cell metabolism

Whilst general gene expression in cells appears to return to normal once the infection front has passed through, it is clear that changes occur to metabolic processes in these cells, resulting in the characteristic symptoms of viral infection such as vein clearing, chlorosis and mosaics. In general there is loss of photosynthetic capacity, increased respiration, changed carbohydrate partitioning and altered starch accumulation. How these changes occur is not clear but it has been postulated that plant signal molecules such as salicylic acid may be involved. Soluble sugars may also be involved, as these accumulate at the infection site and have been shown to repress genes involved in photosynthesis, and induce genes involved in synthesis of storage compounds, and some defence-related genes. In *Cucumber mosaic virus*-infected cucurbit, there is an initial localised increase in photosynthesis, NADP$^+$-dependent malic acid enzyme activity and oxidative pentose phosphate pathway activity in cells in which the virus is replicating. This is followed by changes in starch levels, and increased rates of respiration and glycolysis. Studies using PVY infection of transgenic tobacco lines expressing cell wall invertases and lacking salicylic acid have suggested that virus infection increases apoplastic sucrose levels, possibly because of the alterations to plasmodesmata caused by viral movement proteins (*Figure 8.4*). This sucrose is then converted to the hexoses, glucose and fructose by cell wall invertases, and these hexoses are then taken up into the mesophyll cells. Elevated levels within these cells result in further increases in invertase activity to re-enforce the conversion, and at the same time this results in the down-regulation of photosynthetic genes and the zones of chlorosis that appear as mottles and mosaics.

8.8.3 Suppression of defence responses

A further consequence of elevated hexose levels is an increase in defence-related genes within the zone through which the virus has passed. However,

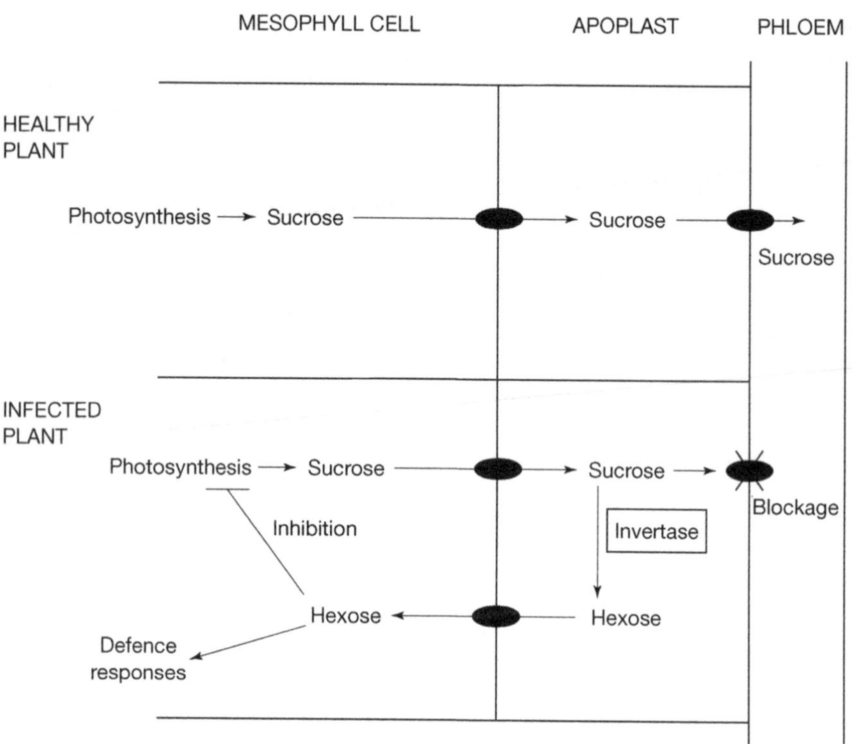

Figure 8.4

A model for the role of soluble sugars in alterations to host cell metabolism during viral infection. In healthy plants, sucrose is transported into the phloem. In infected plants, this transport is blocked, and the apoplastic sucrose is converted to hexoses as described in the text.

this presumably occurs too late to stop the virus moving into the phloem and spreading through the plant, and it may be that they represent a response to the toxic effect of high sugar levels rather than to the virus. Plants possess a number of additional defences against viruses. Gene-for-gene resistance against viruses is discussed in Sections 8.9 and 10.1. However, a further line of defence that has become apparent through studies on transgenic plants, is that of post-transcriptional gene silencing (PTGS). PTGS, which incorporates the phenomena termed co-suppression and virus-induced gene silencing (VIGS) is considered to be the plant equivalent of RNA interference (RNAi) in animals and 'quelling' in fungi. It is the process in which host organisms appear to detect 'foreign' RNAs once they are above a certain threshold level, and then degrade these RNAs by producing short complementary RNAs that anneal to the invading RNA producing a target for RNA-degrading enzymes. The evidence that this is a defence mechanism against plant viruses is fourfold. Firstly, dsRNA has been shown to be a potent inducer of PTGS, and most plant viruses replicate through dsRNA intermediates. Secondly, in *Arabidopsis* mutants in which PTGS is impaired, there is hypersensitivity to viral infection. Thirdly, some viruses have been shown to possess mechanisms to suppress PTGS, and fourthly, PTGS is often restored in cells that recover from virus infections.

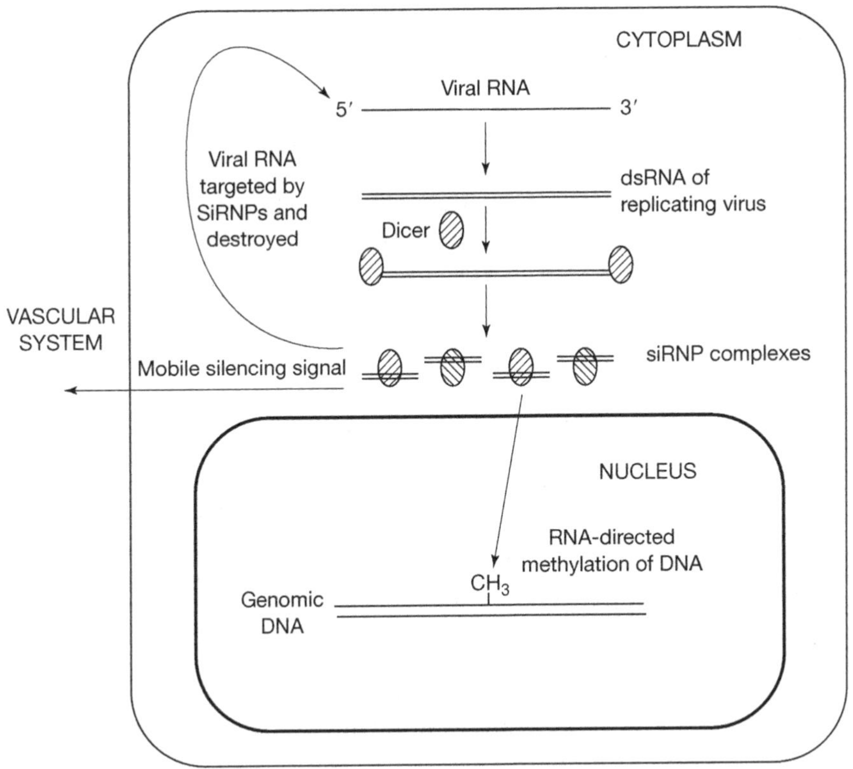

Figure 8.5

Model for post-transcriptional gene silencing against viruses in plants. The dsRNA of the replicating virus is degraded to small (21–25 nucleotide) RNAs by the plant equivalent of the DICER-1 enzyme.

RNAi, the mechanism equivalent to PTGS, has been extensively studied in *Drosophila melanogaster* and nematodes, using radiolabelled dsRNA and target ssRNAs. The process occurs in two stages (*Figure 8.5*). Initially, the dsRNA is degraded into 21–25 nucleotide RNAs (i.e. 2 helix turn) from both ends, by the DICER-1 RNase III-type enzyme. An *Arabidopsis* homologue of this (CAF1, SIN1 or SUS1), has recently been identified. Each of these 21–25 nucleotide fragments is then associated into a dicer-containing small interfering ribonucleoprotein (siRNP) complex, such that the dsRNA becomes un-paired and anneals with its complementary target RNA. The ssRNA is then cleaved in the middle of the 21 bp sequence. How the process is controlled to ensure that enough ribonucleotide complexes are produced is unclear, and it has been suggested that the siRNP complexes act as gatekeepers at the nuclear pores and plasmodesmata to select out and destroy foreign RNAs. It is also believed that the dicer complexes are involved in directing DNA methylation of the transcribed region of the silenced gene, and control transcription of genes through this process, although the role of such control in PTGS resistance to viruses remains unclear.

A further feature of PTGS is that once initiated, it can spread systemically through plants in a manner analogous to viral spread, i.e. through plasmodesmata and then into the phloem. These signals are able to cross graft junctions in plants. Whether this signal is the dicer-dsRNA siRNP complex or something else, and what the role of this signalling is remains unclear. However, it has been suggested that the invading virus and silencing mechanism may enter a race through the vascular system such that if the virus moves faster, systemic infection occurs, whereas if the silencing mechanism catches the virus, the infection is aborted.

Whilst PTGS is a defence mechanism that is capable of degrading viral RNA, it is clear that it is not a very successful strategy for resistance to disease. In part, this may be because it is unable to cope with the rapid rate at which viruses replicate, but there is also now clear evidence that viruses possess mechanisms to suppress PTGS. One mechanism identified in the potyviruses stems from the observations of synergisms between viral infections, in which plants containing PVY are more susceptible to infections with other unrelated viruses such as PVX. This suggests that PVY somehow modifies the plant to help the PVX infection. Further studies using plants expressing the PVY HC-Pro protein from a transient expression system in plants containing a GFP marker transgene, have indicated that HC-Pro blocks both initiation and maintenance of PTGS through blocking accumulation of the small RNAs associated with silencing, but DNA methylation still occurs. Conversely, experiments using the 2b protein from *Cucumber mosaic virus* have shown that this protein suppresses PTGS only in the shoot apex, suggesting that it interferes with PTGS initiation. It therefore appears that viruses have evolved more than one mechanism to evade PTGS to ensure their own replication and spread, and that this can work synergistically to aid infection by other viruses.

8.9 Gene-for-gene interactions with plant viruses

As for fungi and bacteria, plants utilise gene-for-gene resistance mechanisms (see Section 10.1) to recognise viral proteins and elicit defence responses. Since all viral genes are essential for pathogenicity, it is not surprising that these are all targets for gene-for-gene resistance and could be defined as classical avirulence genes. For example, the *N′* resistance gene in *Nicotiana sylvestris* recognises the coat proteins of most tobamoviruses, though not that of TMV, and there are *R* genes in peppers and eggplants that recognise different domain of tobamovirus coat proteins. Replicase genes are also elicitors of gene-for-gene resistance, and the *N* gene from *N. glutinosa* recognises the C-terminal helicase domain of the TMV 126 kDa RdRp protein, as does the tomato *Tm-1* gene, whilst the *Tm-2* and *Tm-2*^{2} resistance genes in tomato recognise the TMV 30 kDa movement protein. More recently, it has been shown that the amino-terminal sequence of the Tav2b suppressor of PTGS from the *Tomato aspermy virus* is recognised by a gene-for-gene mechanism in tobacco, indicating how complex the various defence strategies in plants and counter-defence mechanisms in viruses are.

8.10 Genomic variation in plant viruses

As with fungi and bacteria, viruses are able to mutate and evolve to overcome plant defence strategies such as gene-for-gene resistance. Mutation rates in RNA plant viruses are notoriously high (10^{-4} per nucleotide per round of replication cycle) because of the lack of proofreading activities for RNA replicating enzymes. Viral genomes are also prone to insertions and deletions. However, the efficient use of both coding and non-coding regions shows that viral genomes are highly evolved, and they cannot support many mutational changes. Therefore, even though these high mutation rates and genomic instabilities result in progeny from viral infections composed of individual particles with different genomic sequences, most of these mutations are not passed on to the next generation. In addition, it has been shown that defective interfering (DI) RNAs, which result from deletions of helper viruses, often result in less severe infections and interfere with virus accumulation, supporting the view that such changes reduce the evolutionary fitness of viruses. Thus, plant viruses appear to be highly evolved entities which remain important for furthering our understanding of many biological processes, as well as being significant plant pathogens.

References and further reading

Aranda, M. and Maule, A. (1998) Virus-induced host gene shutoff in animals and plants. *Virology* **243**: 261–267.

Callaway, A., Giesman-Cookmeyer, D., Gillock, E.T., Sit, T.L. and Lommel, S.A. (2001) The multifunctional capsid proteins of plant RNA viruses. *Annual Review of Phytopathology* **39**: 419–460.

Erickson, F.L., Dinesh-Kumar, S.P., Holzberg, S., *et al.* (1999) Interactions between tobacco mosaic virus and the tobacco *N* gene. *Philosophical Transactions of the Royal Society, London B* **354**: 653–658.

García-Arenal, F., Fraile, A. and Malpica, J.M. (2001) Variability and genetic structure of plant virus populations. *Annual Review of Phytopathology* **39**: 157–186.

Hannon, G.J. (2002) RNA interference. *Nature* **418**: 244–251.

Herbers, K., Yakahata, Y., Melzer, M., Mock, H-P., Hajirezaei, M. and Sonnewald, U. (2000) Regulation of carbohydrate partitioning during the interaction of potato virus Y with tobacco. *Molecular Plant Pathology* **1**: 51–59.

Jorgensen, R.A., Atkinson, G.R., Forster, R.L.S. and Lucas, W.J. (1998) An RNA-based information superhighway in plants. *Science* **279**: 1486–1487.

Lee, J-Y. and Lucas, W.J. (2001) Phosphorylation of viral movement proteins – regulation of cell-to-cell trafficking. *Trends in Microbiology* **9**: 5–8.

Lindbo, J.A., Fitzmaurice, W.P. and Della-Cioppa, G. (2001) Virus-mediated reprogramming of gene expression in plants. *Current Opinion in Plant Biology* **4**: 181–185.

Matzke, M.A., Mette, M.F., Aufsatz, W., Jakowitsch, J. and Matzke, A.J.M. (1999) Host defences to parasitic sequences and the evolution of epigenetic control mechanisms. *Genetica* **107**: 271–287.

Maule, A.J. and Wang, D. (1996) Seed transmission of plant viruses: a lesson in biological complexity. *Trends in Microbiology* **4**: 153–158.

Maule, A., Leh, V. and Lederer, C. (2002) The dialogue between viruses and hosts in compatible interactions. *Current Opinion in Plant Biology* **5**: 279–284.

Power, A.G. (2000) Insect transmission of plant viruses: a constraint on virus variability. *Current Opinion in Plant Biology* **3**: 336–340.

Rhee, Y., Tzfira, T., Chen, M.H., Waigmann, E. and Citovsky, V. (2000) Cell-to-cell movement of tobacco mosaic virus: enigmas and explanations. *Molecular Plant Pathology* **1:** 33–40.

Santa Cruz, S. (1999) Perspective: phloem transport of viruses and macromolecules – what goes in must come out. *Trends in Microbiology* **7**: 237–241.

Vance, V. and Vaucheret, H. (2001) RNA silencing in plants – defense and counterdefense. *Science* **292**: 2277–2280.

Waterhouse, P.M., Wang, M-B. and Lough, T. (2001) Gene silencing as an adaptive defence against viruses. *Nature* **411**: 834–842.

Resistance mechanisms in plants

9

Just as pathogens have evolved to colonise living plants, so plants have developed means to prevent or tolerate their presence. Because plants are unable to move to escape these challenges, they have developed many diverse and unique strategies, and as each mechanism is studied in greater detail, so new layers of complexity are uncovered. This chapter gives an overview of the range of mechanisms that are known to exist and how they relate to each other, before we look in detail at the genetics of resistance and structure of resistance genes in Chapter 10, and the signalling and cross-talk between resistance mechanisms in Chapter 11.

9.1 Classical concepts of resistance

Most plants are resistant to most microbes, and it is generally only specialist organisms that have evolved the capacity to overcome these defences, so that many pathogens have narrow host ranges. Classically, two levels of resistance have been defined. Non-host resistance is where the entire plant species or genus is resistant and therefore not a host for the particular pathogen. For example, wheat is not infected by the potato late blight oomycete, *P. infestans,* so is a non-host. Host resistance is where individuals within a species have developed genetically inherited ways of defending themselves against an organism that causes disease on other individuals within that plant species. For example, wheat is infected by yellow (stripe) rust, *P. striiformis,* but certain genotypes will be resistant to specific races of the fungus.

Whilst useful as concepts, these classical definitions do little to help with understanding the underlying mechanisms, which will involve both the lack of capacity for the potential pathogen to colonise this plant species, combined with constitutive and inducible defences in the plant. Perhaps the vector for the pathogen does not feed on the particular species, or the organism does not have the necessary pathogenicity factors to penetrate the host and/or access nutrients from it. Perhaps the cuticle of the plant is too thick, or the innate immune response in the plant (analogous to similar non-self recognition systems present in all eukaryotes), is able to recognise and respond to non-specific elicitors from the pathogen. Against some necrotrophic fungi, such as *Cochliobolus carbonum,* in which host-specific toxins are the essential pathogenicity factors, some plants may also possess toxin-degrading enzymes to prevent pathogen invasion.

Mechanisms of resistance in plants can be subdivided into two categories, passive (constitutive) and active (induced) (*Figure 9.1*). Passive mechanisms involve both structural elements, such as the cuticle and root border cells, and pre-formed antimicrobial chemical compounds within the plant termed phytoanticipins (see Section 9.2). These form the initial layers of protection

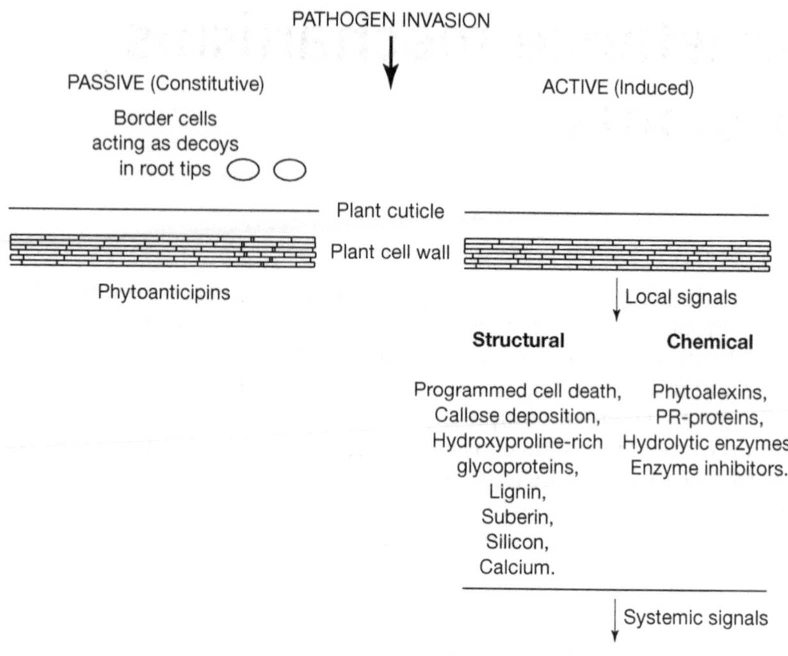

Figure 9.1

Examples of passive (left-hand side) and active (right-hand side) defence mechanisms in plants.

against microbial attack. Should these be breached, plants have a number of active or inducible defence mechanisms (see Section 9.3), which include the hypersensitive response (local plant cell death) and induction of specific gene expression within the plant, including genes involved in cell wall strengthening and/or repairing structural defences, genes for biosynthesis of additional antimicrobial compounds, and localised induction of genes encoding hydrolytic enzymes and other defence-related proteins.

In addition to the localised induction of defence responses, there are mechanisms that induce resistance in other parts of the plant through systemic signals, such as systemic acquired resistance (SAR) and induced systemic resistance (ISR) (see Sections 9.4 and 11.10). Furthermore plants can signal to neighbouring plants through volatile compounds to enhance resistance in these plants (see Section 9.5). It is the nature of these various mechanisms that we shall now consider.

9.2 Preformed defences

9.2.1 Structural barriers

The first passive barrier that defends plants against microbes is the wax layer present on the cuticle of leaves and fruits (see *Figure 2.7* in Chapter 2). This forms a water-repellent surface that prevents formation of water droplets necessary for bacterial ingress and fungal spore germination. Surface hairs on leaves possess a similar function. The thick cuticle, composed of cutin, cellulose and pectin, in combination with the tough walls on epidermal cells form an additional barrier against all pathogens apart from those fungi that have the necessary pathogenicity factors for direct penetration (see Section 2.5), and vectors that are potentially carrying pathogens. This leaves natural openings such as stomata as the main weakness in structural defences, and whilst some fungi are able to force their way through closed stomata, others, such as the cereal rust fungi, have to wait for stomata to open. This can form an additional line of defence in plants and some resistant wheat varieties delay stomatal opening until late in the day so that any rust germ-tubes that are on the leaf surface following germination of urediospores the previous night desiccate.

9.2.2 Root border cells

Root tips that are rapidly elongating as they move towards nutrients and water in the soil are a major potential weakness in the structural defences of plants. This rapid growth of newly synthesised tissue through an environment with a high capacity to cause abrasive damage that is rich in potentially pathogenic microbes makes them extremely vulnerable to attack. In these cells, inducible defences such as programmed cell death of a few plant cells (as occurs in other parts of the plant including behind the root tips) would not be an effective defence since this would terminate root growth. Instead, root tips have developed a mechanism through which they surround themselves with large numbers of detached somatic cells termed root 'border' cells, that effectively guard the root tip from pathogen attack. Production of these border cells is tightly regulated and may range from 12 per day in some plant species such as tobacco, to 10 000 per day for others such as cotton. Once detached from the root, the metabolic activity of these cells increases and gene expression undergoes a global switch to produce anthocyanins and antimicrobial antibiotics. This, in combination with the production of a mucilage layer repels bacteria. The border cells also appear to produce chemical signals that attract fungal zoospores, essentially acting as decoys, so that the border cells become infected rather than the root tips, and there is evidence that they may perform a similar decoy role against nematode infestations.

9.2.3 Phytoanticipins

Chemical barriers in plants have generally been classified as phytoanticipins or phytoalexins depending on whether they are preformed inhibitors of pathogens or synthesised *de novo* following pathogen attack. However, this distinction is somewhat arbitrary, since there are examples of compounds that may be preformed inhibitors in one plant species but induced in another. In other cases, it has not been possible to confirm that the compounds are at appropriate concentrations or locations in plants to be antimicrobial until induced, since many of the studies have been performed *in vitro* using non-physiological levels of the compound. The nature of these chemicals will be discussed in more detail in Section 9.3.4.

9.3 Induced defences

Defence responses in plants can be induced by a number of factors and the exact nature of the response varies depending on what these are. There is evidence that some defences are induced by the physical presence of fungal spores on the leaf surface, whilst others require the pathogen to penetrate the surface before induction. Induction may be in response to non-specific elicitors (the innate immune response that discriminates between self and non-self), or may follow the classical gene-for-gene resistance model originally described by Oort and Flor in the 1940s, which is the mechanism underlying host resistance (see Chapter 10). This is essentially a highly evolved form of inducible resistance, in which the product of a specific resistance gene in the plant is involved in recognition of a specific elicitor from the pathogen. These induced resistance responses have many similarities to the responses that occur when plants are wounded, but they are not identical. Also, natural processes such as abscission induce similar responses, indicating that localised strengthening of structural defences and the production of antimicrobials are fundamental processes in plants.

9.3.1 Local signals

One of the first responses activated in many incompatible interactions prior to the induction of gene expression and protein synthesis is the production of ion fluxes, reactive oxygen species (ROS) (also known as active oxygen species (AOS) or reactive oxygen intermediates (ROI)), production of nitric oxide (NO) and phosphorylation cascades. These are recognised as important signalling mechanisms in many organisms, and the means by which these local signals are generated in plants and their roles in defence responses are discussed in detail in Chapter 11.

9.3.2 Programmed cell death (PCD)

One of the most visible responses of plants to pathogen attack is the hypersensitive response (HR), which is the localised death of plant cells. This is a temporally and spatially co-ordinated mechanism to limit the amount of host tissue lost to the pathogen, and one that restricts the ingress of biotrophic pathogens that require living cells as their source of nutrients. Programmed cell death (PCD) is a general term used to describe the induced cell suicide that occurs in organisms, and amongst the events that occur in the dying cell as part of PCD are membrane blebbing, chromatin condensation and DNA cleavage. These phenomena have been shown to occur as part of the hypersensitive response in plants, and the significance of this is discussed further in Section 11.9.

9.3.3 Induced structural barriers

Plants have a number of inducible structural defences. The first of these is cytoskeleton-based to fend off attack from potential pathogens prior to penetration, involving sensing of the developing pathogen on the surface. These mechanisms have been shown to occur with barley powdery mildew, flax and cowpea rusts, rice blast and *Botrytis* spp. In the case of rice blast (*M. grisea*), it has been shown that these responses occur against *mps1* mutants that are unable to form penetration pegs and colonise plants, indicating that the signal must come from the surface contact and not as a wound response to penetration. At the site of contact, there is accumulation of cytoplasm and sometimes movement of the nucleus in plant cells, along with precise rearrangements of the plant microtubules in the cytoskeleton. Whether the actin filaments are involved in passing signals to the nucleus to induce gene expression is unclear, but there is evidence that plant defence proteins such as osmotin and chitinase associate with the actin cytoskeleton, possibly in preparation for potential invasion, or to stabilise the actin filaments. Formation of appositions referred to as papillae, consisting of callose (a β-1,3-glucan polymer) and phenolics, on the inner surface of cell walls, along with deposition of hydroxyproline-rich glycoproteins such as extensin, phenolic compounds such as lignin and suberin, and minerals such as silicon and calcium also occurs, and these become cross-linked to form insoluble defensive structures and an additional barrier against pathogen invasion and ingress. Blockage of plasmodesmata by callose is also believed to be a key defence against viruses inhibiting their movement. These same structural defences are induced by wounding of plants and in abscission although the signalling mechanisms in these cases are different.

Peroxidase enzymes also have a major role in cell wall strengthening and are induced by pathogen infection and elicitor treatment. These enzymes oxidise a range of substrates, utilising H_2O_2 as an electron donor (see Section 11.4). Peroxidases represent a highly conserved superfamily of proteins that are widespread throughout the plant kingdom, and a single species may have up to 40 different isoforms. Treatment of plant cells with various fungal elicitors has been shown to induce expression of certain peroxidases such as the PR-9 family (see Section 9.3.5), and these peroxidases modify plant cell walls through oxidative cross-linking of two soluble proteins, a hydroxyproline-rich glycoprotein (HRGP) and a proline-rich protein (PRP). These are believed to be important structural proteins in forming foci for lignification. Peroxidases and secondary wall thickening in xylem vessel walls and lumen have also been found to be a significant defence against vascular wilt pathogens such as *Xanthomonas oryzae*.

9.3.4 Phytoalexins

There has been much debate over the years as to whether or not phytoalexins and phytoanticipins have a significant role in plant defence. Recently, genetic studies involving targeted mutagenesis, along with experiments using transgenic plants have yielded more conclusive evidence for the involvement of particular compounds as antimicrobials. Tobacco plants have been engineered to produce resveratrol through introduction of the terminal biosynthetic gene stilbene synthase, and these plants exhibit enhanced resistance to *B. cinerea*. Many genes in the pathways to produce other phytoalexins, such as flavonoids, have also been shown to be regulated in a co-ordinate fashion in response to pathogen invasion, indicative of an important role in defence.

Antimicrobial compounds include a diverse array of low-molecular-weight secondary metabolites (i.e. compounds that are not essential for basic metabolic processes in plants), and these generally act against a broad range of pathogens. They include terpenoid derivatives (e.g. sesquiterpenes); saponins; aliphatic acid derivatives; phenolics and phenylpropanoids (e.g. isoflavonoids); nitrogen-containing organic compounds (e.g. alkaloids); and sulphur-containing compounds including inorganic elemental sulphur (*Figure 9.2a*). Most of these compounds are derived from the isoprenoid, phenylpropanoid, alkaloid or fatty acid/polyketide pathways (*Figure 9.2b*), and some of the genes encoding enzymes involved in these pathways, such as phenylalanine ammonia-lyases (PAL) and chalcone synthases (CHS) have

Terpenoids: 1. Sesquiterpene (rishitin), 2. Diterpine
Saponins: 3. Avenacin
Aliphatic acid derivatives: 4. Butyrolactone, 5. Furanoacetylene (wyerone)
Phenolics and Phenylpropanoids: 6. Flavanone, 7. Pterocarpan (maacklain)
8. Pterocarpan (medicarpin), 9. Stilbene (resveratrol)
Nitrogren and sulphur compounds: 10. Benzophenanthridine alkaloid,
11. Indole (camalexin), 12. Anthranilamide, 13. Benzoxazinone (DIMBOA)
14. Elemental sulphur

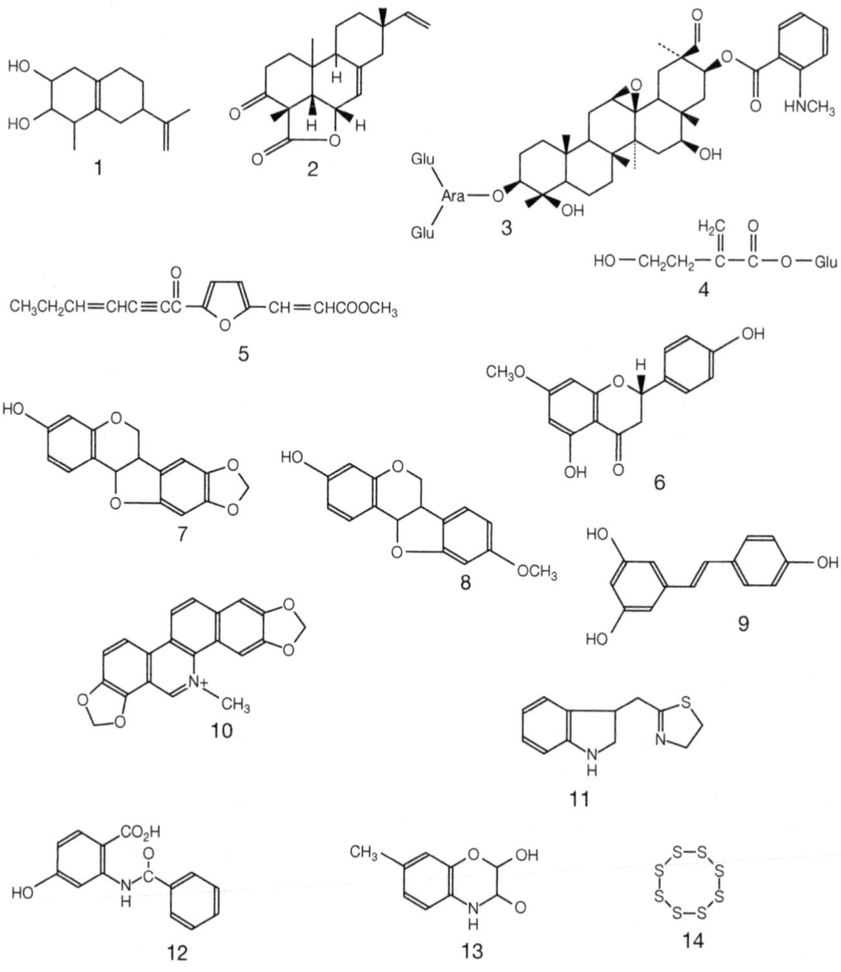

Figure 9.2a

Chemical structures of some examples of the major antimicrobial groups produced by plants.

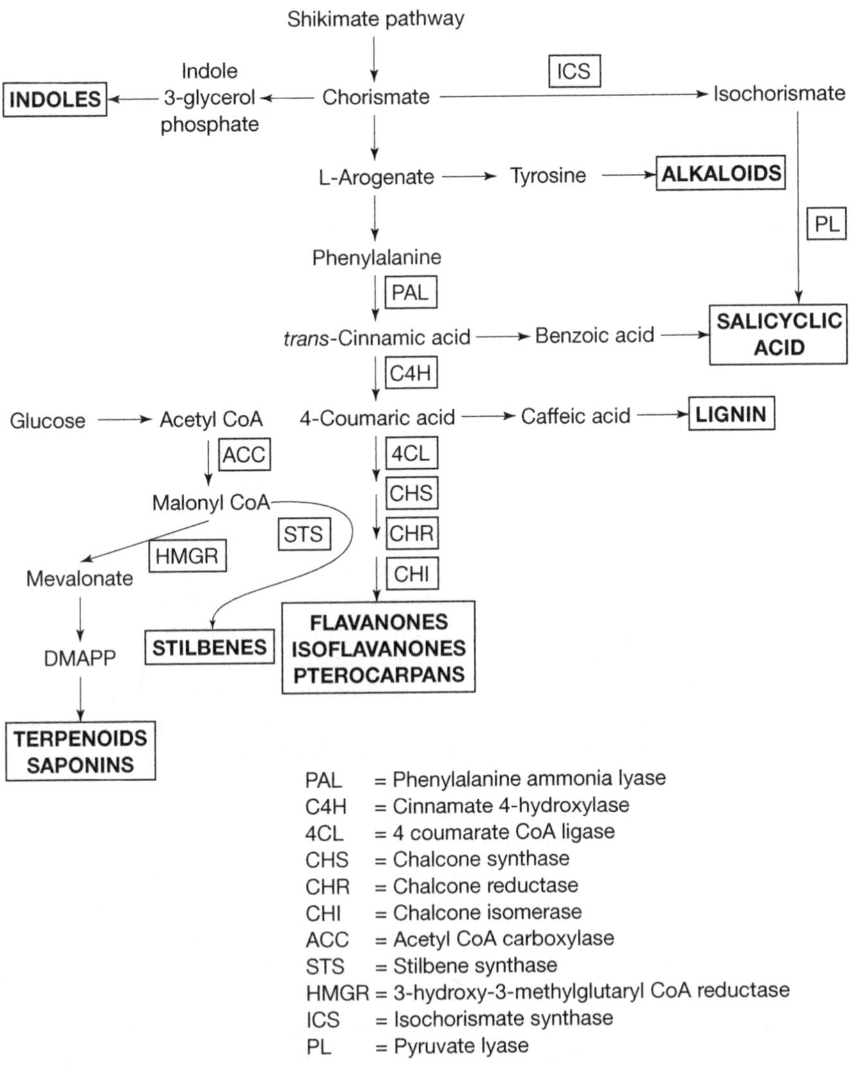

Figure 9.2b

Biosynthetic pathways for the main plant antimicrobials. The main phytoalexin groups are shown in bold, and the enzymes involved in their biosynthesis are boxed.

been shown to be induced in defence responses in association with the hypersensitive response. Others are currently being identified, particularly through EST (expressed sequence tags) projects and DNA-microarray work as discussed in Section 1.5.7, so that transcript expression patterns can be ascertained. Interestingly, plant species within the same families will tend to utilise the same chemical structures as antimicrobials, for example the sesquiterpenes are widely used in the *Solanaceae*.

Many antimicrobial compounds are sequestered within plant cells as inactive precursors or are synthesised and accumulate in specialised vesicles that are then delivered to the site of microbial infection, as occurs for deoxyanthocyanidin in sorghum. Studies on whether sufficient quantities are produced and localised have been difficult to undertake, particularly as some antimicrobials may act synergistically with others. For example, it has been suggested that the soap-like saponins such as avenacin and α-tomatine not only cause membrane permeability and electrolyte leakage directly on invading pathogens, but also have a role in rendering them more permeable to other antimicrobial compounds. Some antimicrobials are secreted out of cells, and ABC transporters have been shown to secrete some antifungal secondary metabolites.

The location of preformed compounds is important in considering the ways that pathogens may evade them. This is particularly appropriate for post-harvest quiescent pathogens (see Section 3.3.1) when they encounter high levels of antimicrobials in the epidermal layers of unripe fruits, such as α-tomatine in tomatoes, resorcinol in mangoes, or 1-acetoxy-2-hydroxy-4-oxo-heneicosa-12,15-diene in avocadoes. To avoid these chemicals, spores of fungi such as *Colletotrichum gloeosporioides* (anthracnose) germinate and produce appressoria, which then remain quiescent until the levels drop at which point invasive growth is initiated. The decrease in concentrations is related to physiological changes that occur during ripening. For example, the diene in avocadoes is oxidised by lipoxygenases, levels of which increase as the amount of the endogenous inhibitor epicatechin decreases in avocado peel. Other means of evading chemical inhibitors include those used by biotrophic pathogens, which may avoid stimulating their production until the fungus has already progressed far enough to survive the levels of antimicrobials produced. However, precise *in vivo* studies on levels of compounds required to preclude ingress of microbes at different stages of development are complex.

A further way of avoiding antimicrobials for which good evidence has been obtained is for the pathogen to produce enzymes that degrade the compound (see Section 3.3.2). Avenacin, present in epidermal cells of oat roots, renders these plants resistant to isolates of *Gaeumannomyces graminis* (take-all) unless the isolates possess the detoxification enzyme avenacinase. This β-glucosidase enzyme removes the β-1,2- and β-1,4-linked D-glucose molecules from the glycoside moiety of avenacin. Mutants in which the avenacinase gene is disrupted are unable to infect oats but are still pathogenic on wheat, and conversely, avenacin-deficient mutants of oats are susceptible to the wheat-attacking isolates of the fungus that lack avenacinase. Further evidence that the relationship between phytoalexin production and disease resistance is complex and highly pathogen-dependent comes from studies on phytoalexin-deficient (*pad*) mutants of *Arabidopsis* that have reduced levels of camalexin. For example, mutants in the cytochrome P450 *pad3* gene are more susceptible to *C. carbonum* and *Alternaria brassicicola*, but not to *Botrytis cinerea, Peronospora parasitica* (downy mildew), *Erysiphe orontii* (powdery mildew) or *Pseudomonas syringae*. Conversely, mutations in *pad-1* and *pad-2* are more susceptible to *P. syringae*.

9.3.5 Pathogenesis-related proteins

In addition to programmed cell death, strengthening of structural defences and production of antimicrobial phytoalexins, various novel proteins are induced during pathogen attack, known collectively as the pathogenesis-related (PR) proteins. These proteins are expressed at low levels in healthy plants, but certain isozymes are induced during pathogen attack both

Table 9.1 The families of pathogenesis-related proteins

Family	Type member	Properties
PR-1	Tobacco PR-1a	Unknown
PR-2	Tobacco PR-2	Endo β-1,3-glucanase
PR-3	Tobacco P,Q	Type I, II, IV, V, VI, VII chitinases
PR-4	Tobacco 'R'	Type I, II chitinases
PR-5	Tobacco S	Thaumatin-like
PR-6	Tomato Inhibitor I	Protease inhibitor
PR-7	Tomato P_{69}	Endoprotease
PR-8	Cucumber chitinase	Type III chitinase
PR-9	Tobacco 'lignin-forming peroxidase'	Peroxidase
PR-10	Parsley 'PR1'	Ribonuclease
PR-11	Tobacco class V chitinase	Type I chitinase
PR-12	Radish Rs-AFP3	Defensin
PR-13	*Arabidopsis* THI2.1	Thionin
PR-14	Barley LTP4	Lipid-transfer protein

locally and systemically, and there is evidence that specific sequences in the promoter regions of these genes are important for the induction. The proteins induced have been grouped into 14 PR classes, according to serology and homology (*Table 9.1*), though not all are induced in all interactions nor in all plant species.

The biochemical role for many of these proteins has been determined. For example, the chitinases (PR-3, PR-4, PR-8, PR-11), which are classified according to their specific activities on different substrates, are presumed to hydrolyse chitin in fungal cell walls. There is experimental evidence that supports the role for induced chitinase enzymes in degrading chitin exposed at the hyphal tips of fungi. This will interfere with fungal growth and also release small oligosaccharide elicitors, which may be involved in inducing and/or amplifying other plant defence responses. Glucanases (PR-2), proteinases (PR-7) and RNases (PR-10) presumably have similar roles as hydrolytic enzymes that will also have activity against bacteria and oomycetes, and in the case of PR-7 and PR-10 against viruses as well. Glucanases may have an alternative role in viral infection through removal of virus-induced callose plugs in the plasmodesmata. Some PR-1 proteins have been shown to inhibit the growth of oomycetes, although their exact biochemical role has yet to be elucidated. Further experiments with these hydrolytic-enzyme-type PR proteins have shown that compartmentalisation of the proteins is important and that if basic isozymes are targeted to vacuoles, they are more effective than acidic isozymes and untargeted enzymes. This suggests that these proteins are more likely to be a secondary line of defence against invading pathogens, with a role in degrading the invading organism once it has already been contained.

Some PR proteins have putative roles in combating pathogenicity factors. Protease inhibitors (e.g. PR-6) are produced that inhibit insect and microbial protease enzymes. Polygalacturonase inhibitor proteins (PGIPs), although not classified as PR proteins, are induced in interactions such as between bean and *Colletotrichum lindemuthianum*, where it has been shown that they accumulate more rapidly and intensely in incompatible compared with compatible interactions. PGIPs are widespread in dicotyledenous plants and pectin-rich monocotyledonous plants such as onions and leeks. Since polygalacturonase (PG) enzymes are used by many cell-wall-softening pathogens (see Section 2.9) it is believed that the PGIPs with specific recognition capabilities may be produced to counter these, and that this may be particularly important in defence of fruit. For example, PGIPs purified from pear fruit were shown to inhibit the *B. cinerea*-encoded PG activity, but not that of the endogenous fruit PGs required for fruit ripening and softening. Furthermore, levels of PGIP mRNAs were much higher in fruit compared to flowers and leaves. The structural similarities between PGIPs and certain classes of plant disease resistance genes, in that they comprise a leucine-rich repeat motif (see Section 10.3.1) and a signal peptide for translocation to the endoplasmic reticulum, has led to speculation that they may also have a role in eliciting further downstream defence responses in plants, or that there may be an evolutionary link between PGIPs and resistance genes.

The roles of the other PR-proteins are more diverse. The PR-9 family are peroxidases, presumably involved in cell wall strengthening (see section 9.3.3). The PR-5 family of thaumatin-like proteins have homology to permatins that permeabilise fungal membranes, whilst the PR-12-type defensins and PR-13-type thionins are classified because of their homologies to similar antimicrobial compounds present in other organisms such as the insect and mammalian defensins. The role of the lipid transfer proteins (PR-14) in defence is as yet unclear.

9.3.6 Other defence-related proteins

Other proteins that may have roles in defence include lectins, and it has been postulated that the chitin-binding lectins common in cereal grains may bind to chitin in the cell walls of fungi and insects and retard their growth *in planta* allowing time for other more active defences to take effect. Similarly, ribosome-inactivating proteins (RIPs) such as the pokeweed antiviral protein (PAP) reduce viral replication in plant species that contain them by inhibiting protein synthesis at affected ribosomes.

For other proteins that are induced as part of the defence response, the specific roles are less obvious. In parsley cell suspension, a number of *Eli* (elicitor activated) genes have been identified following treatment with elicitors from *Phytophthora sojae*, that correspond to genes induced by fungi attempting penetration of intact plants. Some of these appear to have roles in primary metabolism, indicating that a wide range of metabolic changes are induced as part of pathogen attack. As further evidence for the role of primary metabolism, it has been found that environmental conditions such as addition of exogenous nitrogen, can influence whether plants become diseased. Varieties of wheat that are normally susceptible to yellow rust can be made resistant by growing them in the absence of supplementary nitrogen, although the molecular mechanisms for this resistance are not known. A further group of enzymes induced in plants as part of HR are the lipoxygenases. These may contribute to defence both through the production of volatile and non-volatile fatty-acid-derived secondary metabolites that are toxic to invading pathogens, and through induction of the signal molecules jasmonic acid and methyl jasmonates (see Section 11.7).

9.3.7 Post-transcriptional gene silencing (PTGS)

Plants have a specific cytoplasmic defence mechanism (PTGS) that targets dsRNA for destruction, and since most plant viruses replicate in the cytoplasm via dsRNA intermediates, these are candidates for degradation by such a defence mechanism. The mechanism itself has been described in Section 8.8.3. However, whilst there is clear evidence that this mechanism exists, it is unclear how effective it is for preventing viral infections, since a number of viruses have been found to suppress the mechanism. What is perhaps more intriguing is whether the mechanism itself is part of a systemic signalling mechanism (see Section 11.8).

9.4 Systemic resistance mechanisms

As well as activation of defence genes by pathogen invasion, it has been found that the formation of a hypersensitive response, either as part of pathogen invasion or through the action of necrosis-inducing pathogens, results in systemic resistance responses (*Figure 9.3*). Systemic acquired resistance (SAR) is one form of inducible resistance that is activated throughout the whole plant and this resistance has some similarities with animal immune responses, in that it is long-lasting and can be boosted by repeat infections, but unlike animal immunity it can be effective against organisms other than the one used to stimulate the initial response, conferring broad-spectrum resistance.

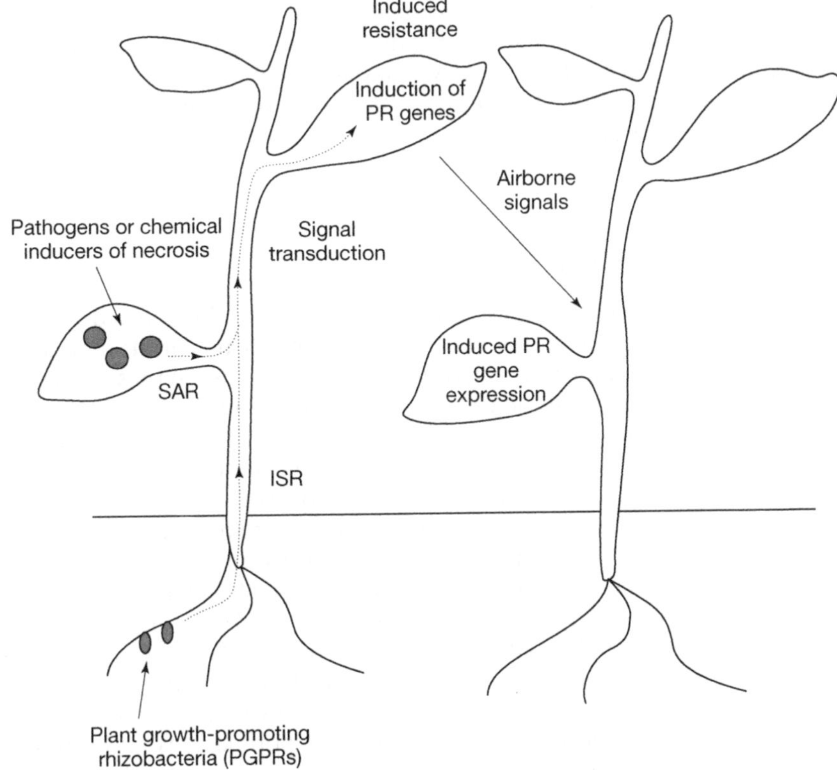

Figure 9.3

Systemic and airborne resistance mechanisms in plants. See text for details of SAR (systemic acquired resistance), ISR (induced systemic resistance) and volatile signals that are transmitted to neighbouring plants to induce resistance.

SAR has been demonstrated in many plant species, including cucurbits, bean, tomato and *Arabidopsis*, upon induction by bacterial, fungal and viral pathogens. The spectrum of pathogens against which systemic resistance is effective remains constant for each plant species, irrespective of the nature of the inducing pathogen. However, this spectrum varies between species. SAR can therefore be considered to provide a characteristic 'fingerprint of protection', which has been useful in discriminating SAR from other resistance mechanisms. However, SAR is not effective against all pathogens. For example, tobacco cannot be protected against challenge by *B. cinerea* and *P. syringae* pv. *tomato*, and cucurbits cannot be protected from powdery mildew. The nature of the resistance induced has led to the concept that SAR acts to prime host cells for a more rapid future deployment of defences, and the mechanism through which this occurs is discussed in Section 11.10.

The phenomenon of induced systemic resistance (ISR) is similar to SAR, in giving protection to normally virulent pathogens. However, rather than induction by phytopathogens, ISR is activated by plant growth-promoting rhizobacteria. It also appears to provide protection against pathogens such as *B. cinerea* and *Alternaria brassicicola*, for which SAR is ineffective and vice versa, but the nature of the genes induced is unclear. Nevertheless, evidence exists for overlap and also antagonism between the mechanisms regulating resistance in SAR and ISR, as discussed in Section 11.10. In addition, some necrotrophic pathogens induce a further form of systemic resistance, also via jasmonic acid and ethylene but independent of SAR and ISR. Whilst induction of the *PR-1*, *PR-2* and *PR-5* genes are taken as markers for SAR in *Arabidopsis*, so *PR-3*, *PR-4*, *Thi2.1* and *PDF1.2* are the equivalent local and systemic markers of this necrosis-induced resistance as discussed in Sections 11.7 and 11.10.

9.5 'Communal' resistance

In addition to systemic signalling within plants and the induction of defences, recent evidence has shown that plants can communicate with their neighbouring plants and induce the activation of defence genes in these. Volatile signals such as methyl jasmonate and methyl salicylate (see Section 11.7) are produced from insect- and pathogen-infested plants, and in laboratory experiments at least, have been shown to induce defence gene expression. For example, methyl jasmonates released from big sagebrush (*Artemisia tridentate*) following insect feeding were able to induce production of proteinase inhibitors in adjacent tomato plants, and reduce the numbers of insects feeding. Similarly, volatiles from lima bean leaves released following spider mite damage induced glucanase, chitinase, lipoxygenase and PAL gene expression in adjacent plants. Methyl salicylate released from TMV-infected tobacco plants containing the *N* resistance gene induced *PR-1* expression in adjacent plants and increased resistance to disease. However, the role of such signalling and gene induction in natural plant populations and crops is unclear and studies are currently in progress to determine the significance of volatile signals in disease resistance. One feature that has been noted is that the volatiles produced when herbivorous insects feed on plants can act as attractants for predatory insects that feed on these herbivorous species, and this is being developed into a means of controlling insect infestations on plants.

References and further reading

Baldwin, I.T., Kessler, A. and Halitschke, R. (2002) Volatile signalling in plant–plant–herbivore interactions: what is real? *Current Opinion in Plant Biology* **5**: 351–354.

De Lorenzo, G. and Ferrari, S. (2002) Polgalacturonase-inhibiting proteins in defense against phytopathogenic fungi. *Current Opinion in Plant Biology* **5**: 295–299.

Dixon, R.A. (2001) Natural products and plant disease resistance. *Nature* **411**: 843–847.

Farmer, E.E. (2001) Surface-to-air signals. *Nature* **411**: 854–856.

Hammond-Kosack, K.E. and Jones, J.D.G. (1996) Resistance gene-dependent plant defence responses. *The Plant Cell* **8**: 1773–1791.

Hawes, M.C., Gunawardena, U., Miyasaka, S. and Zhao, X. (2000) The role of root border cells in plant defense. *Trends in Plant Science* **5**: 128–133.

Heath, M.C. (1998) Apoptosis, programmed cell death and the hypersensitive response. *European Journal of Plant Pathology* **104**: 117–124.

Prusky, D. (1996) Pathogen quiescence in postharvest diseases. *Annual Review of Phytopathology* **34**: 413–434.

Van Loon, L.C. and van Strien, E.A. (1999) The families of pathogenesis-related proteins, their activities, and comparative analysis of PR-1 type proteins. *Physiological and Molecular Plant Pathology* **55**: 85–97.

Wojtaszek, P. (1997) Oxidative burst: an early plant response to pathogen infection. *Biochemical Journal* **322**: 681–692.

Resistance genes

10

Induced resistance mechanisms are of major importance in the defence of plants against pathogens, as discussed in Chapter 9. Why are these effective in some plants against some pathogens, but not in others? The answer is complex and presumably reflects how and when the induction occurs. In non-host plants, induction of the innate immune response by non-specific elicitors from microbes may be sufficient for resistance. In host plants, the pathogen may have evolved to evade these innate responses or to trigger them later in infection so that they are ineffective, and specific resistance genes for recognising specific elicitors from the pathogen may be required for resistance. These may be either one or a few genes whose individual effects can be easily detected (gene-for-gene or vertical resistance) or numerous genes with small additive effects (quantitative or horizontal resistance) (*Figure 10.1*). It is the nature of these resistance genes that we focus on in this chapter.

10.1 Gene-for-gene resistance

The simple model for how the host resistance mechanism works is through a dominant resistance (*R*) gene in the plant encoding a product that recognises a pathogenicity factor (product of a dominant gene) in the pathogen to confer resistance. If the plant does not have this *R* gene or loses it, it becomes susceptible, or if the pathogen loses or modifies this pathogenicity gene to avoid recognition, it will overcome the resistance (although loss of

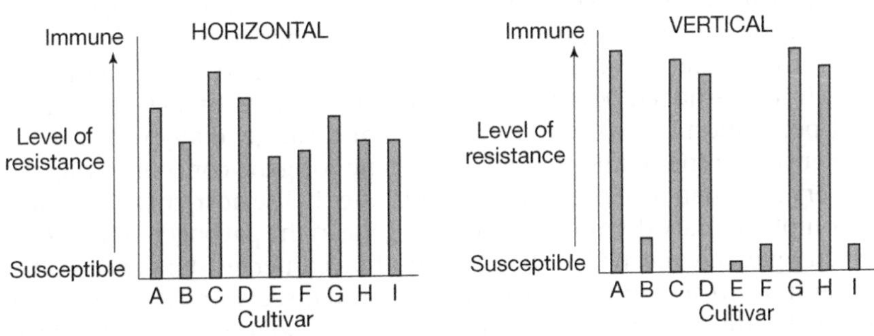

Figure 10.1

Horizontal versus vertical (gene-for-gene) resistance. In horizontal resistance, numerous genes have small additive effects so that the resistance varies by small amounts between cultivars. In vertical resistance, controlled by single genes, resistance is either close to complete immunity if the gene is present, or complete susceptibility if it is absent.

Resistance genes in plants

Avirulence genes in pathogens	CULTIVAR 1 (*R1R1, R2R2*) (*R1r1, R2R2*) (*R1R1, R2r2*) (*R1r1, R2r2*)	CULTIVAR 2 (*R1R1, r2r2*) (*R1r1, r2r2*)	CULTIVAR 3 (*r1r1, R2R2*) (*r1r1, R2r2*)	CULTIVAR 4 (*r1r1, r2r2*)
RACE 1 (*AVR1, AVR2*)	No disease	No disease	No disease	Disease
RACE 2 (*AVR1, avr2*)	No disease	No disease	Disease	Disease
RACE 3 (*avr1, AVR2*)	No disease	Disease	No disease	Disease
RACE 4 (*avr1, avr2*)	Disease	Disease	Disease	Disease

Figure 10.2

The genetics of gene-for-gene resistance. Based on two resistance genes (*R1* and *R2)* in the plant species and two avirulence genes (*AVR1* and *AVR2*) in the pathogen. For each cultivar, the genotypes that could give this phenotype are noted, and for each race of the pathogen, the possible avirulence genotypes for a 'haploid' genome are noted.

its pathogenicity factor may also render the pathogen ineffective). If this occurs, there will be selection pressure on the plant population for individuals that recognise other pathogenicity factors in the pathogen so that they can resist it. Thus an evolutionary 'arms race' will develop with complementary changes occurring in the plant and pathogen populations. The result in natural plant/ pathogen ecosystems would be a balance and evolutionary stability between plant and pathogen populations, as discussed in Section 10.9.

Support for this evolutionary model was originally developed from the studies of Farrer in the 1890s, who described resistance in wheat against yellow/stripe rust (*Puccinia striiformis)*, as following Mendelian genetics, followed by the work of Biffen in the early 1900s who demonstrated that resistance could be a monogenic trait. In the early 1940s, Flor, working on flax/flax rust (*Melampsora lini*) in the USA, and Oort, working independently on wheat/wheat smut (*Ustilago tritici*) in The Netherlands, made the major breakthrough that it was possible to discriminate between different genotypes in the pathogen population by using different resistance genes in the plant population. It was found that virulence was generally recessive and avirulence dominant, and this led to the gene-for-gene concept stating that for every dominant gene determining resistance in the host, there is a matching complementary dominant avirulence gene in the pathogen (*Figure 10.2*).

Such genes have been identified in plants that confer resistance against bacteria, fungi, viruses, oomycetes, nematodes and insects, and a simple model to explain this concept is the elicitor/receptor model (*Figure 10.3*). In this, recognition of the elicitor derived from the functional avirulence gene in the pathogen by the product of the *R* gene in the plant activates a signal transduction pathway leading to the hypersensitive response and resistance. This model has been an important framework for establishing the underlying mechanisms of resistance, although there is increasing evidence that *R* genes do not act in isolation from other genes and pathways in plants,

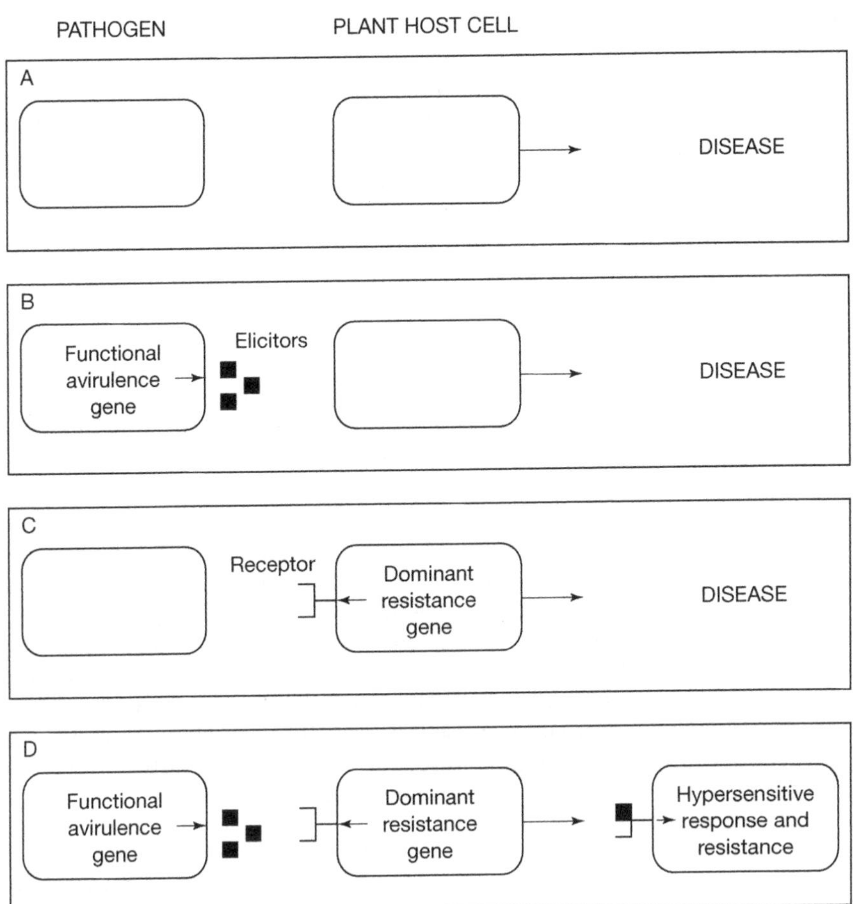

Figure 10.3

The dual functions of plant resistance genes. For any particular resistance gene/avirulence gene combination, there are four possible alternatives (A–D) depending on whether functional genes are present or absent as shown.

and that the outcome of any plant–pathogen interaction is determined by many genetic factors in both plant and pathogen. It should also be noted that some *R* genes are only semi-dominant, for example some of the *Cf (Cladosporium fulvum)* resistance genes in tomato and the *Mla* genes in barley against *Blumeria graminis*, allowing a greater degree of fungal ingress in the heterozygous state, and other *R* genes are influenced in their effectiveness by environmental conditions (e.g. temperature), and also by the age of the plant.

10.2 Features of cloned resistance genes

Through the 1980s and early 1990s, there was a concerted effort worldwide to clone *R* genes from plants as a step towards determining how they function. The two main approaches used were map-based cloning and transposon tagging (see Section 1.5.4), which culminated initially in identification of the *Hm1* gene from maize conferring resistance to *Cochliobolus carbonum*.

However, this pathogen is a necrotroph that causes disease through toxin production, and identification of the resistance gene as a toxin reductase indicated that this gene was unlikely to be a model for gene-for-gene resistance genes. In particular, non-host close relatives of maize such as sorghum and rice were shown to contain functional *Hm* loci, and it is only in maize plants where this gene is disrupted by transposon insertions that susceptibility and gene-for-gene-like resistance is apparent.

In 1993, the *Pto* gene was cloned from tomato, conferring resistance to bacterial speck (*Pseudomonas syringae*). This was cloned by mapping the *R* gene to a location between two molecular markers. A yeast artificial chromosome (YAC) clone was then found with both these flanking markers and therefore the *R* gene, and this YAC was used to probe a leaf cDNA library. The positive cDNA clones were then transformed independently into susceptible plants to identify the one that complemented *Pto* resistance. Subsequently, a large number of *R* genes have been cloned from many different plant species, conferring resistance against viruses, bacteria, fungi, oomycetes and nematodes (*Table 10.1*). In some cases, such as for the tobacco *N* gene and the tomato *Cf-9* gene, elegant selection strategies have been used. Both these genes were cloned by transposon tagging, but rather than using pathogen screens to select for progeny plants in which the gene was tagged, alternative approaches were adopted. For *Cf-9,* the corresponding avirulence gene *Avr9* was cloned into tomato, so that all the plants died from a lethal hypersensitive response apart from those in which the resistance gene had been inactivated. For the *N* gene, a heat shock approach was used to create the same effect. Above 28°C, the *N* gene is ineffective, so plants were grown at this temperature after virus infection to allow the virus to spread. The temperature was then lowered to a permissive temperature, which activated the *N* gene and resulted in a lethal hypersensitive response in all apart from those plants where the gene was tagged. The similarity in sequence between many *R* genes has now made it possible to clone resistance gene analogues (RGAs) by PCR-based techniques, although in most cases it has not been confirmed that these represent functional *R* genes. Indeed, *Arabidopsis* has more than 400 *R* gene candidates (approximately 2% of it genes) and other plant genomes have even larger proportions of their genomes as RGAs. In addition, synteny between closely related genomes such as rice and sorghum has facilitated the cloning and identification of resistance genes.

The *R* genes cloned to date have been grouped into a number of classes based on sequence relationships (*Table 10.1* and *Figure 10.4*). The largest class encodes a nucleotide-binding site plus a carboxy-terminal leucine-rich repeat domain (NBS-LRR), and this group makes up at least half (more than 200) of the *R* gene candidates in *Arabidopsis*. The group can be subdivided into a group that possess a Toll/Interleukin-1-receptor homology (TIR) region at their amino terminal, and a group that lack this, though some of these may contain coiled coil (CC) domains (also sometimes referred to as leucine zippers, LZ). Interestingly, the TIR subgroup is very rare in the grasses but is predominant in *Arabidopsis*, where some genes have been identified that encode only TIR domains, although whether these have a role in disease resistance is unclear. The other classes of *R* genes are more diverse. The tomato *Cf*-gene products consist almost entirely of a predicted extracellular LRR attached to a short transmembrane anchor at the carboxy terminal. The *Ve* genes for resistance in tomato to *Verticillium albo-atrum* encode cell-surface glycoproteins in which the LRRs are also presumed to be extracellular, and have PEST

Table 10.1 Classification of some cloned resistance genes (see text for details)

Class	Predicted protein structure	Gene	Host	Pathogen
1	TIR-NBS-LRR	*L*	Flax	*Melampsora lini* (fungus)
		M	Flax	*Melampsora lini* (fungus)
		P	Flax	*Melampsora lini* (fungus)
		N	Tobacco	*Tobacco mosaic virus*
		RPP1	*Arabidopsis*	*Peronospora parasitica* (oomycete)
		RPP5	*Arabidopsis*	*Peronospora parasitica* (oomycete)
		RPS4	*Arabidopsis*	*Pseudomonas syringae* (bacterium)
	CC-NBS-LRR	*Prf*	Tomato	*Pseudomonas syringae* (bacterium)
		Mi	Tomato	*Melodogyne incognita* (nematode)
		Gpa2/Rx1	Potato	*Globodera* (nematode) & *Potato virus X*
		RPS2	*Arabidopsis*	*Pseudomonas syringae* (bacterium)
		RPS5	*Arabidopsis*	*Pseudomonas syringae* (bacterium)
		RPM1	*Arabidopsis*	*Pseudomonas syringae* (bacterium)
		RPP8/HRT	*Arabidopsis*	*Peronospora* & *Turnip crinkle virus*
	NBS-LRR	*Bs2*	Pepper	*Xanthomonas campestris* (bacterium)
		Dm3	Lettuce	*Bremia lactuca* (oomycete)
		I2	Tomato	*Fusarium oxysporum* (fungus)
		Cre3	Wheat	*Heterodera avenae* (nematode)
		Xa1	Rice	*Xanthomonas oryzae* (bacterium)
		Pib	Rice	*Magnaporthe grisea* (fungus)
		Pi-ta	Rice	*Magnaporthe grisea* (fungus)
		Rp1	Maize	*Puccinia sorghi* (fungus)
		Mla	Barley	*Blumeria graminis* (fungus)
	TIR-NBS-LRR-NLS-WRKY	*RRS1-R*	*Arabidopsis*	*Ralstonia solanacearum* (bacterium)
2	LRR-TM	*Cf-2, Cf-4 Cf-5, Cf-9*	Tomato	*Cladosporium fulvum* (fungus)
3	Kinase	*Pto*	Tomato	*Pseudomonas syringae* (bacterium)
		PBS1	*Arabidopsis*	*Pseudomonas syringae* (bacterium)
	Kinase-kinase	*Rpg1*	Barley	*Puccinia graminis* (fungus)
4	LRR-TM-Kinase	*Xa21*	Rice	*Xanthomonas oryzae* (bacterium)
		FLS2	*Arabidopsis*	Innate immunity (flagellin)
5	Unique	*HS1$^{pro\text{-}1}$*	Sugar beet	*Heterodera schachtii* (nematode)
6	Unique	*RPW8*	*Arabidopsis*	*Erysiphe* (fungus)
7	Membrane protein	*mlo*	Barley	*Blumeria graminis* (fungus)
8	Cell-surface glycoprotein	*Ve1*	Tomato	*Verticillium albo-atrum* (fungus)
9	Toxin reductase	*Hm1*	Maize	*Cochliobolus carbonum* (fungus)

For the predicted protein structures, TIR = Toll interleukin receptor; LRR = leucine-rich repeat; NBS = nucleotide binding site; CC = coiled coil; NLS = nuclear localisation signal; WRKY = transcription factor; and TM = transmembrane.

(Pro-Glu-Ser-Thr)-like sequences in the C-terminal domain which may be involved in ubiquitinisation and protein turnover (see Section 11.4). The *Pto* gene encodes a cytoplasmic serine/threonine protein kinase, barley *Rpg1* is a tandem repeat of kinase domains, whilst the *Xa21* gene from rice is a combination of the *Cf* and *Pto* types, having an LRR attached to an intracellular protein kinase. Products of other *R* genes, such as the *Hs1$^{pro\text{-}1}$* conferring resistance in sugar beet to the cyst nematode *Heterodera schachtii*, and the

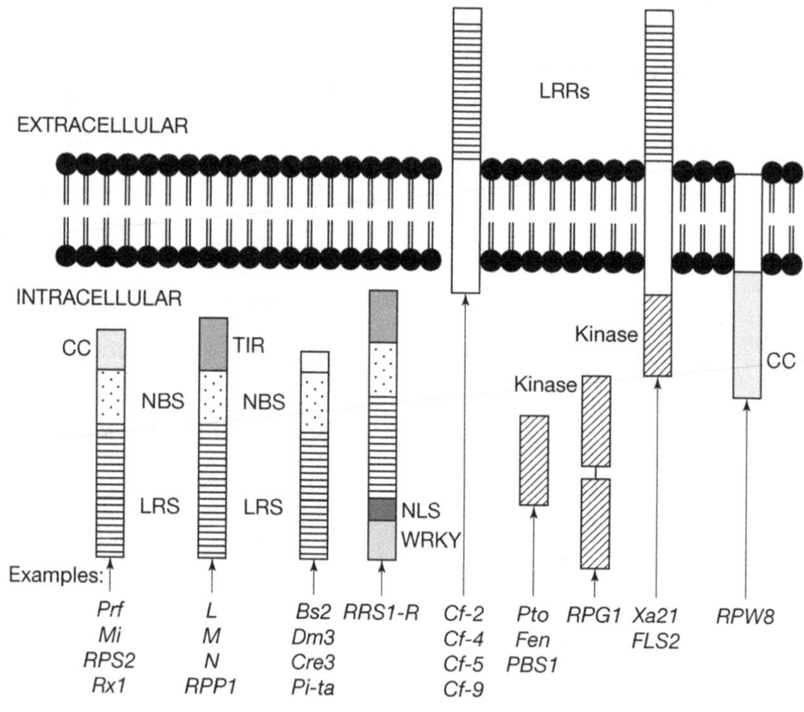

Figure 10.4

Structures and putative locations in plant cells of disease resistance genes. NBS (nucleotide binding site), LRR (leucine rich repeat), CC (coiled coil), TIR (Toll interleukin receptor), NLS (nuclear localisation signal) and WRKY (transcription factor) domains are shown. The amino end of the proteins is at the top.

RPW8 gene against mildew in *Arabidopsis* have little or no homology to other known proteins.

Expression of resistance genes in plants is believed in most cases to be constitutive but at a low level, not induced by pathogen invasion and not tissue-specific. However, some experiments using the functional promoter of the *I2 Fusarium oxysporum* resistance gene from tomato fused to the *GUS* reporter gene suggest that transcripts of this gene are more abundant in tissues surrounding the vascular system than in other tissues, implying some enhanced tissue-specific expression.

10.3 *R* gene specificity

10.3.1 Leucine-rich repeats (LRRs)

Leucine-rich repeats (LRRs) are multiple, serial amino acid repeats (normally around 24 amino acids long) that contain leucines or other hydrophobic residues at regular intervals, along with regularly spaced prolines and asparagines. Because LRR domains are often involved in protein–protein interactions in mammalian cells, such as in hormone receptors that recognise glycoprotein ligands, or enzyme inhibitors (e.g. the porcine RNase inhibitor) that recognise enzymes, the identification of these in plant resistance genes led to the hypothesis that these are the recognition domains. Structural studies from the porcine RNase inhibitor protein suggest that the hydrophobic residues are orientated internally to provide a curved spring tertiary structure, and that the intervening amino acids in the β-strand-β-turn structural motifs are exposed on the surface to confer functional specificity.

The role of resistance gene LRRs in recognition is supported by analysis of the *Cf-4* and *Cf-9* genes from tomato, in which the majority of amino acid substitutions between the two genes occur in these regions (see Section 10.8). Domain swaps between *Cf-4* and *Cf-9* LRRs have confirmed the importance of these in determining specificity, and there is evidence that the numbers of LRR repeat units may also be important. Furthermore, in the *L* alleles of flax, domain swaps in which the *L2* LRR was combined with the *L6* or *L10* N-terminal TIR and NBS domains, resulted in *L2* rust resistance specificity. However, evidence from the *L6* and *L7* alleles indicates that the N-terminal region of the TIR domain can be an alternative source of specificity, since amino acid changes between these two alleles only occur in this region. In the case of the *Pto* protein kinase resistance gene, the absence of LRRs was initially an anomaly. However, subsequent analysis revealed the presence of a genetically linked NBS-LRR type gene, *Prf* (*Pseudomonas r*esistance and *f*enthion sensitivity), that is required along with *Pto* for resistance to occur. The possible role of this is discussed in Section 10.3.3.

10.3.2 Cellular location of recognition

The receptor-ligand model for *R* genes (*Figure 10.3*) initially predicted that they would encode extracellular receptor-like proteins to detect the pathogen as it was attempting ingress into plant cells. Sequence analysis suggests that some (e.g. the *Cf* and *Ve* genes) do have LRRs located extracellularly and tagging experiments have confirmed the plasma-membrane location for Cf-9. This is consistent with the fungus *C. fulvum* having an entirely extracellular lifestyle growing within the plant apoplast. However, other *R*-gene products have LRR domains located within cells. For example, the RPM1 protein for resistance to *P. syringae* in *Arabidopsis* lacks signal peptides or transmembrane domains and appears to be linked to the plasma membrane via a second protein identified through yeast two-hybrid screens. Since it is now known that bacteria secrete elicitors via the *hrp* secretion mechanism into cells (see Chapter 6), that most viruses replicate cytoplasmically (see Chapter 7), and that many fungi form haustorial invaginations into cells and probably secrete peptides from these (see Section 3.7), the location of LRRs for many *R* genes within cells is not surprising and presumably reflects where detection of the elicitor occurs (see *Figure 10.4*).

The location of the LRR may also influence the timing of detection of the invading pathogen, and this in turn may affect the outcome of the resistance response and account for some of the variation in degrees of hypersensitive reactions and pathogen colonisation in different *R* gene/*Avr* gene-dependent interactions. Resistance responses to the cereal rust fungi, for example, range from small hypersensitive flecks to medium-sized uredia pustules surrounded by necrosis depending upon the resistance gene involved, and this may reflect when the resistance gene detects the elicitor combined with when the elicitor is produced by the pathogen. Small flecks may be a consequence of rapid detection, with the *Rx* gene in potato, in which no HR is discernible, being the most extreme example of rapid detection.

10.3.3 Does the R gene interact directly with the pathogen elicitor?

The receptor-ligand model predicts that the *R* gene product will interact directly with the elicitor derived from the pathogen *Avr* gene to result in the signal transduction mechanism. However, despite many attempts, evidence for direct binding between elicitors and R proteins have been difficult to obtain. The AVR9 peptide from *C. fulvum* for example binds as effectively to plasma membranes in plants without *Cf-9* as it does to membranes in plants with this resistance gene. However, an AVR9 high-affinity binding site (HABS) has been identified in plasma membranes of tomato and the binding affinity of mutant AVR9 peptides to this correlates with their ability to induce HR. In the interaction of *Arabidopsis* with AvrRpt2 and AvrB from *P. syringae*, the RPS2 peptide was found to bind to both avirulence peptides *in vivo*, but an additional plant protein of 75 kDa was bound in the interaction with AvrRpt2, the peptide to which this resistance gene confers gene-for-gene resistance. In the case of the AVR-Pita peptide from *M. grisea*, there is evidence from yeast two-hybrid studies of a direct interaction with the C-terminal leucine-rich-repeat domain of the Pi-ta NBS-LRR protein. Interestingly, in susceptible rice varieties, there is a single amino acid difference, serine instead of alanine, at position 918 in the Pi-ta resistance protein LRR domain.

More significant perhaps are the studies from the Pto/Prf/AvrPto interaction. Here it has been shown by yeast two-hybrid analysis that AvrPto and Pto interact directly, and that there is a threonine residue at position 204 that is crucial for this to occur. Whether this interaction alone is sufficient to elicit HR is unclear, particularly as Prf appears to be necessary. A general model to account for this and other data in which *Avr* and *R* gene products have been shown not to interact directly is the 'Guard Hypothesis' (*Figure 10.5*). Although

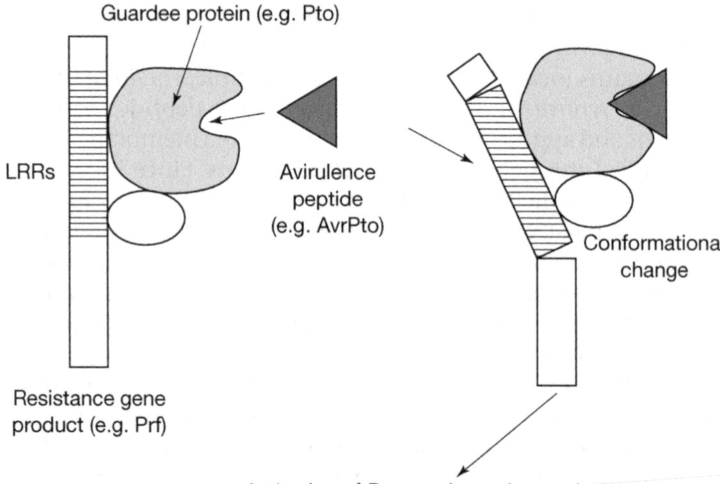

Figure 10.5

The Guard Hypothesis. In this model, Pto is a general component of host defence that is targeted by AvrPto. AvrPto is produced by the pathogen to suppress this defence pathway. The LRR regions of *R* gene products (or Prf in the case of Pto) are guards for these general components of host defence mechanisms.

this guard hypothesis is not yet proven and may not apply to all *R* genes, further support for it comes from the *Rpm1* gene in *Arabidopsis,* that is able to respond to two completely unrelated avirulence elicitors from *P. syringae,* and a protein RIN4 has been identified as a potential guardee that could account for this. In addition, the *Mi* resistance gene in tomato, that confers resistance to the *Meloidogyne incognita* root knot nematode (an obligate endoparasite of roots), also confers resistance to the *Macrosiphum euphorbiae* potato aphid (an insect that feeds from phloem in leaves), and it may be that rather than both organisms producing identical avirulence elicitors, they target and modify the same host protein which is guarded by this *R* gene.

10.4 The TIR domain

The TIR domain is so called because it contains similarity to the cytoplasmic signalling domain of the *Drosophila melanogaster* Toll protein and the interleukin-1 receptors of birds and animals. These proteins are key components of the innate immune system which is presumed to exist in all multicellular organisms and relies on a set of germ-line encoded receptors that are expressed in a wide variety of cells, particularly those at host/environment boundaries. In animals, the receptors trigger a variety of host defences such as complement activation, phagocytosis and expression of antifungal and antibacterial peptides such as defensins. The activation involves recognition of conserved microbial structures referred to as pathogen-associated molecular patterns (PAMPs). These are factors that are produced by microbes and not host cells and are essential for survival or pathogenicity of the microbe. They often provide a 'molecular signature' of the invading organism, and include such factors as lipopolysaccharides. The role of TIR domains is discussed in more detail in Section 11.2.

10.5 The NBS (NB) domain

Nucleotide-binding sites (referred to in the literature as NBS or NB domains) occur in diverse proteins with ATP- or GTP-binding activity, such as adenylate kinases, ATP synthase β-subunits, and the apoptosis proteins, APAF-1 (apoptotic protease activating factor) in humans, and CED-4 (*Caenorhabditis elegans* death regulating protein 4) in the nematode *C. elegans*. NBS domains consist of a nucleotide-binding P-loop (kinase 1a) and two distally located domains, kinase 2 and kinase 3. The homology between NBS domains in *R* genes and those in apoptosis proteins has led to the proposed roles for these domains in plant defence, particularly as this sequence homology extends beyond the NBS domains. This role is discussed in detail in Section 11.9.

10.6 Other R gene domains

10.6.1 Protein kinases

The *Pto-* and *Xa21*-encoded proteins possess serine-threonine protein kinase (PK) domains. Pto has been shown to interact with several proteins in yeast two-hybrid studies, including two that are phosphorylated by Pto, Pti1 (another serine-threonine kinase) and Pti4 (a DNA-binding protein). Whether Pti1 is part of a kinase-signalling cascade that leads to activation of specific genes through the Pti4 transcription factor has not been proven. However, Pti4 is similar to the ethylene response element-binding proteins (EREBPs) that bind the GCC box *cis* element in the promoter regions of many PR protein genes, and it has been shown that phosphorylation of Pti4 by the Pto kinase enhances the ability of Pti4 to activate expression of GCC-box PR-protein genes in tomato (see Section 11.2).

10.6.2 Coiled coil (leucine zipper) domains

Coiled coil (formerly referred to as leucine zipper) motifs have the amino acid pattern (hxxhxxx)n where h is a hydrophobic residue (A, F, I, L, M or V) and x is normally a polar or charged residue. These motifs are present in a number of *R* genes (e.g. *RPM1, RPP8, Rx1, Mi* and *Prf*) and the patterns are consistent with them forming an amphipathic alpha-helical domain into a coiled coil structure. Coiled coils are often involved in protein homo/hetero dimerisation (particularly of eukaryotic transcription factors) but their role in resistance gene function is as yet unclear, although their presence suggests that dimerisation may be important for many *R* gene-encoded proteins.

10.7 Genetic organisation of resistance genes

The classical gene-for-gene 'arms race' model implies that the product of a single resistance gene in the plant interacts with a specific pathogen elicitor. If the pathogen evolves to lose this elicitor, selection pressure in the plant will result in another resistance gene capable of recognising another elicitor. Evidence from genetics and cloning of *R* genes indicates that they are organised in plant genomes in one of two ways. Some, such as *RPS2, RPS4* and *RPM1* in *Arabidopsis* and *Brassica napus*, appear to occur at simple loci comprising a single gene with minor allelic variation in gene sequence between plant lines. In *Arabidopsis* lines that lack *RPM1*, the gene is replaced by an unrelated 98 bp sequence, and a similar situation occurs for the *RPM1* homologue in *B. napus*. Conversely, the *L* locus for rust resistance in flax, which also appears to comprise a single gene, has 13 different specificities identified in different alleles at this locus. Most *R* genes however seem to be members of gene families residing at complex loci. These loci may include apparently inactive resistance gene analogues, sequences disrupted by frameshifts, mutations and/or transposon insertions, and *R* genes capable of recognising more than one pathogen, such as the *Mi* (nematode resistance) and *Cf-2/Cf-5* locus on chromosome 6 of tomato, or the *HRT* gene in *Arabidopsis* conferring resistance to *Turnip crinkle virus,* which is adjacent to the *RPP8* gene conferring resistance to *P. parasitica*, and has arisen from this gene through duplication (i.e. is a paralogue). Conversely, genes recognising different avirulence determinants from the same pathogen are often dispersed throughout the genome, for example the *Cf-4* and *Cf-9* genes conferring resistance to other races of tomato leaf mould are on chromosome 1 of tomato.

These *R* gene loci may be large and diverse between plant lines. The *Dm3* locus in lettuce for resistance to *Bremia lactucae* (downy mildew) contains 10–35 sequences related to *Dm3* spanning over 3.5 Mbp, whilst the *M* locus for *Melampsora lini* (rust) resistance in flax consists of 15 similar genes spanning 300–1000 kb. The *Mla* locus in barley spans 240 kb and has at least 11 NBS-LRR type sequences within it. Over 30 different specificities have been detected at this locus (*Mla-1–Mla-32*), many of them bred into commercial cultivars from wild barley, *Hordeum spontoneum*. One of the best-studied complex loci has been the *RPP5* resistance locus on chromosome 4 in *Arabidopsis* conferring resistance to *Peronospora parasitica* (downy mildew). Complete sequencing of this locus in two ecotypes, Landsberg *erecta* (L*er*) and Columbia (Col), has revealed ten homologues in L*er* compared with only eight in Col. Of the eight homologues in Col however, only two are intact with the others disrupted through frameshifts and/or mutations.

10.8 Mechanisms for generating new *R* gene specificities

It is generally recognised that many *R* genes are unstable and often give rise spontaneously to susceptible mutants. There have been parallels drawn between the evolution of *R* genes and that of the major histocompatability complex (MHC) of vertebrates. In this complex, there is a birth and death process within the MHC complex that involves the expansion or contraction of the cluster (i.e. gene duplication or gene loss) through unequal crossing over, in combination with the evolution of individual genes by diversifying selection.

Since the basis for *R* gene specificity is believed to reside mainly in the LRR regions, research has focused on looking at variation in these between alleles and plant lines. At the *RPP5* locus, there is clear evidence from DNA sequencing that changes occur within LRRs between alleles. These changes appear to take two forms, mutations and rearrangements. Evidence for the role of mutations has come from a test often used to analyse selection for diversity, which is to compare the K_a/K_s ratios between nucleotide sequences. K_s is the frequency of synonymous substitutions (i.e. in nucleotides that do not alter the encoded amino acid), whilst K_a is the frequency of asynonymous substitutions (i.e. in nucleotides that do change amino acids). If this ratio is low ($K_a < K_s$), it indicates that the selection favours conservation of that particular amino acid, whilst if it is high ($K_a > K_s$), it implies that selection favours amino acid changes. In the case of the *RPP5* locus, the ratio is low in the TIR, NBS and conserved amino acids of the LRR, but in the hypervariable solvent-exposed amino acid residues of the LRR, this ratio is high (averaging 2.8), indicating that selection pressure is favouring diversity in these amino acids. Similar results have been obtained for the *L* rust resistance locus of flax although in this case there were also two regions in the TIR domain that had high ratios, suggesting that these regions may also play a role in ligand binding specificity.

As well as selection for amino acid diversity within domains, there is evidence of shuffling of polymorphic sites through recombination to result in different numbers of LRRs. For example at the *RPP5* locus, a variant has been found containing a complete in-frame deletion of an LRR repeat unit that has no effect on *R* gene function. However, other members of the *RPP5* cluster have between 13 and 23 LRRs and it is believed that these have evolved from a progenitor gene with eight LRRs probably through unequal recombination events (*Figure 10.6*). Several alleles at the flax rust *L* locus have also been found to have internal deletions and expansions within the LRR, and the *Cf-2* and *Cf-5* genes in tomato also differ from each other by a precise deletion of six LRRs.

Unequal recombination has been documented in other parts of *R* genes giving rise to chimaeric genes. At the *Rp1* locus for rust resistance in maize,

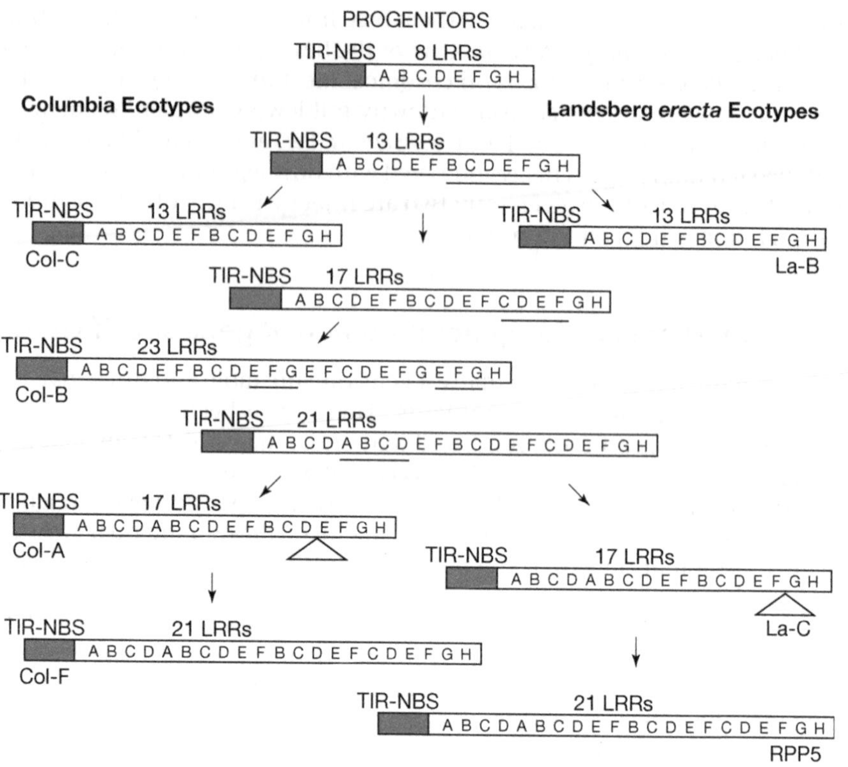

Figure 10.6

Model for the evolution of the leucine rich repeats (LRRs) at the *RPP5* locus in *Arabidopsis* by shuffling of the leucine rich repeats. The LRRs are denoted by letters. Regions of duplication are underlined and deletion marked by triangles. Names of the ecotypes are noted under the relevant protein structures.

selection for *Rp1-D*-susceptible mutants showed at least nine deletions of various *Rp1* gene family members. Recombinants with altered phenotypes, including disease lesion mimics and susceptible recombinants with flanking alleles of both possible combinations were generated from this locus. However, whether these gross recombination events are responsible for new specificities of *R* genes is unclear, and it is believed that the more subtle changes within the LRR and other domains through random nucleotide changes, inter-allelic recombination and gene conversion are more likely to be responsible for generating new specificities. An interesting recent observation has been that fungal extracts and defence-related stresses such as methyl jasmonate treatments activate the *Tnt1* and *Tto1* transposons in tobacco. This suggests that stresses in plants may result in genetic plasticity in these organisms (as also appears to occur in microbes, see Section 4.8), and may be responsible in part for the co-evolutionary dynamics. In addition, recent analysis of *R* genes from a number of plant species has suggested that they can be broadly divided into two types. Type 1 are fast-evolving and possibly involved in detecting avirulence elicitors with little fitness penalty in the pathogen, whilst type 2 are slow-evolving and presumably recognise avirulence elicitors with important functions in the pathogen.

10.9 Coevolution of resistance genes

Much information has been gained from molecular studies of *R* genes about their functional domains and organisation within host genomes. However, it is important to remember that the significance of these genes is in protecting plants from disease, and that this is a dynamic co-evolutionary process involving both plant and pathogen. The 'arms-race' concept (in which *R* gene effectiveness is lost by pathogen *Avr* gene mutation, and new resistance specificities then evolve) is a gross over-simplification, since it is now apparent that many *Avr* genes have a positive function in virulence and that their loss has a fitness cost on the pathogen. As a result, it appears that rather than *Avr* and *R* genes being lost once they are no longer effective, both *R* genes in plants and *Avr* genes in pathogens are maintained in populations for long periods by balancing selection.

This relationship is supported by population genetic studies in natural host/pathogen populations, such as that between Australian native flax (*Linum marginale*) and flax rust (*M. lini*), and between groundsel (*Senecio vulgaris*) and powdery mildew (*Erysiphe fischeri*). Here it has been shown that within a locality, there are plants with a high degree of variation in resistance phenotypes, and pathogens with a high variation in virulence phenotypes. Amongst the plants, none are found to be resistant to all pathotypes of the fungus, and those with the broadest range of resistance genes are the rarest. Amongst the pathogen population, those that can overcome most resistance genes are the rarest, and those that can overcome only a few resistance genes are the most common. Over time, there are fluctuations in the frequencies of different genotypes in host and pathogen populations, but the overall level of disease in the population remains constant, often referred to as an endemic balance.

A model to account for this balance is one in which the overall plant population (metapopulation) is comprised of many micropopulations. Each micropopulation is susceptible to different pathotypes of the fungus to different degrees, so that at any one time, one micropopulation may be under severe disease pressure whilst another is not. The micropopulation that is under severe disease pressure will subsequently produce fewer seed and less progeny in the following season resulting in selection against the pathotype that infected it and predominance of a different pathotype. At any given time, each micropopulation will be in a different phase of this boom-and-bust cycle resulting in an overall balance within the metapopulation. Instead of the 'arms race' scenario, a 'trench warfare' hypothesis has been invoked, in which there are advances and retreats of resistance and avirulence allele frequencies to maintain dynamic and stable polymorphisms. Furthermore, co-evolutionary dynamics are influenced by the nature of the plant (whether it is annual or perennial) and the nature of the pathogen (for example how it is transmitted). It is when crops are grown in 'artificial' agricultural environments as genetically uniform monocultures that co-evolutionary relationships are disturbed and the 'arms race' develops.

10.10 Recessive resistance genes

Whilst there are numerous *R* genes that follow the classical gene-for-gene model of dominant resistance genes, there are some, most notably the *Mlo* resistance gene in barley, in which monogenic resistance is conferred by recessive (*mlo*) alleles. Originally described in barley samples collected from

Ethiopia in 1937/38, and subsequently produced by mutagen treatment of *Mlo* cultivars, this resistance is particularly significant because it confers resistance against nearly all known isolates and genotypes of barley powdery mildew, and has been an effective long-lasting, or 'durable' resistance in the field, although it does increase susceptibility to some other pathogens such as rice blast, *M. grisea*.

Using a map-based cloning approach, the *Mlo* gene has been isolated, and encodes a 60 kDa integral plasma-membrane protein anchored by seven transmembrane helices. Whilst the protein sequence yields no direct clues to function through comparisons with databases, it has some similar properties to metazoan G-protein coupled receptors (GPCRs), which are proteins involved in relaying extracellular signals through activation of the α-subunit of heterotrimeric G proteins. However, it has recently been shown that the MLO protein functions independently of G-proteins and that signalling activity is enhanced through a Ca^{2+}-dependent interaction with calmodulin encoded within the C-terminal cytoplasmic domain of the protein.

Transient expression studies in which *Mlo* has been expressed in single epidermal cells of *mlo* genotypes has shown that the gene is cell-autonomous. The fungus is able to penetrate and form haustoria in these cells, but not in adjacent cells, although it can channel enough nutrients from adjacent cells to complete its life cycle and sporulate from a single epidermal cell. MLO is believed to operate as a negative regulator of a resistance response, and there is genetic evidence from mutations in two downstream genes, *Ror1* and *Ror2*, that these are involved in the resistance pathway. The defence response that is suppressed in *Mlo* plants but which occurs in *mlo* plants involves the formation of cell-wall appositions (CWAs) comprising phenolics, callose, peroxidases and cell wall material that are deposited directly under fungal penetration pegs, and these CWAs resist penetration (see Section 9.3.1). How MLO prevents this happening is yet to be resolved, and *mlo* resistance is only effective against powdery mildew and not other pathogens. However, searches to find homologues of the *Mlo* gene in other plant species as a means of developing durable resistance in these have resulted in identification of a homologue in wheat that complements barley *mlo*.

A further gene that shows recessive inheritance is the *RRS1-R* gene for resistance in *Arabidopsis* to *Ralstonia solanacearum*. In this case, the allele in the susceptible parent (*RRS1-S*) appears to suppress the *RRS1-R*-mediated defence responses, and it has been proposed that the *RRS1-R* and *RRS1-S* gene products compete for factors that are essential for perception of the pathogen.

10.11 Quantitative resistance

The cloning and characterisation of gene-for-gene resistance genes has rapidly increased our understanding of how these work to trigger defence responses. However, a large amount of resistance in plants is controlled quantitatively through polygenic resistance (see *Figure 10.1*). Identification of the genes involved in this will be a major challenge and one that will involve mapping QTLs (quantitative trait loci) onto complete genome sequences of plant species. In *Arabidopsis*, some progress has been made through the finding that resistance to some species of powdery mildew appears to be polygenic. Whether the *R* genes involved are resistance gene analogues that work co-operatively to confer polygenic resistance, or completely different sequences remains to be determined.

References and further reading

Bergelson, J., Kreitman, M., Stahl, E.A. and Tian, D. (2001) Evolutionary dynamics of plant *R*-genes. Science **292**: 2281–2284.

Bonas, U. and Lahaye, T. (2002) Plant disease resistance triggered by pathogen-derived molecules: refined models of specific recognition. *Current Opinion in Microbiology* **5**: 44–50.

Collins, N.C., Webb, C.A., Seah, S., Ellis, J.G., Hulbert, S.H. and Pryor, A. (1998) The isolation and mapping of disease resistance gene analogs in maize. *Molecular Plant–Microbe Interactions* **11**: 968–978.

Dangl, J.L. and Jones, J.D.G. (2001) Plant pathogens and integrated defence responses to infection. *Nature* **411**: 826–833.

Hulbert, S.H., Webb, C.A., Smith, S.M. and Sun, Q. (2001) Resistance gene complexes: evolution and utilization. *Annual Review of Phytopathology* **39**: 285–312.

Kajava, A.V. (1998) Structural diversity of leucine-rich repeat proteins. *Journal of Molecular Biology* **277**: 519–527.

Martin, G.B. (1999) Functional analysis of plant disease resistance genes and their downstream effectors. *Current Opinion in Plant Biology* **2**: 273–279.

Medzhitov, R. and Janeway, Jr. C. (2000) The Toll receptor family and microbial recognition. *Trends in Microbiology* **8**: 452–456.

Michelmore, R.W. and Meyers, B.C. (1998) Clusters of resistance genes in plant evolve by divergent selection and a birth-and-death process. *Genome Research* **8**: 1113–1130.

Mitchell-Olds, T. and Bergelson, J. (2000) Biotic interactions; genomics and coevolution. *Current Opinion in Plant Biology* **3**: 273–277.

Schneider, D.S. (2002) Plant immunity and film noir: what gumshoe detectives can teach us about plant–pathogen interactions. *Cell* **109**: 537–540.

Song, F. and Goodman, R.M. (2001) Molecular biology of disease resistance in rice. *Physiological and Molecular Plant Pathology* **59**: 1–11.

Signalling in plant disease resistance mechanisms

11

In this chapter, we will examine the diverse and complex signalling pathways that transmit detection of pathogens into resistance. The earliest events to occur in these pathways are activation of protein kinases, fluxes in calcium and other ions, production of reactive oxygen species (ROS) and nitric oxide (NO), and changes in gene expression (e.g. those encoding pathogenesis-related (PR) proteins) in the plant (*Figure 11.1*). Many of these mechanisms have parallels in mammalian systems, and a number of comparisons have been made between defence pathways in plants and programmed cell death in mammals.

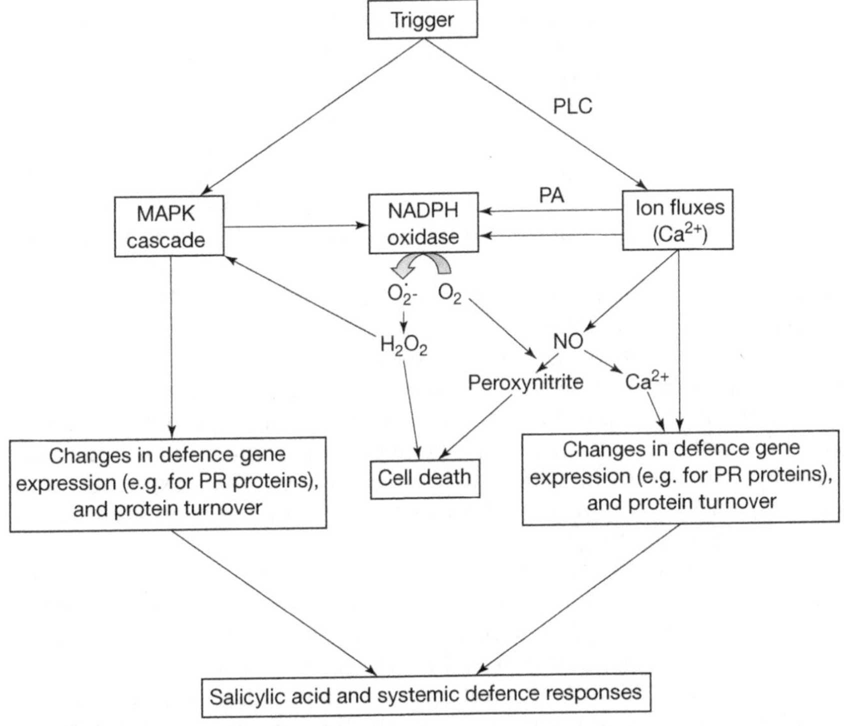

Figure 11.1

A general scheme indicating some of the signalling pathways that are known to be triggered through resistance gene recognition of pathogen elicitors. PLC = phospholipase C; PA = phosphatidic acid; NO = nitric oxide. See text for details.

Amongst the consequences of local resistance responses are the biosynthesis of salicylic acid (SA) and other potent signalling molecules that induce systemic forms of resistance in distal tissues. Wounding and attack by necrotrophic pathogens may also result in jasmonic acid (JA)/ethylene-based signalling pathways, that can act antagonistically to the SA pathways, and there is evidence of volatile signals being transmitted to adjacent plants (see Section 9.5). It is through a combination of pharmacological studies (using inhibitors of signalling pathways), mutational analysis and gene expression profiling, particularly in *Arabidopsis*, that many of these pathways are being dissected, and the means through which they are co-regulated determined. What is emerging is a highly branched signalling network with multiple signal amplification loops. Superimposed on this signalling machinery are mechanisms that establish the hierarchy of pathways, depending on the nature of the signal input.

11.1 Genetic analyses

Following the cloning and identification of resistance genes (see Chapter 10), a range of molecular genetic approaches have been targeted at identifying components of signalling pathways, including analysis of lines with combinations of mutants to obtain information about signalling co-ordination and hierarchies. The approach has often been to mutagenise seed of resistant lines and then identify mutants that have alterations to their resistance phenotypes when challenged by pathogens, such as increased susceptibility. Through this approach, the *ndr1* (*n*on-specific *d*isease *r*esistance), *eds1-5* (*e*nhanced *d*isease *s*usceptibility) and *pbs1*, *pbs2* and *pbs3* (avr*P*ph*B* *s*usceptible) mutations have been identified in *Arabidopsis*, the *rcr1*, *rcr2* and *rcr3* (*r*equired for *C*ladosporium *r*esistance) mutations in tomato, and *rar1* and *rar2* (*r*equired for Ml*a*12 *r*esistance) in barley. The majority of these genes appear to be involved in pathways from multiple resistance genes, although other resistance genes such as *Mla-1* and *Mla-7* in barley act independently through as yet unidentified signalling pathways. Other mutants, such as the *Arabidopsis pad* (*p*hytoalexin *a*ccumulation *d*eficient) loci *pad1*, *pad2*, *pad3* and *pad4* have been identified by screening for defects in pathogen-induced accumulation of the phytoalexin, camalexin, whilst a further group of mutants, such as *npr1* (*n*onexpressor of *PR* genes) have been obtained by searching for defects in salicylic acid (SA) signalling pathways. In addition, genes such as *cpr* (*c*onstitutive expresser of *PR* proteins) and *ssi* (*s*uppressor of *SA* *i*nsensitivity) have been identified as resistance up-regulating mutants.

Further approaches for identifying genes involved in downstream signalling have included the yeast two-hybrid screen, which has been used to isolate gene products that physically interact with *R* gene products, such as the *relA/SpoT* homologues as discussed in Section 11.6. In addition, there have been searches to identify homologues of mammalian genes known to be involved in defence responses in plants, for example to identify homologues of the Bax family of proteins involved in apoptotic cell death, and also of the anti-apoptotic proteins Bcl-2 and Bcl-x_L that are at the heart of mitochondrion-mediated cell death in mammals (see Section 11.9). *Table 11.1* lists some of the genes identified and what they encode, and some of the putative roles of these are discussed in Section 11.10.

Table 11.1 Some of the genes that have been identified as involved in *R*-gene-specified disease resistance through mutational and other screens

Gene	Plant species	How identified	Putative structure of gene product
PRF	Tomato	Required for fenthion resistance	CC-NBS-LRR resistance gene
RCR1	Tomato	Required for *Cladosporium* resistance	
RCR2	Tomato	Required for *Cladosporium* resistance	
RCR3	Tomato	Required for *Cladosporium* resistance	
RAR1	Barley	Required for Mla12 resistance	Zinc-binding CHORD protein
RAR2	Barley	Required for Mla12 resistance	
NDR1	*Arabidopsis*	Non-specific disease resistance	Transmembrane protein
EDS1	*Arabidopsis*	Enhanced disease susceptibility	Cytoplasmic protein with lipase motifs
EDS5 (SID1)	*Arabidopsis*	Enhanced disease susceptibility	Multidrug and toxin extrusion (MATE) transporter
PAD1	*Arabidopsis*	Phytoalexin accumulation deficient	
PAD2	*Arabidopsis*	Phytoalexin accumulation deficient	
PAD3	*Arabidopsis*	Phytoalexin accumulation deficient	Cytochrome P450 monooxygenase
PAD4	*Arabidopsis*	Phytoalexin accumulation deficient	Lipase-like protein
PBS1	*Arabidopsis*	AvrPphB susceptible	Serine-threonine kinase
PBS2 (RAR1)	*Arabidopsis*	Orthologue of barley *RAR1*	Zinc-binding CHORD protein
PBS3	*Arabidopsis*	AvrPphB susceptible	
NPR1 (NIM1)	*Arabidopsis*	Nonexpressor of PR genes	Ankyrin repeat protein
CPR1	*Arabidopsis*	Constitutive expressor of PR proteins	
CPR5	*Arabidopsis*	Constitutive expressor of PR proteins	Transmembrane protein
CPR6	*Arabidopsis*	Constitutive expressor of PR proteins	
SSI1	*Arabidopsis*	Suppressor of SA insensitivity	
SID1	*Arabidopsis*	SA induction deficient	
SID2 (ICS1)	*Arabidopsis*	SA induction deficient	Isochorismate synthase
SGT1b	*Arabidopsis*	Homologue of yeast *SGT1*	Ubiquitin ligase – regulator of cell cycle
DND1	*Arabidopsis*	Defence, no death	Cyclic nucleotide-gated ion channel
DND2	*Arabidopsis*	Defence, no death	
JAR1	*Arabidopsis*	Jasmonic acid resistant	
ETR1	*Arabidopsis*	Ethylene receptor	Ethylene binding histidine kinase
COI1	*Arabidopsis*	Coronatine insensitive	F-box protein
MPK4	*Arabidopsis*	Mitogen activated protein kinase 4	

11.2 MAP kinases (MAPK)

As shown in *Figure 11.1*, a key signalling pathway for disease resistance involves MAP kinases (MAPK). MAPK signalling pathways are common in eukaryotes as a means of transducing external signals perceived by cell surface receptors into downstream cellular responses. They involve the activation of a serine/threonine MAP kinase enzyme through dual phosphorylation of specific tyrosine and threonine residues by an upstream MAP kinase kinase (as discussed in Section 2.7.1). Early experiments in which pharmacological inhibitors of phosphatases and kinases were used to inhibit pathogenesis-related protein production and the hypersensitive response, indicated that protein kinases might have a role in defence, and the cloning of the *Pto* resistance gene (see Section 10.2) supported this. The presence of TIR domains in many *R* genes increased the evidence for kinase signal-ling pathways in disease resistance, since TIR-domain proteins activate gene expression through signalling pathways that are essentially conserved between mammals and *Drosophila*. In mammals for example, a protein MyD88 that contains an N-terminal death domain (so-called because it regulates cell death) is recruited by the TIR protein. The MyD88 is associated with a serine/threonine protein kinase (IRAK) via the death domain protein, and IRAK is autophosphorylated and acts through a kinase signalling cascade to activate $I_{\kappa}B$, which is an ankyrin-repeat-containing inhibitor of the $NF_{\kappa}B$ transcription factor. This results in degradation of $I_{\kappa}B$ through the proteasome pathway and release of the $NF_{\kappa}B$ transcription factor, which then translocates to the nucleus and co-operates with other transcription factors to activate target gene expression.

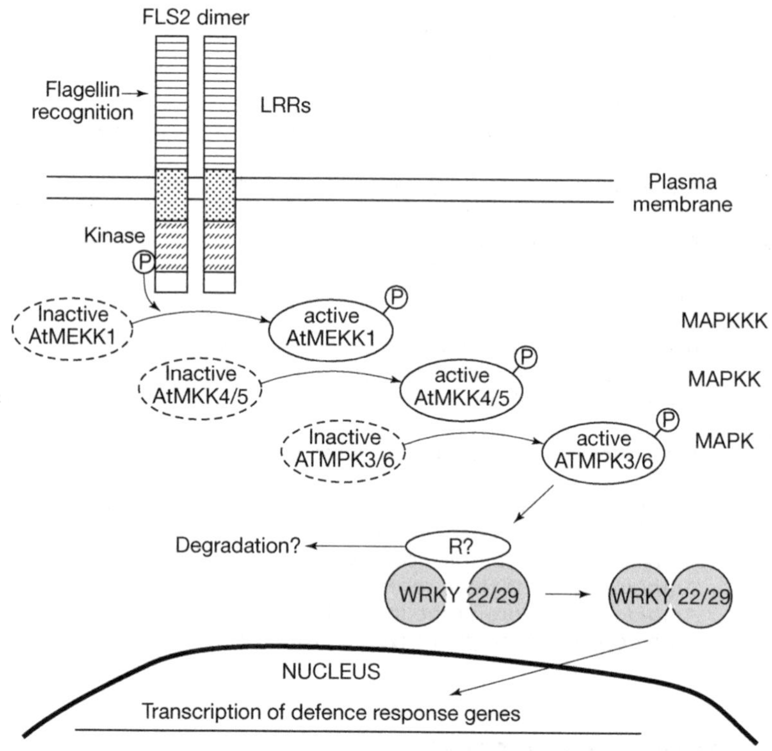

Figure 11.2

The FLS2 flagellin receptor in *Arabidopsis*. The action of the flagellin receptor is analogous to that of Toll-like receptors, in which the LRR domain recognises the pathogen associated molecular patterns (PAMP) setting off a kinase signalling cascade which culminates in release of the NF-κB transcription factor and defence gene expression. In *Arabidopsis*, the MAPKKK, MAPKK and MAPK proteins are as shown and the WRKY 22/29 transcription factor is the equivalent of NF-κB.

In addition to this proposed role for the TIR domains of *R* genes, an analogous MAP kinase signalling cascade has recently been identified in the innate immune response in *Arabidopsis* (*Figure 11.2*). Flaggelin, which is a pathogen-associated molecular pattern (PAMP) in mammalian systems was used to identify a receptor gene *FLS2*, which encodes a leucine-rich repeat-transmembrane domain-kinase protein similar in structure to Xa21 from rice. As indicated in *Figure 11.2*, specific MAPKKK, MAPKK and MAPK proteins have been identified associated with this signalling cascade which results in activation of defence-related genes such as *PAL*, *PR-1* and *PR-5* through the WRKY-type transcription factors. These transcription factors, which possess zinc-finger-like motifs and are unique to plants, interact with W boxes, which are found in the promoter regions of many defence genes and upstream of *NPR1* (see Section 11.10.2). In addition, the *Arabidopsis RRS1-R* gene for resistance to *Ralstonia solanacearum* is composed of TIR-NBS-LRR domains linked through a nuclear localisation signal to a WRKY domain, suggesting that this gene may be targeted to the nucleus and interact directly with W-boxes to induce gene expression following perception of the bacterial avirulence elicitor.

MAPKs have now been identified in many plant species, and the *Arabidopsis* genome contains at least 20 MAPK-like genes. In tobacco, two MAPKs, one of which had been previously identified as salicylic acid induced (SIPK, an orthologue of MPK6 in *Arabidopsis*) and the other as wound induced (WIPK, an orthologue of MPK3), are both activated during TMV infection when the *N* resistance gene is present, though with different activation kinetics. In addition, the tobacco *SIPK* gene positively regulates programmed cell death. For the *Cf-9* gene in tomato, which has no apparent kinase-signalling domain, SIPK and WIPK activities are stimulated in the interaction with Avr9 through Ca^{2+} influx and phosphorylation events, but not the oxidative burst. Also, blocking SIPK and WIPK activation does not abolish *Cf-9-Avr9*-dependent ROS production, suggesting that the MAP kinase pathway can operate independently of the oxidative burst.

It is apparent from these analyses that there are aspects of both innate immunity (involving recognition of PAMPs) and gene-for-gene resistance (involving specific *Avr* genes), as well as wound and mechanical stress responses, that operate through MAPK-signalling pathways. How these multifunctional kinases discriminate between the diverse stimuli to result in defence responses is unclear. It may be that signal specificity is determined by selective recruitment of particular kinases into signalling complexes or that the extent and duration of MAP kinase activation is important rather than a simple on/off setting.

11.3 Ion fluxes and calcium homeostasis

When plant cell cultures are treated with pathogen-derived elicitors to induce defence-related responses, some of the most rapid measurable changes that occur are large transient ion fluxes across the plasma membrane. These include efflux of K^+ ions, and influx of Ca^{2+}. This precedes the oxidative burst, and it is the elevated cytosolic Ca^{2+} that appears to be of particular significance, since in some cases omission of Ca^{2+} from culture media prevents these elicitor-inducible defence responses from occurring.

Using aequorin-based luminescence and Ca^{2+} reporter dye monitoring techniques in conjunction with confocal laser scanning microscopy, it has been possible to visualise fluxes in intracellular calcium levels *in situ*. In the rust (*Uromyces vignae*)/cowpea interaction, Ca^{2+} levels increase in resistant cultivars following inoculation but not in susceptible cultivars, and this occurs as the fungus attempts penetration of the plant cell wall. In *Pseudomonas syringae*/*Arabidopsis* interactions involving the *AvrRpm1/RPM1* genes, elevation of calcium levels occurs in a biphasic fashion. Initial increases appear to be non-specific in that they are observed in both compatible and incompatible interactions, and in interactions with other bacteria such as *E. coli*. However, the subsequent sustained increase in Ca^{2+} levels only occurs in incompatible interactions and correlates with delivery of AvrRpm1 to the plasma membranes, where RPM1 is located as a peripheral membrane-associated protein (see Section 10.3.2). Yeast two-hybrid analysis has shown that RPM1 itself is associated with a putative membrane protein, which may be involved in calcium channelling through the membrane. In the tomato/*C. fulvum* interaction, patch-clamp analysis has shown that calcium channels are activated in both *Cf-5/Avr5* and *Cf-9/Avr9* interactions. This activation (which occurs from *R* genes that are extracellularly located), is thought to be modulated through a G-protein-dependent mechanism. Interestingly, the activation of these plasma membrane calcium channels leads to inhibition of plasma membrane Ca^{2+}-ATPase, which may reflect an additional mechanism

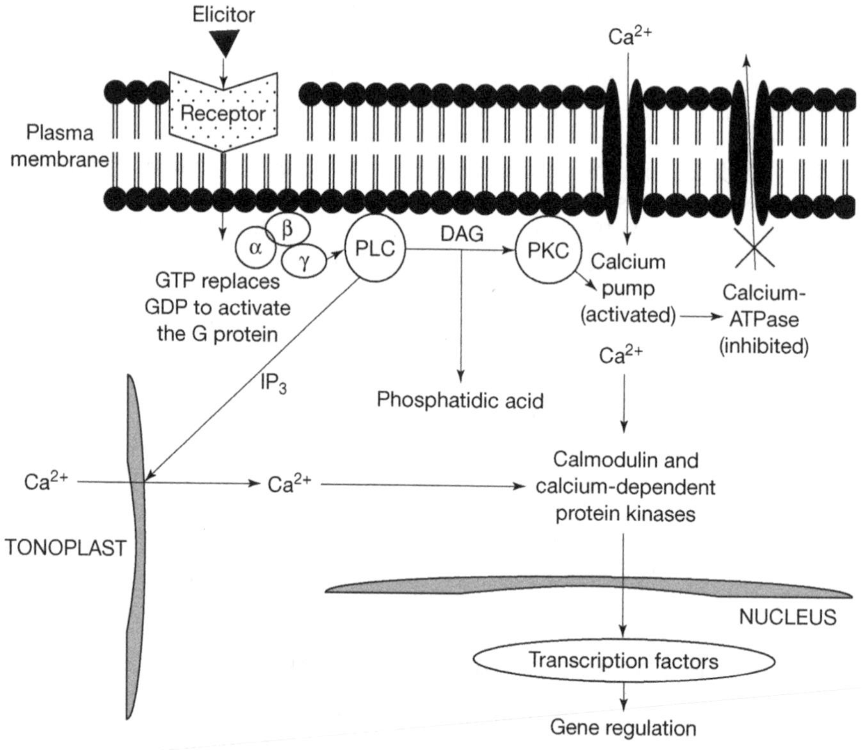

Figure 11.3

Possible role of G-proteins in stimulating calcium influx in response to pathogen elicitors. Elicitor binding activates G-proteins to release calcium from the tonoplast, open plasma membrane calcium channels, and inhibit the calcium-ATPase. The result is an influx of calcium and blockage of its efflux (see text for details).

to prevent leakage back to the apoplast and so maintain the high levels of cytosolic calcium.

In animals, G-protein signalling cascades operate through phospholipase C, diacylglycerol (DAG), and inositol 1,4,5-triphosphate (IP_3) to increase calcium levels, with calmodulin, calcium-dependent protein kinases (CDPKs) and protein kinase C (PKCs) as modulators of Ca^{2+} (see *Figure 2.4* in Section 2.7.1). There is recent evidence for phospholipase C working through DAG in plants to activate PKCs and calcium influx (*Figure 11.3*), although DAG is also readily converted to phosphatidic acid (PA), which is itself a signal in plant defence responses. IP_3 is also generated in plants and is believed to release Ca^{2+} from the tonoplast. Downstream, CDPKs have been found such as the 68 kDa CDPK in tomatoes that is activated in *Cf-9/Avr9* interactions prior to the oxidative burst. Calcium is also involved in regulating the oxidative burst (see Section 11.4) and nitric oxide (NO) production (see Section 11.5).

11.4 The oxidative burst

Like calcium fluxes, reactive oxygen species (ROSs) are weakly induced in both compatible and incompatible interactions, but in the incompatible interactions there is a later more prolonged and extensive induction. ROSs are the products of the sequential reduction of molecular oxygen (*Figure 11.4*), primarily the superoxide radical ($O_2^{\cdot -}$), hydrogen peroxide (H_2O_2) and the hydroxyl radical ($OH^{\cdot}$). They are produced at a low level in all plants, and there are mechanisms involving glutathiones, superoxide dismutates and catalases to remove them so that they cause minimal damage to the plant. However, within 2–3 minutes of pathogen attack or elicitor treatment, ROSs are induced to much higher levels so that they overwhelm the removal mechanisms. This burst appears to occur after the calcium and ion fluxes, and PA production.

The superoxide radical itself can be generated through a number of mechanisms in plants. The main two identified in plant–pathogen interactions are via NADPH/NADH oxidases, or via apoplastic peroxidases (*Figure 11.5*). Evidence for the first (which is analogous to the mechanism used in mammalian neutrophils as part of the respiratory burst to phagocytose and kill invading bacteria), comes from the identification of analogues of the mammalian NADPH-oxidase *gp91^phox^* subunit gene in plant species such as rice, *Arabidopsis* and parsley. Interestingly, these genes appear to have an extended N-terminus containing two Ca^{2+}-binding motifs, suggesting that calcium has a major role in regulation of activity. The *Arabidopsis* analogue ATRBOH also has features such as a putative haem-binding site and homology to the superfamily of plasma membrane flavocytochromes, suggesting that it may have a role in iron uptake. However, functional equivalents of other parts of the mammalian NADPH oxidase complex have been less easy to find in plants, presumably reflecting some fundamental differences in the ways plants and animals combat invading pathogens.

Apoplastic peroxidases have also been shown to generate ROS, and co-localisation of peroxidases that are able to perform this function under the alkaline pHs achieved through the movement of ions (H^+ and Ca^{2+} influx; K^+ and Cl^- efflux) during pathogen invasion have been reported. At the same time, the acidification of the cytoplasm may be a stimulus for activating pathogenesis-related gene expression. The demonstration of enhanced susceptibility to disease in transgenic plants in which peroxidase genes have

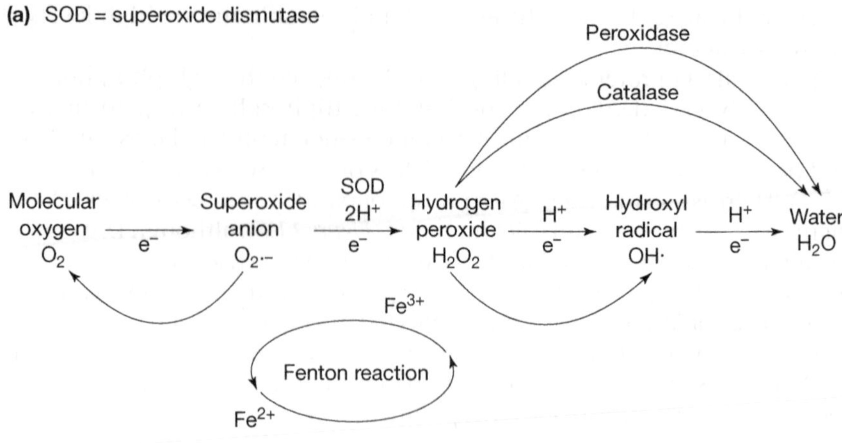

(b) The Fenton reaction

$$H_2O_2 + Fe^{2+} (Cu^+) \longrightarrow Fe^{3+} (Cu^{2+}) + OH\cdot + OH^-$$

$$O_2{\cdot}^- + Fe^{3+} (Cu^{2+}) \longrightarrow O_2 + Fe^{2+} (Cu^+)$$

Figure 11.4

General scheme for the production of reactive oxygen species by sequential reduction of molecular oxygen. (a) SOD = superoxide dismutase. (b) The Fenton reaction.

been down-regulated through antisensing supports the role of peroxidases as a means of ROS generation.

The superoxide radical itself rapidly dismutates in cells to form hydrogen peroxide, either spontaneously or through the action of superoxide dismutase (SOD). Hydrogen peroxide is a relatively stable, unreactive ROS, with a half-life of <1 second in plant cell walls, but it is able to pass through membranes and reach cell locations away from where it is generated. Hydrogen peroxide is directly toxic to microbes at the levels produced in plants. However, since plant cells lack the ability to surround and engulf invading pathogens, it is not clear whether this has a significant role in defence. In addition, hydrogen peroxide has been implicated as the most important ROS for signalling. Defence genes, such as those encoding phenylalanine ammonia lyase (PAL) and glutathione-S-transferase are activated when hydrogen peroxide is added to plants and also in transgenic plants that produce H_2O_2 constitutively. Since PAL is the first enzyme in one pathway for salicylic acid biosynthesis (see Section 11.7), elevated levels of hydrogen peroxide may also work through SA as part of the systemic signalling discussed in Sections 11.7 and 11.10, and there is evidence in *Arabidopsis* that the primary oxidative burst at the site of pathogen inoculation induces systemic 'microbursts' that appear to be an essential component of systemic immunity. Mitogen-activated protein kinase (MAPK) cascades (see Section 11.2) are also activated by hydrogen peroxide, and there is good evidence for hydrogen peroxide forming an integral part of signal amplification along with nitric oxide (NO).

(a) Via NADPH/NADH oxidases (gp$^{91\ phox}$). Following ligand binding and calcium influx, the Rac GTP-binding protein is activated and the membrane-spanning oxidase converts oxygen to the superoxide radical.

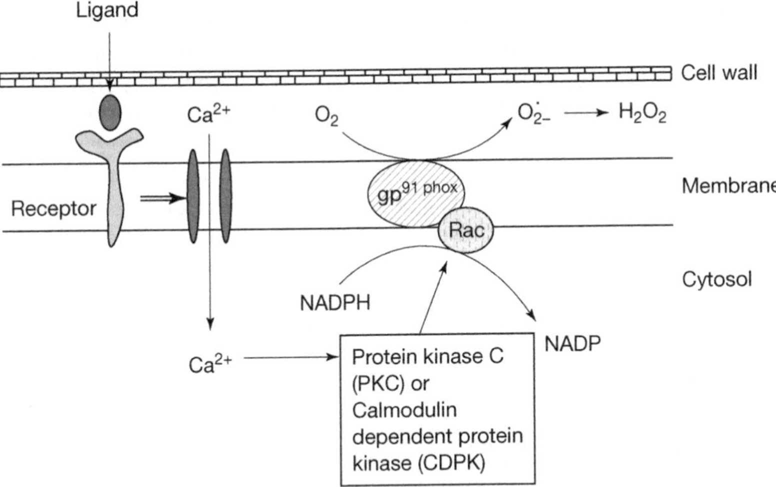

(b) Via apoplastic peroxidases. Ligand binding results in ion fluxes through the membrane that cause alkalinisation of the apoplast. This activates the peroxidase to convert oxygen into hydrogen peroxide.

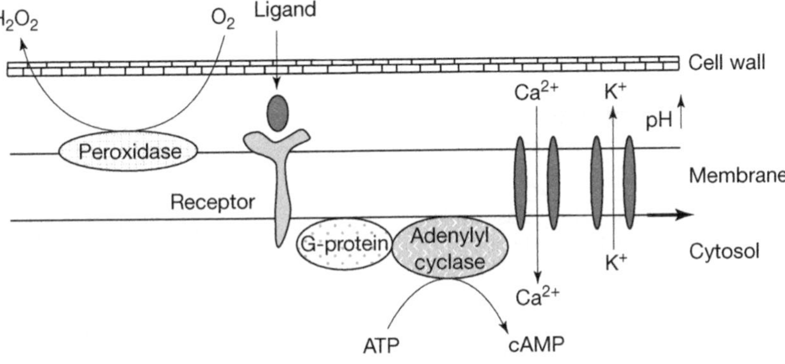

Figure 11.5

The main mechanisms identified for the production of the superoxide radical in plants.

The most damaging ROS is the hydroxyl radical which is believed to be formed from hydrogen peroxide readily in plant cells through the Fenton cycle of reactions, involving oxidation of transition metals such as Fe^{2+} or Cu^{+} (*Figure 11.4b*). It has been proposed that if hydrogen peroxide were to pass to the pathogen or plant nucleus, it would react there with intracellular metal ions to produce the hydroxyl radical, and that this in turn would cause DNA fragmentation and cell damage. This would result in both pathogen death and also plant cell death, which would provide a barrier against further ingress of any biotrophic pathogens.

In addition the oxidative burst may regulate protein turnover. In barley, the RAR1 protein (see Section 11.1) is involved in the prolonged and extensive second burst of ROS production. RAR1 has homology to eukaryotic zinc-binding proteins, and has two 60 amino acid *C*ysteine and *H*istidine *R*ich *D*omains, ('CHORD' domains), suggesting a role in signal cascades and induction of gene expression. Evidence from yeast two-hybrid screens suggests that this protein interacts with a second protein SGT1, encoded by the *SGT1b* gene in *Arabidopsis*. This protein has homology to the yeast SGT1 protein which is an ubiquitin ligase involved in targeting proteins for degradation by the proteasome protein complex. It has been proposed that the role of RAR1 and SGT1 may be as regulators of defence responses, co-ordinating the degradation and turnover of proteins involved in defence signalling (see Section 11.10). Such a role for resistance genes has also been predicted from the PEST (Pro-Glu-Ser-Thr)-like sequence in the C-terminal domain of the *Ve* gene, which may be involved in ubiquitinisation and protein turnover.

11.5 Nitric oxide (NO)

In some defence induction experiments using plant cell cultures, physiological levels of ROS are not sufficient to cause plant cell death, indicating that other stimuli are required. In mammalian macrophages, ROSs are known to

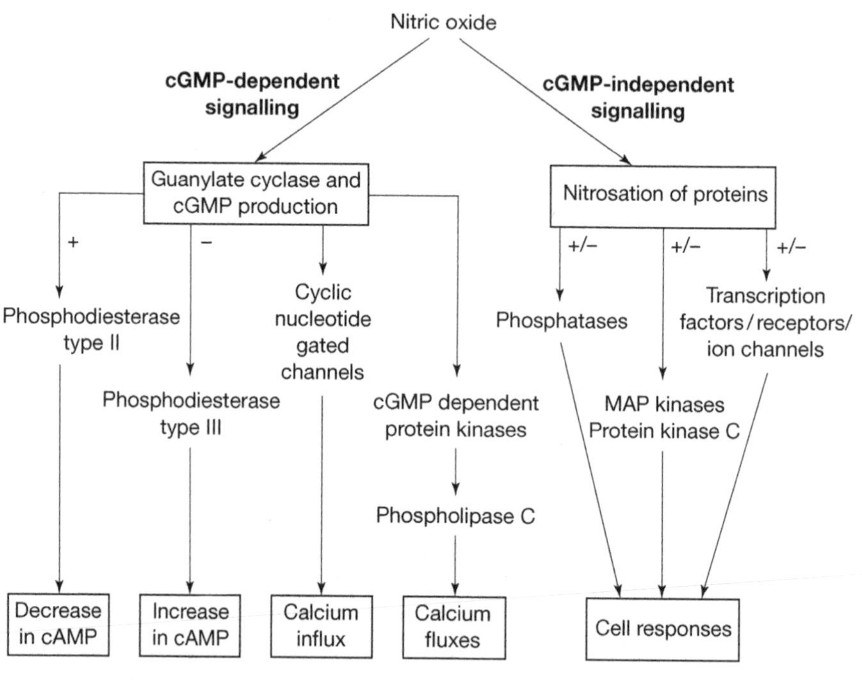

Figure 11.6

Nitric oxide signalling in animal cells. This occurs through two pathways, one involving cGMP that culminates in changes in cAMP levels and calcium fluxes, and the second which is cGMP-independent involving nitrosation of proteins.

work with nitric oxide (NO) in defence responses. Biosynthesis of NO in animals is primarily catalysed by the nitric oxide synthase (NOS) enzyme, of which there are three isoforms, two of which depend on levels of intracellular free Ca^{2+} for their activation. The NO produced is highly diffusible and is an important signalling molecule. It works through two types of signalling pathway in animals, those that involve cyclic GMP (cGMP) and those that do not (*Figure 11.6*). One consequence of these signalling networks can be the triggering of apoptosis, and in animals, NO can be both an inducer and a repressor of apoptosis depending on the cell type and other poorly characterised factors.

In plants, NO is known to be produced both non-enzymatically and through the action of NAD(P)H-dependent nitrate reductase. Evidence from the use of nitric oxide synthase (NOS) inhibitors, and antibodies raised against mammalian NOSs, has indicated that Ca^{2+}-dependent NOS-like proteins are also present in plants. Moreover, synthesis of NO is induced in incompatible interactions in soybean and tobacco experiments, but not in compatible interactions, and the addition of NO donors or recombinant mammalian NOS to tobacco plants or cell suspensions induced expression of the *PR-1* and *PAL* genes. These genes were also induced by a cGMP analogue, and this activation could be suppressed by guanylate cyclase inhibitors, strengthening similarities between NO signalling pathways in animals and those that might occur in plants, although as yet no guanylate cyclase genes have been cloned from plants. The induction of both *PR-1* and *PAL* gene expression is particularly intriguing, since *PR-1* expression is salicylic acid-dependent, whilst expression of *PAL* is salicylic acid-independent. The implication is that NO induction of *PAL* results in SA biosynthesis which in turn activates *PR-1*, providing additional evidence for the inter-relationships between ROS, NO and SA signalling molecules (*Figure 11.7*). At the same time, there is evidence that

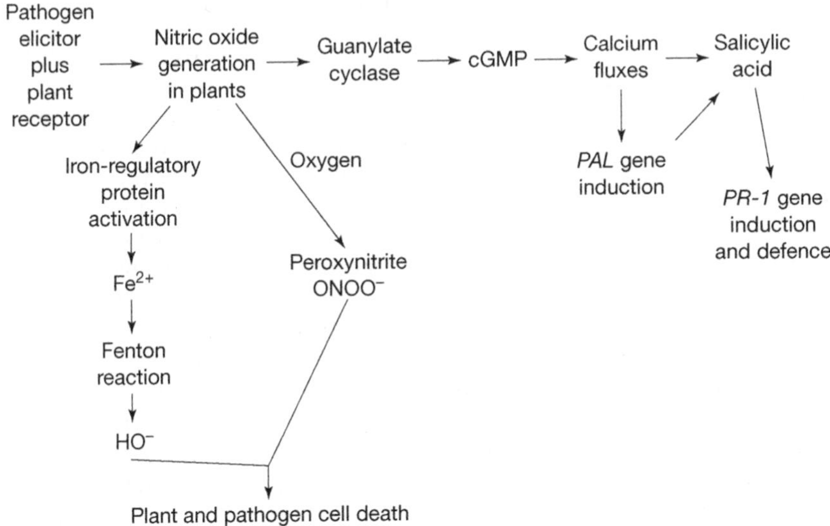

Figure 11.7

Proposed role for nitric oxide in defence responses in plants, both as an inducer of pathogenesis-related proteins (horizontal pathway) and as an inducer of plant and pathogen cell death (vertical pathway).

NO inhibits ethylene production, which suggests a key role for NO in the systemic signalling pathway as discussed in Section 11.10. Along with the increasing evidence of a role for NO in plant cell apoptosis and signalling, it may have an additional role as an antimicrobial. Production of peroxynitrite ($ONOO^-$) occurs through reaction between NO and oxygen, and peroxynitrite has been shown to have direct antimicrobial affects.

11.6 (p)ppGpp signalling

In experiments using the yeast two-hybrid system to identify proteins that interact with resistance genes, the *At-RSH1* gene was identified in *Arabidopsis* downstream of the *RPP5* resistance gene. The At-RSH1 protein, which is predicted to be a plasma-membrane-anchored cytoplasmic molecule, has significant homology to the RelA and SpoT proteins found in bacteria such as *E. coli* and *Streptomyces coelicolor.* Further homologues have subsequently been identified in *Arabidopsis.* In *E. coli,* these proteins regulate the levels of guanosine tetraphosphate (ppGpp) and guanosine pentaphosphate (pppGpp), and are activated by sudden nutritional and environmental stresses such as amino acid, carbon, nitrogen or phosphate starvation, or changes in temperature and osmolarity. The resultant (p)ppGpps that accumulate act as signalling nucleotides to elicit the bacterial 'stringent response' in which nucleic acid and protein synthesis is down-regulated whilst protein degradation is up-regulated. The *Arabidopsis* homologues have been shown to complement the function of RelA and SpoT in *E. coli* and *S. coelicolor,* implicating (p)ppGpp as a signalling molecule in mediating a stress-induced defence response in plants.

11.7 Low-molecular-weight signalling molecules

Early experiments on systemic acquired resistance (see Section 9.4) in the late 1970s indicated that addition of salicylic acid (SA) to plants induced the same pathogenesis-related (PR) protein expression fingerprint as occurs in SAR, and gave the same spectrum of resistance. Subsequently, structurally related chemicals such as 2,6-dichloroisonicotinic acid (INA) and benzo-(1,2,3)-thiadiazole carbothioic acid-S-methyl ester (BTH) were shown to have the same effects (*Figure 11.8*). SA and its conjugated glycoside derivative (SAG) have repeatedly been shown to accumulate to high levels, in both local and systemic tissue, after primary infection of several plant species, and accumulation has also been observed in the phloem. They regulate cell growth by specifically affecting cell enlargement. Two alternative pathways have been proposed for the biosynthesis of SA in plants (see *Figure 9.2*), and there is evidence that both of these occur in *Arabidopsis.* In one pathway synthesis is from phenylalanine involving phenylalanine ammonia lyase (PAL), which is often part of the hypersensitive response and systemic resistance. However, SA can be produced when this enzyme is inhibited, and recent cloning of an isochorismate synthase (*ICS2*) gene from *Arabidopsis* has shown that a second pathway analogous to that used by some bacteria, also occurs in plants. By use of mutant screens, it has been confirmed that the SA produced by this pathway is involved in both local and systemic resistance responses.

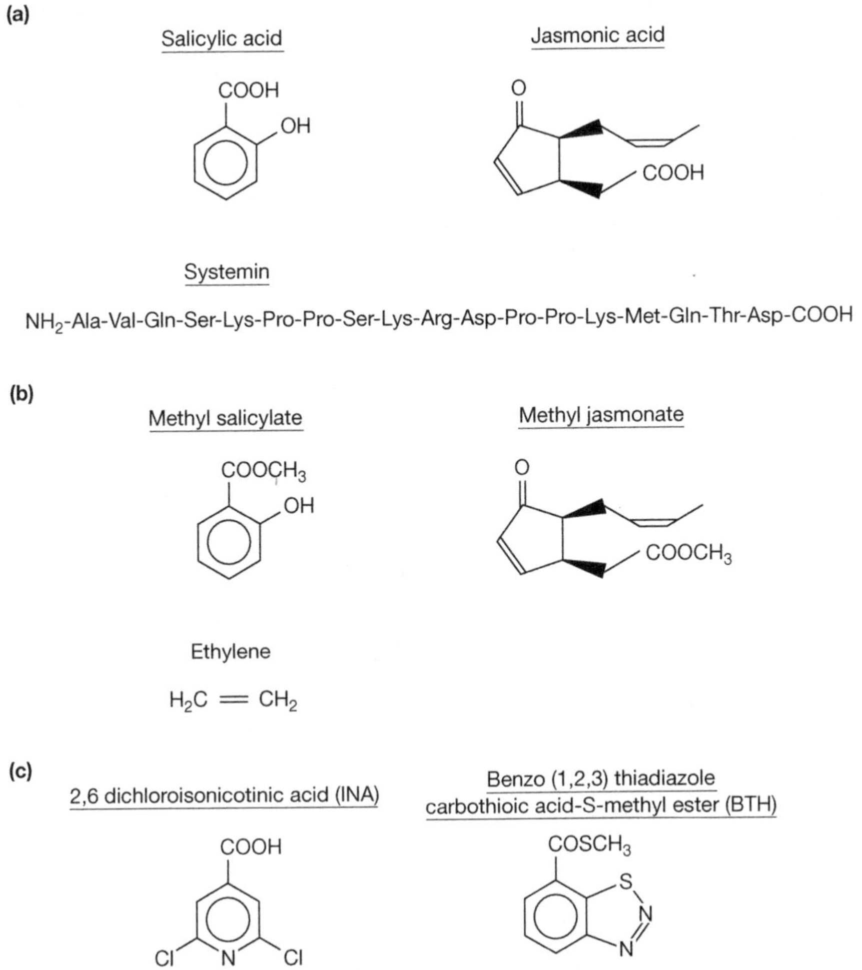

(a)

Salicylic acid

COOH
OH

Jasmonic acid

O
COOH

Systemin

NH_2-Ala-Val-Gln-Ser-Lys-Pro-Pro-Ser-Lys-Arg-Asp-Pro-Pro-Lys-Met-Gln-Thr-Asp-COOH

(b)

Methyl salicylate

$COOCH_3$
OH

Methyl jasmonate

O
$COOCH_3$

Ethylene

$H_2C = CH_2$

(c)

2,6 dichloroisonicotinic acid (INA)

COOH
Cl N Cl

Benzo (1,2,3) thiadiazole carbothioic acid-S-methyl ester (BTH)

$COSCH_3$
S N N

Figure 11.8

The major (a) systemic and (b) airborne signalling compounds, along with (c) chemical activators of induced resistance

There are a number of systemic resistance responses in which *PR* gene activation does not correspond to enhanced levels of SA, and also in which the set of *PR* genes activated are different. For example, whilst SA accumulation and activity generally induces expression of *PR-1*, *PR-2* and *PR-5*, infection of tobacco with *Erwinia carotovora* induces expression of a basic β-1,3-glucanase (*PR-2*) and a basic chitinase (*PR-3*) with no *PR-1* induction. Evidence suggests that jasmonic acid, methyl jasmonate and ethylene play an important role in these additional systemic defence responses. Exogenous application of these signalling molecules induces these genes in addition to plant defensins and thionins, which are small cysteine-rich basic proteins that have antimicrobial activity, similar to the antifungal peptide drosomycin induced in *Drosophila*. In *Arabidopsis*, the *Thi2.1* thionin gene, and the *PDF1.2* defensin genes are activated by methyl jasmonate and infection with certain pathogens that cause necrotic lesions, such as *Alternaria brassicicola*, but not by addition of salicylic acid, and these responses also occur in *NahG* transgenic plants. *NahG* encodes the enzyme salicylate hydroxylase, that converts SA into inactive catechol. However, in barley, thionin genes are induced by SA and JA, and there are other thionins in *Arabidopsis* that are not induced by JA.

Jasmonates are fatty acid derivatives with a 12-carbon backbone, derived from 18-carbon intermediates via the octadecanoid (ODA) pathway (*Figure 11.9*), and the identification of JA-related mutants in *Arabidopsis* has helped elucidate this pathway. This biosynthesis requires a plasma membrane located lipase, which is believed to be induced by calcium. This particular pathway is potently inhibited by salicylic acid. Ethylene biosynthesis involves two enzymatic steps. S-adenosyl methionine is converted to the cyclic amino acid 1-aminocyclopropane-1-carboxylate (ACC) by ACC synthase. ACC is then

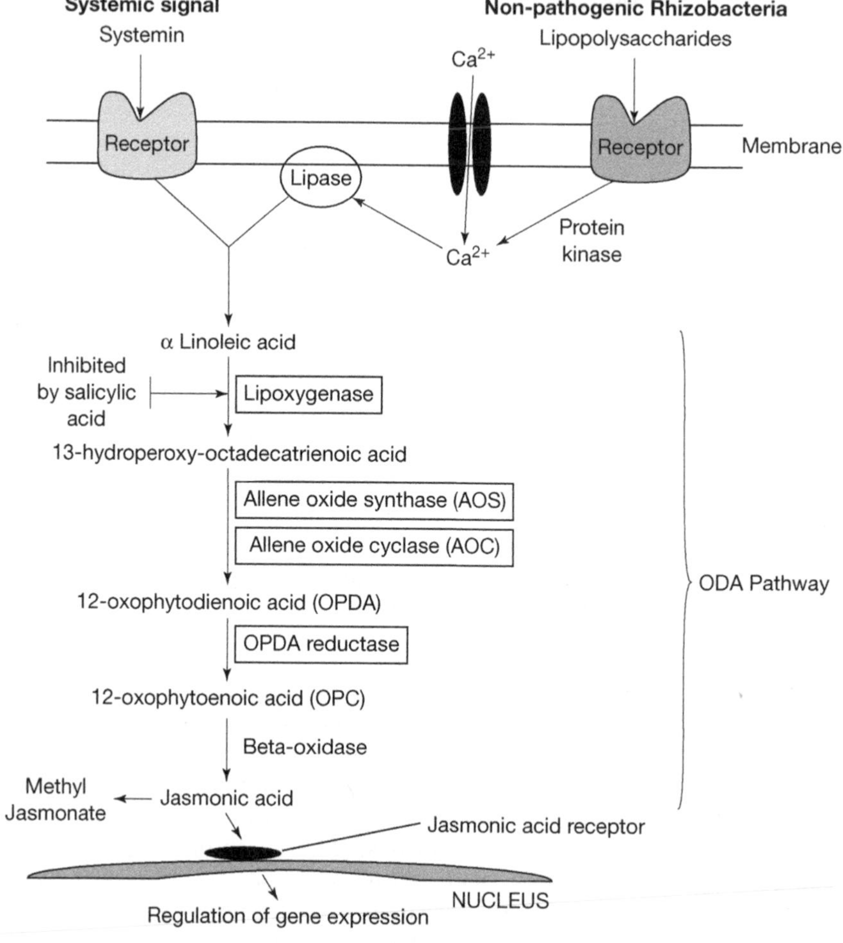

Figure 11.9

The biosynthesis of jasmonic acid. Signals (either through systemin or from microbes) activate a membrane-bound lipase resulting in linoleic acid production. This enters the ODA pathway to produce jasmonic acid, which regulates gene expression. Enzymes involved are boxed.

converted to ethylene by ACC oxidase. The genes encoding these enzymes are generally part of multigene families in plants, with different regulation and expression characteristics.

Jasmonic acid and ethylene also appear to be involved in signal transduction of wound responses and responses to non-pathogenic rhizobacteria (bacteria that colonise the roots of plants living off root exudates, and which sometimes promote plant growth, such as *Pseudomonas fluorescens* and *P. aeruginosa*). This is referred to as induced systemic resistance (ISR) (see Section 9.4), and it is often PAMPs such as lipopolysaccharides in the bacterial cell wall that are the elicitors of this response. There is also a wound response that is activated by feeding insects in which proteinase inhibitor proteins are induced, and a key signalling molecule has been identified as the 18 amino acid polypeptide systemin (*Figure 11.8*). This polypeptide, which is believed to be derived from a cytoplasmically located precursor, prosystemin, is activated during insect feeding, and is then believed to function like polypeptide hormones in animals and yeast, but passing through the phloem of the plant. The signal is transduced via the octadecanoid (ODA) pathway, linking into jasmonic acid production, and acts through hydrogen peroxide as a secondary messenger to activate defence gene expression in mesophyll cells. Whilst systemin-like signalling molecules have so far only been detected in the *Solanaceae*, there is no reason to suppose that they are not widespread in other plant families.

11.8 RNA as a signal

Evidence from grafting experiments has revealed that the post-transcriptional gene silencing mechanism (see Section 8.8.3) that is involved in defence against RNA viruses may also be transmitted through plants systemically. This mechanism, referred to as systemic acquired silencing (SAS), has been identified in experiments in which non-suppressed scions were grafted onto co-suppressed rootstocks, and co-suppression was then found to occur in the scions. It has been postulated that this mechanism may be responsible for the formation of patches of virus-resistant cells within infected plants (the dark areas between mosaics for example), and also in part for systemic resistance that occurs in virally infected plants. Whether the signals are naked RNA molecules that result from viral replication, or RNA molecules associated with plant proteins is as yet unknown.

11.9 Co-ordination of cell death responses

In Chapter 9, we discussed the nature of the hypersensitive response and its similarities to programmed cell death (PCD) in animals. The nature of many resistance genes combined with genetic and pharmacological dissection of signalling pathways in plants supports the notion that forms of PCD occur in plants. However, whilst there are many similarities in the mechanisms between plants and animals, there are also clear differences. One of the key regulatory components of stress responses and PCD in animals is the mitochondrion, which is involved in interpreting and amplifying stress signals from different receptors and sub-cellular compartments as shown in *Figure 11.10*. These include detection of calcium fluxes, ionic fluxes and pH changes, reactive oxygen species, nitric oxide, specific proteinaceous intracellular

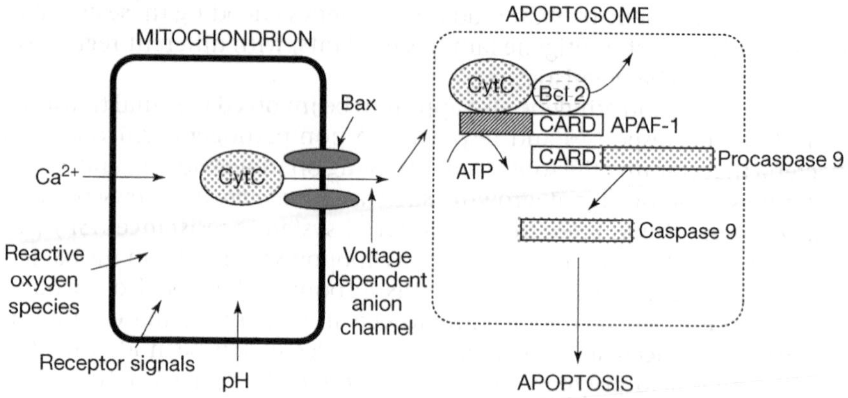

Figure 11.10

The role of the mitochondrion in co-ordinating apoptosis in human cells. APAF-1 is held in an inactive form by Bcl-2 in the apoptosome. When the mitochondria receives a stress signal, cell-death regulators e.g. cytochrome C are released. These, in combination with ATP, release APAF-1, which activates procaspase-9 and initiates cell death.

death signals, including Bax, and changes in the levels of metabolites that reflect the energy status of the cell, such as ATP, ADP, NADH, NADPH and creatine phosphate. The result is release from the mitochondria of cell death activators, inhibitors and inhibition derepressors, including cytochrome c, which regulate apoptosis proteins. These apoptosis proteins in humans and *C. elegans* work by activating a cysteine protease (caspase) cascade leading to cell suicide. The adapter proteins, APAF-1 (CED-4 in *C. elegans*) are held in inactive conformations by Bcl-2 (CED-9) in an 'apoptosome' until the death inducing stimulus in combination with ATP or dATP releases the APAF-1 (CED-4) to activate the pro-caspase-9 (CED-3) and initiate cell death.

In plants, there is increasing evidence for a similar role for mitochondria in regulating PCD, and many of the induction signals are similar to those in animals. For example, the TIR and NBS domains of *R* genes resemble those of receptor signals in mammalian systems, with motifs similar to that in the mammalian PCD mediator APAF-1. Evidence that supports the importance of NBS domains in plant *R* genes comes from studies on the *P2* flax rust-resistance gene. Here, a spontaneous susceptible mutant line of flax has been identified with a single G to E amino acid substitution in the GLPL motif that is conserved between R gene products, mammalian APAF-1 and *C. elegans* CED4.

There are however clear differences in the responses that occur between plants and animals. Evidence for cytochrome c release in plants is limited, and the caspase-like proteins that have been identified in *Arabidopsis* (the metacaspases) are not direct homologues of these in animals. Furthermore, no Bcl homologues have been found in plants, although a putative homologue of Bcl-2 has been identified in tobacco using anti-Bcl-2 antibodies. In yeasts, the functional equivalent of Bcl-2 is belived to be the BI-1 (Bax inhibitor 1) protein, and homologues of this have been isolated from *Arabidopsis* (*AtBI-1, 2* and *3*) and rice, and these genes were able to suppress Bax-mediated cell death in yeast.

It has been postulated that lack of conservation of these PCD regulators in plants may reflect the greater stringency of regulation that is required in animals, where it is necessary to ensure that cell death is rapid and that the dead cells are then rapidly removed to prevent inflammatory responses in adjacent cells. In animal cells, PCD mechanisms of many types occur, normally categorised into three morphotypes. Apoptosis (morphotype 1) is viewed as the important active programme, involving caspases (cysteine proteases), and results in cell death through chromatin condensation, nuclear DNA fragmentation and formation of membrane-bound apoptotic bodies containing intact organelles, that eventually become phagocytosed by neighbouring cells or macrophages. Morphotype 2 is autophagic or cytoplasmic degenerative PCD, in which cytoplasm is consumed by autophagic organelles, not normally involving macrophages. This is the kind of death in which cells die *en masse*, for example in degeneration of tadpole tails, or in the embryonic suspensor of plants. The third type of PCD (oncosis) is generally considered as 'accidental' death in animals. Here, the cells and organelles become swollen and the plasma membrane breaks. In animals, this involves leakage of cellular components likely to result in inflammation of neighbouring tissue, and so it has not been favoured in animal evolution. However, in plants, these same evolutionary constraints do not apply, and there is growing support for the notion that the principal PCD mechanism in plants is a form of 'programmed oncosis' with apoptosis as a secondary mechanism. Spread of PCD from this oncosis to adjacent cells could be regarded as a selective advantage in plants forming an additional defence barrier against pathogen ingress. Thus the controls on this process may not need to be as stringent as in animals, and the dead cells remain in place as a structural barrier and as the visible necrotic symptoms of the hypersensitive response.

11.10 Interplay of downstream signalling pathways

As outlined above, there are many diverse pathways and signals that culminate in disease resistance responses involving PCD. In addition, there are numerous downstream signalling pathways that lead to systemic defence responses as well as responses in neighbouring plants. These involve responses to pathogens, wounding and non-pathogenic rhizobacteria, and operate through ROS, SA, JA, ethylene, systemin and possibly other signals such as RNA in the case of viruses. How these pathways are integrated and how they result in induction of the different fingerprints of PR proteins, defensins, thionins and proteinase inhibitors that are characteristic of the different mechanisms is complex and not fully understood, but mutational analysis, principally in *Arabidopsis* is helping to position specific genes in signalling cascades and hierarchies (see Section 11.1 and *Table 11.1*).

11.10.1 The EDS1 and NDR1 pathways

Mutations in one gene, *EDS1* have been shown to suppress resistance conditioned by several *RPP* genes against *P. parasitica* as well as *RPS4* (resistance to *P. syringae*), but not *RPM1*-, *RPS2*- or *RPS5*-specified resistance to *P. syringae*. Resistance can be rescued in *eds1* plants by application of SA, allowing *EDS1* to be placed upstream of SA-dependent processes that involve the *NPR1* gene (see Section 11.10.2). In contrast to *eds1*, the *ndr1* mutation strongly suppresses resistance conferred by *RPM1*, *RPS2* and *RPS5* but has a negligible effect on *EDS1*-dependent resistance responses. Thus, it appears that TIR-NBS-LRR type R proteins are strongly dependent on *EDS1* in *Arabidopsis*, whereas most CC-NBS-LRR type R proteins are dependent on *NDR1* (*Figure 11.11*).

NDR1 has been identified as a transmembrane protein, whilst part of EDS1 has some homology in databases to the catalytic site of eukaryotic lipases, and lipases are known to be involved in the biosynthesis of the plant fatty acid signalling molecule jasmonic acid (JA) (see Section 11.7). However, applications of JA failed to rescue disease resistance in *eds1* plants, suggesting that

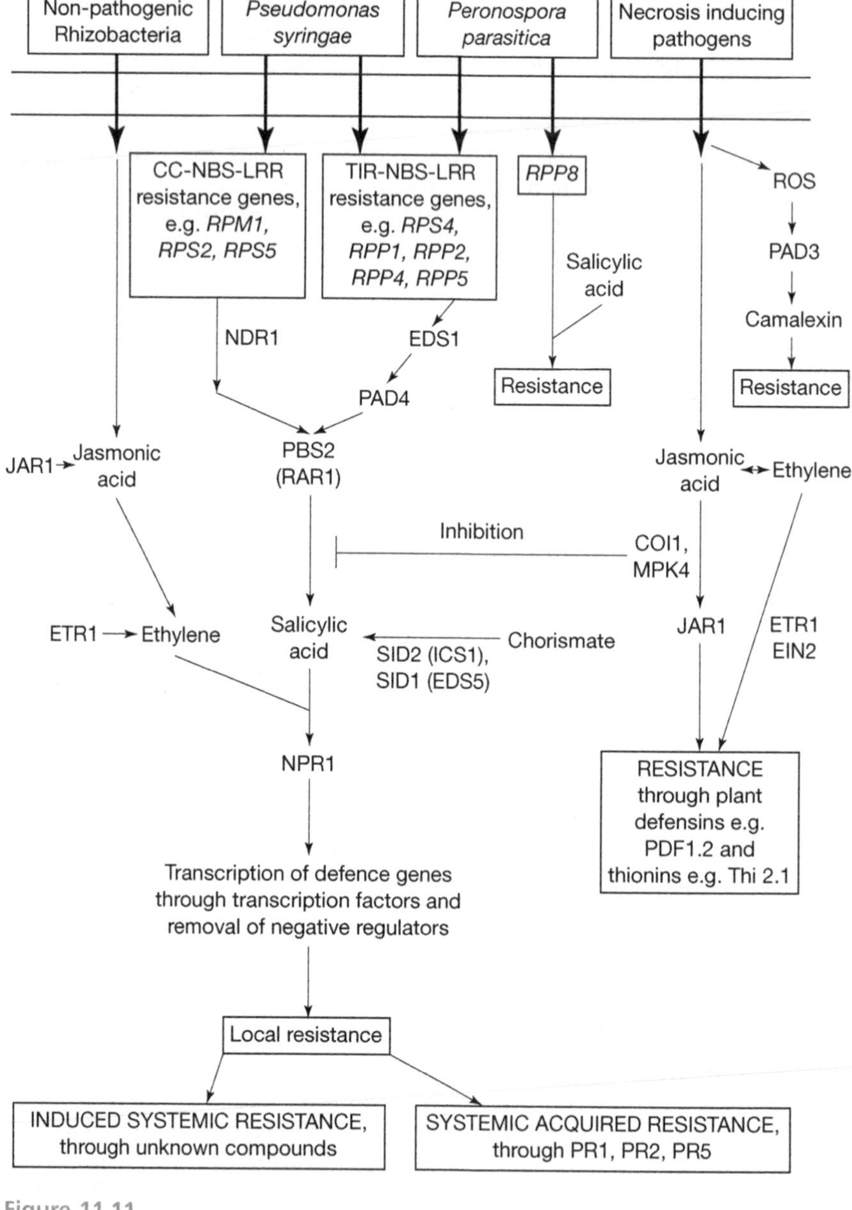

Figure 11.11

The putative positioning of some proteins involved in signalling pathways in *Arabidopsis,* based on analysis of mutants.

if EDS1 is a lipase it must process a different lipid metabolite. Further evidence has indicated that EDS1, in conjunction with PAD4 and acting through reactive oxygen species, controls runaway cell death and the extent of the hypersensitive response in surrounding cells depending on the resistance gene activated. Both pathways converge at PBS2, which is the *Arabidopsis* equivalent of barley RAR1, and therefore probably act through the degradation and turnover of proteins involved in defence signalling (see Section 11.4). There is also evidence that RAR1 is involved in some non-host resistance signalling pathways in plants.

11.10.2 The role of NPR1

Studies to dissect the role of SA and JA in systemic signalling pathways point to a central modulating role for NPR1 in both SA-dependent and SA-independent processes that is clearly influenced by the nature of the input signal, as shown in *Figure 11.11*. How this role is achieved is unclear but it might rely on selective association with other signalling components after sensing a particular pathogen-derived stimulus. Some clues to NPR1 function come from its predicted protein structure, its localisation and potential associations within the plant cell. *NPR1* encodes a protein with ankyrin repeats that have been shown to mediate protein–protein interactions in several prokaryotic and eukaryotic proteins (see Section 11.2). Recent studies reveal that NPR1 interacts in a yeast two-hybrid assay with two transcription factors of the TGA family of basic leucine zippers. Specificity of this interaction was demonstrated using two *npr1* single amino acid exchange mutations that both destroy the protein interactions and NPR1 activity in the plant. Interestingly, electrophoretic mobility shift assays showed that NPR1 binding to one of the transcription factors, TGA2, increased binding of this factor to its cognate promoter element and to an SA-inducible element (*LS7*) plus a negative element (*LS5*) in the promoter of the *Arabidopsis PR-1* gene, suggesting a direct link between NPR1 activity and downstream *PR-1* activation. This model is supported by cellular localisation studies of NPR1-GFP reporter fusions in which NPR1 protein becomes mobilised from the cytoplasm to the nucleus after pathogen or SA treatment, and it has been shown that this nuclear localisation is essential for its activity in inducing *PR* genes. A further gene product identified in the induction of *PR* genes is SNI1. This is a negative regulator of transcription, and it appears that NPR1 also works by removing this transcriptional blocker in order to stimulate gene expression.

Induced systemic resistance (ISR) through non-pathogenic rhizobacteria also appears to require functional NPR1, although this mechanism requires JA and ethylene and is independent of SA. Effective resistance appears to engage JA and ethylene sequentially. However, the NPR1-mediated process is not associated with transcriptional induction of SAR-type genes nor the JA/ethylene resistance response genes generated through necrosis-inducing pathogens, suggesting that either accumulation of JA and ethylene is below the threshold needed to induce expression of these marker genes, or that other as yet unknown genes are induced.

11.10.3 Pathways that are independent of NPR1

Necrosis-inducing pathogens, such as *Alternaria brassicicola, Botrytis cinerea* and *Erwinia amylovora* appear to use alternative NPR1-independent pathways to induce local and systemic resistance via different response proteins (*Figure 11.11*). In one branch of this pathway, the production of reactive oxygen species (ROS) appears to operate through PAD3 to result in camalexin production, which is a phytoalexin with antifungal activity against *A. brassicicola*. PAD3 has homology to maize cytochrome P450 monooxygenases known to direct biosynthesis of these indole-based phytoalexins (see Section 9.3.4). Alternative pathway(s) from necrosis use ethylene and jasmonic acid and operate through plant defensins and thionins. These pathways involve a number of genes such as *MPK4* (encoding a MAP kinase) and *Coi1* (encoding an F-box protein), which negatively regulate SA-mediated defence responses whilst positively regulating JA-mediated responses. A mutation in the *MPK4* gene has recently been identified that confers severe dwarfism in *Arabidopsis* combined with greatly elevated levels of salicylic acid.

The extent of positive and negative cross talk between SA and JA/ethylene signalling pathways remains a controversial topic although several studies demonstrate a clear antagonism between SA-associated plant-pathogen resistance and JA-mediated wound responses to insect feeding. Also, mutual antagonism was observed between SA-dependent *PR* gene expression and defence responses induced by *Erwinia carotovora*-derived elicitors in tobacco plants. It may be that opposing or synergistic interactions between SA and JA/ethylene signals reflect the plant's ability to prioritise responses very effectively through subtle differences in the kinetics of accumulation or distribution of particular signalling species. A 'turnable dial' model has been proposed in which different pathogen-derived input signals are responded to by fine tuning of the SA/JA/ethylene balance.

The existence of mutants deficient in SA, JA and ethylene biosynthesis or perception, and lines genetically engineered to deplete SA have been crucial in assessing the relative contributions of these diverse signalling molecules in defence pathways in *Arabidopis*. However, there is still a paucity of information about signalling systems in other plant species such as in the cereals, and whether the same mechanisms are important. There is also some evidence that other signals, such as abscisic acid (ABA), electrical signals and hydraulic pressure changes may be involved in some systemic responses.

References and further reading

Asai, T., Tena, G., Plotnikova, J., *et al.* (2002) MAP kinase signalling cascade in *Arabidopsis* innate immunity. *Nature* **415**: 977–983.

Berger, S. (2002) Jasmonate-related mutants of *Arabidopsis* as tools for studying stress signalling. *Planta* **214**: 497–504.

Blumwald, E., Aharon, G.S. and Lam, B.C-H. (1998) Early signal transduction pathways in plant–pathogen interactions. *Trends in Plant Science* **3**: 342–346.

Bolwell, G.P. (1999) Role of active oxygen species and NO in plant defence responses. *Current Opinion in Plant Biology* **2**: 287–294.

Feys, B.J. and Parker, J.E. (2000) Interplay of signalling pathways in plant disease resistance. *Trends in Genetics* **16**: 449–455.

Gómez-Gómez, L. and Boller, T. (2002) Flagellin perception: a paradigm for innate immunity. *Trends in Plant Science* **7**: 251–256.

Grant, J.J. and Loake, G.J. (2000) Role of reactive oxygen intermediates and cognate redox signalling in disease resistance. *Plant Physiology* **124**: 21–29.

Grant, M. and Mansfield, J. (1999) Early events in host–pathogen interactions. *Current Opinion in Plant Biology* **2**: 312–319.

Innes, R.W. (2001) Mapping out the roles of MAP kinases in plant defense. *Trends in Plant Science* **6**: 392–394.

Jones, A. (2000) Does the plant mitochondrion integrate cellular stress and regulate programmed cell death? *Trends in Plant Science* **5**: 225–230.

Kunkel, B.N. and Brooks, D.M. (2002) Cross talk between signalling pathways in pathogen defense. *Current Opinion in Plant Biology* **5**: 325–331.

Lam, E., Kato, N. and Lawton, M. (2001) Programmed cell death, mitochondria and the plant hypersensitive response. *Nature* **411**: 848–853.

Laxalt, A.M. and Munnik, T. (2002) Phospholipid signalling in plant defence. *Current Opinion in Plant Biology* **5**: 332–338.

May, M.J. and Ghosh, S. (1999) $I_{\kappa}B$ kinases: kinsmen with different crafts. *Science* **284**: 271–273.

Nishimura, M. and Somerville, S. (2002) Resisting attack. *Science* **295**: 2032–2033.

Nürnberger, T. and Brunner, F. (2002) Innate immunity in plants and animals: emerging parallels between the recognition of general elicitors and pathogen-associated molecular patterns. *Current Opinion in Plant Biology* **5**: 318–324.

Nürnberger, T. and Scheel, D. (2001) Signal transmission in the plant immune response. *Trends in Plant Science* **6**: 372–378.

Pieterse, C.M.J. and van Loon, L.C. (1999) Salicylic acid-independent plant defence pathways. *Trends in Plant Science* **4**: 52–58.

Singh, K.B., Foley, R.C. and Oñate-Sánchez, L. (2002) Transcription factors in plant defense and stress responses. *Current Opinion in Plant Biology* **5**: 430–436.

Wendehenne, D., Pugin, A., Klessig, D.F. and Durner, J. (2001) Nitric oxide: comparative synthesis and signalling in animal and plant cells. *Trends in Plant Science* **6**: 177–183.

Wildermuth, M.C., Dewdney, J., Wu, G. and Ausubel, F.M. (2001) Isochorismate synthase is required to synthesize salicylic acid for plant defence. *Nature* **414**: 562–565.

Molecular diagnostics

12

A key application for plant pathology in agriculture, horticulture and forestry is the development and use of methods to rapidly and accurately diagnose disease. In human medicine, correct diagnosis of the causal agent can ensure that appropriate control methods are implemented and that unnecessary methods are avoided, and the same is true in plant pathology. Diagnostics are particularly important with the increase in global trade of plants and plant produce, and the associated risk of movement of pathogens and their vectors from one country to another. Tests that can reliably, quickly and cheaply detect the presence of plant pathogens are essential for preventing new outbreaks and potentially devastating crop diseases. Not only are diagnostics important for detection of organisms, but they are also important for identifying the isolate/race/pathotype/pathovar of the organism, so that pre-emptive breeding/control methods can be adopted based on predictions of likely future outbreaks. Annual surveys of rust populations in Australasia and the USA have been used to determine which resistance genes to deploy in wheat to reduce the likelihood of disease outbreaks, and surveys of antibiotic resistance in strains of bacteria, and of fungicide resistance in fungal populations have been used to improve the methods of deployment of chemical controls, such as use of mixtures and/ or alternating sprays. This chapter will discuss the molecular methods used for diagnosis of disease, and those used for identifying genetic variation within pathogen populations.

12.1 Classical approaches

In many cases, and particularly for the more common foliar pathogens of agriculturally and horticulturally important plants, macroscopic symptoms are clear enough for diagnosis. Advisory leaflets, books and internet resources provide clear descriptions of symptoms for many diseases, making accurate diagnosis possible without the need for extensive training or diagnostic kits. However, care must be taken with all types of diagnosis to establish that the organisms identified are the cause of disease and not secondary infections by saprophytic bacteria and fungi that have colonised damaged and diseased plants, and for this, the skilled plant pathologist who is experienced with the system may be required. For some fungi, microscopic study of the mycelia, fruiting structures and spores are also useful for diagnosis, and selective media/specific culture conditions may be appropriate to induce spore formation and facilitate these approaches. In addition, bacterial disease may be diagnosed by microscopic observations, and the use of selective media and biochemical tests on isolated bacteria, such as Gram staining, fatty acid profiles and nutritional analysis can be used to aid identification. Some viral and phytoplasma diseases can be diagnosed based on visual symptoms, but in most cases electron microscopy is required to see the causal agent. Use of transmission tests to indicator host plants can be used for some viruses and sensitivity to tetracycline antibiotics can be used for spiroplasmas and phytoplasmas.

These tests and methods have been invaluable over the years and will continue to remain so, but there has always been scope for improvements, for example to detect pathogens before symptom development, and to screen larger numbers of samples accurately, reliably, quickly, with greater sensitivity and in ways that provide additional information. For example, information on the genetic make-up of the organism, such as what virulence phenotype it possesses or whether it is resistant to particular chemical controls may be useful, and this has classically been acquired by sending samples to diagnostic laboratories and advisory services for testing. Tests may involve growing the isolate on 'tester' plants or differential cultivars to determine what resistance genes are overcome by it. Resistance to chemicals can also be determined by testing with different dose rates either on plants or in some cases by culturing the organism in media containing different concentrations of the chemical. Over the past 30 years, two major technologies, antibody-based and nucleic acid-based, have been developed to support these classical diagnostic methods, and recent advances in genomics and molecular biosystematics are now revolutionising these approaches further so that they can deliver faster, more accurate and more comprehensive diagnoses.

12.2 The use of antibodies

Some of the most widely used diagnostic systems in plant pathology are based on serology, in which antibodies are raised in animals against specific antigens from the pathogen, and these antibodies are then purified and used in the diagnostic test. The major soluble antibody in vertebrates, IgG is the most commonly used and is a Y-shaped molecule with two antigen-binding sites. Variation in the amino acid composition of the antigen-binding sites determines the specificity, whilst other regions of the IgG can be labelled, for example by attachment of enzymes, for the diagnostic detection system.

12.2.1 Polyclonal antibodies (Pabs)

Success of serological techniques relies on their ability to recognise antigenic sites that are specific to the pathogen that is to be detected. Polyclonal antisera has been routinely used for a number of pathogens, particularly viruses, because of the relative ease with which it can be produced. Essentially, the pathogen or a part of the pathogen (such as the viral coat protein or a particular peptide) has to be purified and/or chemically synthesised. The purity of the antigen is crucial, since antibodies will be raised to any proteins that are injected into the animal, which may result in problems with antibodies detecting contaminants in subsequent analyses. The purified pathogen or peptide is injected into a test animal, such as a rabbit, and following subsequent booster injections, the animal is bled. The resultant antisera will contain a population of antibodies that may react with different determinants of the immunogen, which can be utilised in detection techniques (see Section 12.3). However, there are potential problems of cross-reactions to contaminants, lack of a sustainable source of antibodies (when the test animal dies, so do the antibodies of that specificity), along with problems of lack of specificity. For example, if purified bacteria and fungi are used, antibodies will be raised against a whole range of peptides, many of which may be common to a wide range of bacteria or fungi. This problem may be alleviated by using purified peptides. Alternatively, monoclonal antibodies provide a more reliable source of specific antibodies for detection systems.

12.2.2 Monoclonal antibodies (Mabs)

Monoclonal antibodies (Mabs) are superior to Pabs because they can provide a constant supply of specific antibodies. They are however relatively expensive to produce and maintain because of the specialised cell cultures and low-temperature storage facilities that are required. The production method is shown in *Figure 12.1*. Monoclonal antibodies have been produced and are commercially available against a wide range of viruses, such as common seed-borne and potato viruses, where diagnosis is important to prevent dissemination of virus in seeds and tubers. They have also been produced against specific oligosaccharide side-chains of *Erwinia* bacteria, and

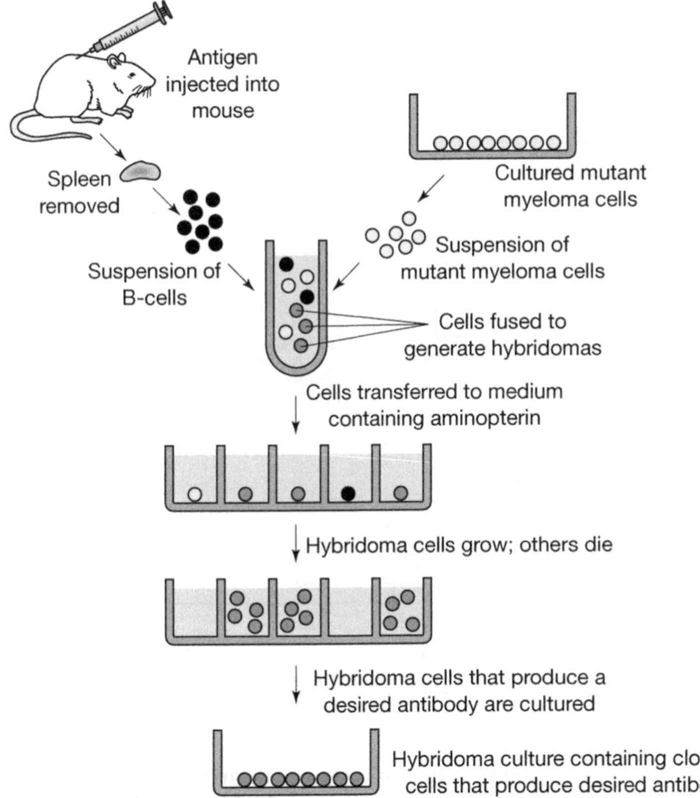

Figure 12.1

The production of monoclonal antibodies.

against proteins from other bacteria and phytoplasmas, as well as to specific proteins from plant-pathogenic fungi, allowing for example, the identification of *Botrytis cinerea* in grape juice, and discrimination between cereal eyespot and other stem-based diseases of wheat that give similar symptoms.

12.2.3 Recombinant DNA techniques

Further sources of antibodies for use in immunological screens are phage display libraries (*Figure 12.2*). This technique uses filamentous bacteriophages, such as M13 or fd, to express foreign peptides in fusion with one of the phage coat proteins. In particular, libraries have been made by cloning genes for human single chain variable fragments (scFv), to result in libraries consisting of phage expressing complete antibody-variable domains. These libraries are commercially available and can be screened with plant

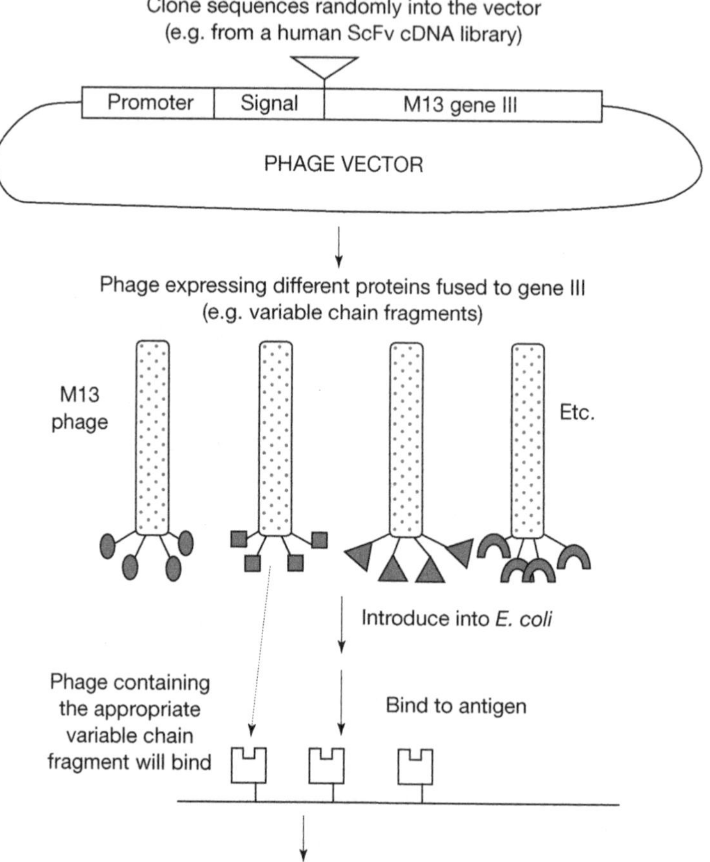

Figure 12.2

The production of antibodies by phage display. Libraries are made of human ScFv cDNAs in a filamentous bacteriophage such as M13, so that the cloned sequence is expressed as a fusion protein with the phage coat protein. When expressed in *E. coli,* phage that recognises a specific antigen can be selected and maintained as a source of the antibody.

pathogens or peptides from plant pathogens to identify clones that contain antibodies capable of recognising the specific plant pathogen antigen, without the need to immunise any animal. Such clones can then be used as a source of antibodies for a diagnostic system. These techniques have been developed for a number of potato viruses, and have the potential to be used for a wide range of other pathogens because of the relative ease and speed with which antibodies with the desired specificity can be identified and developed.

12.3 Serological tests

12.3.1 ELISA (enzyme-linked immunosorbent assay)

The ELISA-based techniques combine the specificity of the antibody/antigen interaction with a detection system involving an enzyme conjugated to an antibody. The activity of this enzyme is measured spectrophotometrically by adding an appropriate substrate that results in a colour change, and the intensity of the colour change can be used as a measure of the quantity of antigen present. As a system, ELISA has many benefits in that it can be easily automated, particularly through the use of multi-well disposable microtitre plates and plate readers that can measure and record colour changes in all of the wells simultaneously. The components of the systems are also stable and safe to use, such that the method can be readily transferred to extension laboratories lacking sophisticated equipment and/or in remote areas.

There are a number of variations in the way ELISA is used as shown in *Figure 12.3*. Because the double-antibody sandwich (DAS) (*Figure 12.3a*) uses antibodies to specifically trap the antigen, it is useful for detecting pathogens in crude samples, such as soil or plant extracts, since the contaminating material is washed away prior to the enzyme detection. The advantage of TAS (triple-antibody sandwich) (*Figure 12.3b*) is that enzyme-conjugated goat anti-rat antibodies are commercially available, removing the need to conjugate an enzyme to the pathogen-specific antibody. In a further modification, known as PTA (plate-trapped antigen) (*Figure 12.3c*), the pathogen antigens are trapped directly onto the plastic surface of the microtitre dish without the need to pre-coat with antibodies.

Antibodies can also be used directly on tissue prints of leaves for detecting some pathogens and there are now numerous examples of diagnostic kits for detecting viruses, bacteria and fungi, and for the detection of pathogen metabolites in produce. For example, ELISA kits are available to detect mycotoxins in food samples and for fungicide residues.

12.3.2 Lateral flow techniques

'Dipstick' methods suitable for field use by crop inspectors and extension workers without the need for any sophisticated equipment have been produced for some plant pathogens such as the potato viruses PVX and PVY, and can provide results within a few minutes. In these equivalents to medical detection systems such as pregnancy kits, the pathogen-specific antibodies are bound to coloured latex beads or gold particles preimmobilised onto membranes in the release pad (*Figure 12.4*). When the sample (for example a diluted sap extract from plant tissue) is applied to the lateral flow strip, any antigen within it will form a complex with the tagged antibodies as

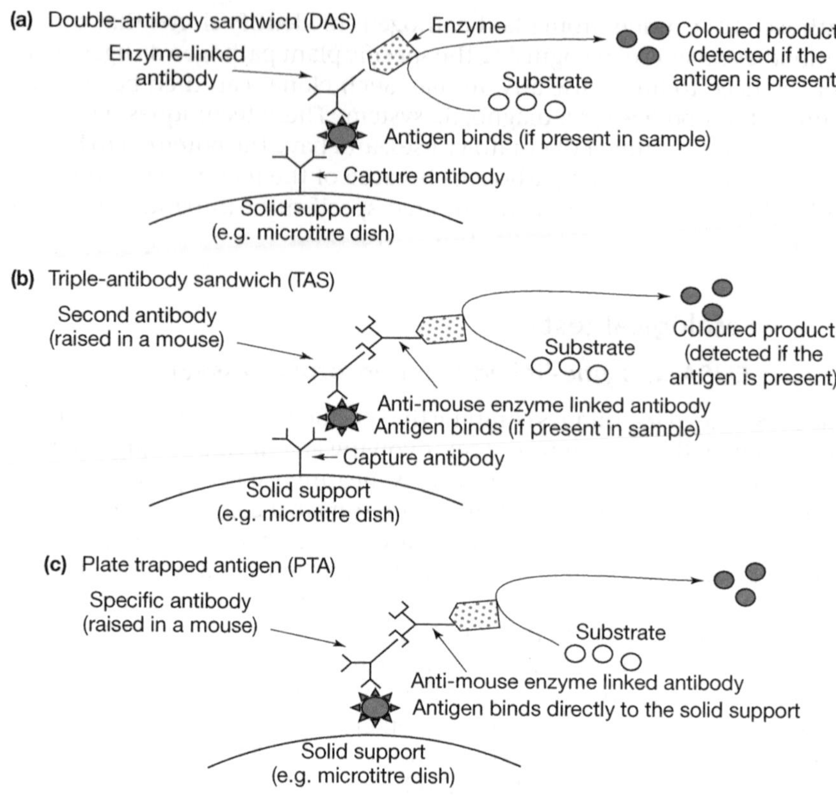

Figure 12.3

ELISA-based methods. (a) In DAS, the unlabelled specific antibody is bound to the microtitre dish. The test sample, enzyme-linked antibody and substrate are then added sequentially, and unbound material is removed after each step by washing. (b) In TAS, the plate is initially coated with a polyclonal antibody that traps the antigen and then a second monoclonal antibody is added. An enzyme-linked antibody is then used to detect the presence of the second antibody. (c) In PTA, the pathogen antigens are trapped directly onto the plastic surface of the microtitre dish without the need to pre-coat with antibodies.

the solution moves through the membrane by capillary action. The bound antigen/antibody complex will continue to move by lateral flow through the membrane until it reaches the capture zones. The first preprinted capture line (containing pathogen-specific antibodies) is designed to trap the bound antigen/antibody complex, and as these are arrested, the line becomes visible as the coloured latex or gold accumulates. The second capture line is the control line, containing species-specific antibodies designed to capture all the unbound antibodies and confirm that the assay has run correctly.

12.3.3 Other uses of antibodies

As well as their use in diagnostics, antibodies can be used to localise pathogens in samples. This generally involves conjugation of an alternative detection system to the antibody. For example, conjugation of the fluorescent dyes fluorescein isothiocyanate (FITC) or rhodamine isothiocyanate

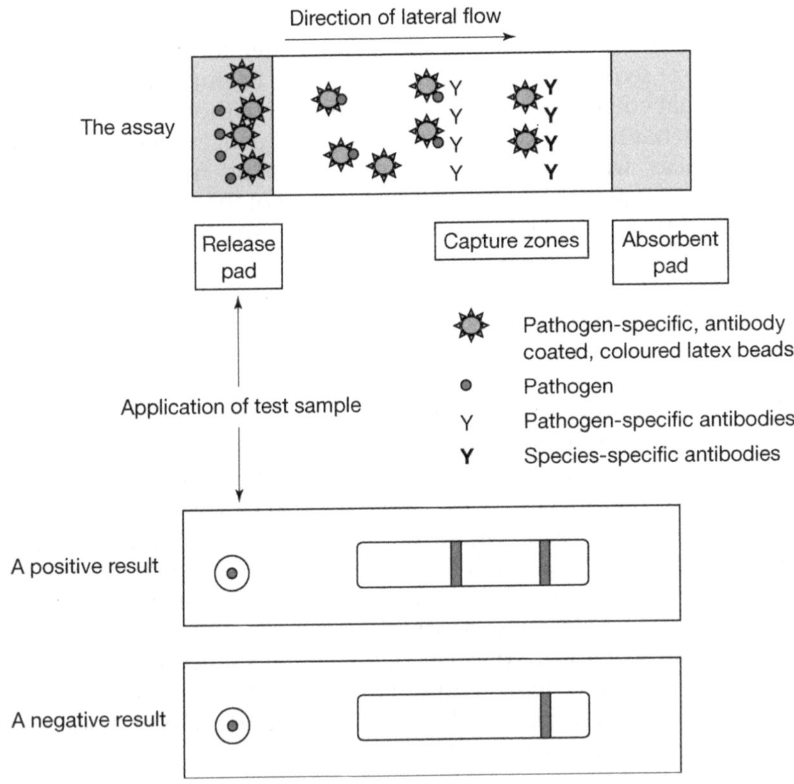

Figure 12.4

Lateral flow 'dipstick' detection. The upper diagram shows the assay principle, whilst the lower diagrams indicate what positive and negative results would look like in a commercial assay strip. See text for detail.

(RITC), followed by application of these antibodies to sections through infected material and fluorescence microscopy can be used to localise bacteria and/or fungi in infected tissue. The same technique can localise pathogens in soil, seed, etc., and also specific proteins onto pathogens during the infection process. In addition, labelling of antibodies through conjugation of colloidal gold can be used for visualisation in electron microscopy. Sections of infected tissue are prepared and incubated with the gold-labelled antibodies, which show up as electron-dense particles in the electron micrographs. This technique is valuable for localising viruses in infected tissue and also for localising specific fungal and bacterial proteins in infected plants.

12.4 Nucleic acid-based techniques

12.4.1 Identification of pathogen-specific markers

The aim for nucleic acid-based diagnostic techniques has been to identify sequences of the pathogen (DNA for bacteria and fungi; RNA for most viruses) that could be used to make pathogen-specific probes. For example, coat protein genes of particular viruses, or pathogenicity genes of bacteria and fungi would be ideal candidates to use for a probe. It may be that these sequences would be species-specific and therefore diagnostic at the species level, or sequences might be identified, for example avirulence gene sequences, that could be diagnostic for individual pathotypes.

Much work has been undertaken on general sequences such as ribosomal DNA sequences, which are relatively easy to obtain from organisms. In fungi, these genes are arranged as repeated clusters of the 18S-5.8S-28S genes, with internal transcribed spacers (ITS) separating the three subunit genes, and IGS (intergenic spacers) separating the repeated units. In general more variation is found in the ITS and IGS regions. In bacteria, the 16S-5S-23S subunit genes may be in a single unit or the individual subunit genes may be widely separated in the genome, and markers have been developed based on the 16S and 23S genes.

In addition, a number of molecular techniques have been developed over the years for identifying random molecular markers that can reveal polymorphisms between isolates of organisms, and many of these have been applied to plant pathogens. These range from RFLP (restriction fragment length polymorphism) analysis, through RAPDs (randomly amplified polymorphic DNA), AFLPs (amplified fragment length polymorphisms), and microsatellites or SSRs (simple sequence, di, tri and tetranucleotide repeats) (see Section 1.5.4). All of these methods have proved invaluable for generating molecular fingerprints of isolates and sequences that have been developed into diagnostic systems. Futhermore, current advances in genomics research is likely to result in development of diagnostic kits with key genes built into them that can identify the virulence profile of any particular pathogen.

12.4.2 Hybridisation techniques

Nucleic acid hybridisations, which depend on the inherent base pairing of nucleotides in complementary sequences, have been used for many years in molecular biology, and there has been a substantial effort to develop these techniques for pathogen diagnosis. Many non-radioactive ways of labelling DNA that can be used in field diagnostic systems have now been developed (see Section 1.5.7), with detection of labelled probe in hybridisation experiments through enzyme-based or fluorescence-based detection systems. Improved methods for extraction of DNA from plant, soil samples, etc., have also been developed to minimise the risk of false-positive results due to non-specific hybridisation. One method has been to use a combination of pathogen-specific antibodies with DNA hybridisation. Antibodies are used to immunocapture the pathogen into microtitre dishes, and fluorescent DNA probes are then used to hybridise to and detect the DNA in captured samples. If hybridisation occurs, due to the presence of pathogen in the sample, this can be detected through fluorescence in the microtitre dish, and the results can be read through the plate reader methods that have been developed for ELISA systems allowing automation of these techniques.

12.4.3 PCR-based techniques

PCR-based methods have also been developed for diagnostics because the technique is rapid and can be used relatively easily in remote laboratories and laboratories with limited facilities. Apart from a PCR machine, gel electrophoresis equipment and a system for visualising results, such as a UV transilluminator and camera, the main resources required are enzymes, chemicals and disposable plasticware. Laboratories that can undertake this kind of work have been established with minimal difficulty in remote field stations in many parts of the world.

The technique utilises specific oligonucleotide primers, which are designed based on nucleic acid sequences that are diagnostic for the pathogen. These may be sequences identified by genomic sequencing or techniques such as DNA fingerprinting of pathogens. To improve the sensitivity of these techniques, it is sometimes an advantage to use nested PCR methods, in which the products from the initial PCR amplification are diluted and re-amplified with a second set of primers internal to the original primer set in the pathogen sequence. These conventional PCR-based techniques have proved to be useful for diagnosis of fungal, bacterial and phytoplasma-associated diseases with a number of good taxon-specific primers developed for example from the rRNA subunit genes. Because of the sequence variation between these regions in different isolates, it has sometimes been possible to identify restriction endonucleases that give different restriction patterns upon digestion of the PCR products depending upon the isolate. This technique has been particularly valuable for organisms that cannot be propagated away from plants, such as phytoplasmas, since the primers are specific for the phytoplasmas and will amplify these sequences out of infected plant samples.

Primers have also been developed based on more specific sequences such as the *argk-tox* gene of *P. syringae* pv. *phaseolicola* which encodes a gene involved in phaseolotoxin biosynthesis and can be used to identify bacteria that possess this trait, or the *aflR* gene of *Aspergillus flavus* which regulates aflatoxin production in these fungi. In addition, PCR-based techniques have proved useful for identifying the vectors for insect-transmitted diseases. For example, DNA extracted from leafhoppers that are potential vectors for phytoplasma diseases can be PCR amplified using phytoplasma-specific primers to identify which species are the true vectors. However, DNA from plant, soil and insect samples can contain potent inhibitors of Taq polymerase, and these have to be removed and appropriate controls undertaken to ensure that negative results are not due to these inhibitors.

Single-strand RNA viruses can be detected by modifying the PCR to include a reverse transcriptase step (RT-PCR). In this, the reverse transcriptase uses a viral specific primer to make a cDNA copy of the viral RNA which is then amplified using Taq polymerase through conventional PCR. RT-PCR has the potential for use in diagnostics of bacterial and fungal pathogens. Through identification of particular genes that are expressed during pathogen growth, or at specific stages of development, it may be possible to use RT-PCR to identify the developmental stage of the pathogen in infected material. This may be a way of showing whether the organism detected is viable or dead, since one of the drawbacks of conventional PCR is that its sensitivity enables it to pick up and amplify DNA from non-viable organisms and organisms such as secondary invaders that may be nothing to do with the disease.

A number of techniques have been developed to improve the reliability, efficiency and cost-effectiveness of PCR-based techniques. Antibodies can be used for 'immunocapture' of virus particles prior to PCR, and dipsticks coated with chemoattractants have been developed to attract zoospores of oomycetes such as *Pythia* and *Phytophthora* prior to diagnostic analysis.

It has also been possible to produce 'multiplex' PCR kits capable of detecting more than one pathogen present in a particular plant or soil sample. Kits are commercially available to unravel the cereal stem-based complex of fungi comprising *Tapesia yallundae* and *T. acuformis* (eyespot fungi), *Fusarium culmorum, F. avenaceum, F. graminearum* and *F. poae* (ear blights), and *Microdochium nivale* (snow-mould of cereals). Also, by using specific primers, multiplex systems have been developed that can determine the mating type of the *Tapesia* present in the DNA sample. These multiplex systems use combinations of primers that give different-size fragments for each of the species, and multiplex systems could be further developed by using primers that fluoresce at different wavelengths for the different pathogens. In mammalian systems, it has been possible to devise successful multiplex systems comprising up to 92 different primer pairs.

12.4.4 Gene-array-based techniques

As more sequences for plant pathogens become available, there is increasing potential to develop diagnostic microarrays that can be used to screen for the presence of multiple pathogens in one assay. Such techniques have already been developed in medicine for HIV testing and genotyping, and for diagnosis of bacterial pathogens. For example, based on the sequence of the 23S rRNA gene, oligonucleotides have been designed and arrayed (currently with a bubble-jet printer on nylon membranes) that can be used to identify more than 90 medically significant bacteria from blood samples. Universal primers are used to amplify a variable region of the 23S ribosomal DNA from any bacteria present in the blood sample, and this DNA is then labelled and used to probe the array to identify the organisms present. Such systems could be developed for plant-pathogenic bacteria and also for fungi, where universal primers are available that can be used to amplify the ITS regions. Samples could be taken from soil, plant surfaces, etc. and screened to determine what organisms are present and these techniques could be further developed to establish the efficacy of control methods in removal of specific species. Some technical limitations still remain, such as the ease of DNA purification from fungal spores, separation of DNA from PCR inhibitors, and the confirmation that specific organisms are the causal agents of disease, but rapid progress is being made in the development of these methodologies.

12.4.5 Quantitative PCR

One of the historical drawbacks of PCR methods compared to hybridisation and ELISA has been the lack of reliable quantification of target populations, since small differences in any of the parameters that affect the efficiency of amplification can affect the outcome of the reaction. A number of approaches have been devised to overcome this problem. One involves titrating the target DNA with known quantities of an internal standard DNA. This internal standard DNA, referred to as competitive DNA, possesses the same primer binding sites as the target DNA and is designed so that it is amplified with the same efficiency as the target in the PCR reaction. By titrating unknown amounts of the target in a heterogeneous DNA sample alongside a dilution series of the competitor, it is possible through image analysis of the relative intensity of products to determine the quantity of starting material in the sample.

An alternative approach that has been developed and is commercially available for *Septoria* and rust diseases of wheat involves the incorporation of a fluorescent dye, PicoGreen as a measure of the amount of DNA present in samples. PCR reactions are performed directly in microtitre dishes, and the dye (which fluoresces specifically upon binding to dsDNA) is then added and the intensity of fluorescence measured using a fluorometer. By comparison of fluorescence with that of known standards, it is possible to determine how much DNA is present in each sample, and this can be related back to the amount of starting material, although not as accurately as the competitive approach.

More sophisticated high-technology approaches of Real-Time PCR/Taqman, based on incorporation of fluorescent dyes into PCR mixes as detailed in Section 1.5.4 have also been developed for a number of plant pathogens. The main advantages of these methods are that they are automated and that relatively cheap portable thermocyclers are now available that can give reliable quantitative data. In addition, some of the newer technologies and machines are able to detect multiple pathogens simultaneously even if the primer annealing and denaturing temperatures are widely different. Primers have been developed for real-time PCR detection of bacterial and fungal plant pathogens such as *Ralstonia solanacearum* in potato tubers, *P. syringae* pv. *phaseolicola* in bean seeds, *Phytophthora* spp. in potatoes and citrus, and Karnal bunt (*Tilletia indica*) in wheat.

Diagnostic systems have been developed and are commercially available for a wide range of pathogens in fast, efficient and reliable kits. There remains considerable potential to improve kits to identify the pathotype of the organism, or the mating type, or perhaps whether it contains resistance to chemical controls such as specific fungicides. As more genome sequences become available and key genes in pathogenicity are identified, these will also be developed into markers. For example, systems are now available for detecting single nucleotide polymorphisms (SNPs) between gene sequences and these could form the basis of diagnostic systems to identify key mutations in specific pathogen genes.

12.5 Phylogenetic analysis

Another field in which polymorphisms in ribosomal sequences and in random molecular markers between isolates has been useful is in phylogenetic classifications. Phylogenetic trees are valuable both for identifying the relative time of origin of particular taxa and also for analysis of the extent of divergence within groups of organisms. It was based on phylogenetic analyses that the oomycetes were recently separated away from the true fungi into the Chromista. Nucleic acid sequence comparisons have added to the classifications previously built on morphological characteristics as a means for building phylogenetic trees. Basepair changes in the ribosomal RNA genes and ITS sequences have been particularly useful, as have DNA fingerprinting techniques such as RAPDs and AFLPs. Sequences for many genes are now available in international databases, and appropriate computer programs and statistical analyses can be accessed particularly through the following website: http://evolution.genetics. washington. edu/phylip/software.html for building phylogenetic trees.

References and further reading

Irving, M.B., Pan, O. and Scott, J.K. (2001) Random-peptide libraries and antigen-fragment libraries for epitope mapping and the development of vaccines and diagnostics. *Current Opinion in Chemical Biology* **5**: 314–324.

Lévesque, C.A. (2001) Molecular methods for detection of plant pathogens – what is the future? *Canadian Journal of Plant Pathology* **24**: 333–336.

Mackay, I.M., Arden, K.E. and Nitsche, A. (2002) Real-time PCR in virology. *Nucleic Acids Research* **30**: 1292–1305.

Martin, R.R., James, D. and Lévesque, C.A. (2000) Impacts of molecular diagnostic techniques on plant disease management. *Annual Review of Phytopathology* **38**: 207–239.

Schaad, N.W. and Frederick, R.D. (2002) Real-time PCR and its applications for rapid plant disease diagnostics. *Canadian Journal of Plant Pathology* **24**: 250–258.

Sicheritz-Pontén, T. and Andersson, S.G.E. (2001) A phylogenomic approach to microbial evolution. *Nucleic Acids Research* **29**: 545–552.

Stowell, L.J. and Gelernter, W.D. (2001) Diagnosis of turfgrass diseases. *Annual Review of Phytopathology* **39**: 135–155.

Tillib, S.V. and Mirzabekov, A.D. (2001) Advances in the analysis of DNA sequence variations using oligonucleotide microchip technology. *Current Opinion in Biotechnology* **12**: 53–58.

Application of molecular biology to conventional disease control strategies

13

13.1 Breeding for resistance

13.1.1 The basis of resistance breeding programmes

Since the recognition in the early 1900s that resistance to disease in plants is genetically inherited, breeding resistant cultivars has been a popular and intensively used approach for crop protection. It has many advantages over alternative approaches, both to the plant breeders, who can gain commercially by producing improved varieties with better resistance characteristics, for the grower who will have stable and uniform varieties that require fewer expensive alternative crop protection measures such as chemical interventions, and for the consumer who will have produce free of chemical residues and mycotoxins produced by infectious fungi. It is regarded as environmentally friendly crop management because of the reduced need for agrochemicals. However, if the presence of resistance is associated with a significant yield penalty, the cultivar may not be commercially viable.

The success of conventional resistance breeding strategies has been at a cost. As a resistance gene becomes more widely used in the field and the commercial cultivars become less genetically diverse, so the selection pressure on the pathogen to overcome the resistance increases. If isolates are present or evolve in the pathogen population to overcome this resistance, the lack of genetic diversity favours the spread of the new pathogen and epidemic development. As this occurs, so new sources of resistance genes are required, and the arms race between plants and pathogens is perpetuated. The source of such genes is often in old abandoned varieties, wild plant relatives and occasionally through induction of mutations within current cultivars. For example, many of the genes for resistance to *Cladosporium fulvum* (tomato leaf mould) that have been incorporated into commercial cultivars of tomato (*Lycopersicon esculentum*) originate from wild accessions of *L. hirsutum* (e.g. *Cf-4*) or *L. pimpinellifolium* (e.g. *Cf-9*), and many of the disease resistance genes present in cereals originate from wild grasses of the Middle East, the centre of origin of these cereals. It is therefore vital that germplasm stocks of wild accessions and old cultivars are maintained for these strategies to continue, to ensure that potentially useful sources of resistance genes are not lost.

13.1.2 The conventional breeding strategy

The technique of breeding resistance genes into commercial cultivars is a complex and time-consuming one (*Figure 13.1*). Progeny from an initial cross (if this is possible) between the elite cultivar and the identified source of the resistance gene will contain large amounts of DNA from the wild relative (alien DNA) that is likely to make it a poor-quality cultivar. Extensive back-crossing to the elite cultivar is then required to gradually reduce the amount of alien DNA (linkage drag) present in the line, and it has been calculated that as many as seven backcrosses will still leave 1% alien DNA in a cultivar. Only when the presence of this alien DNA is reduced to a level in which the resultant cultivar has better agronomic properties than its predecessors, will it be suitable for commercialisation.

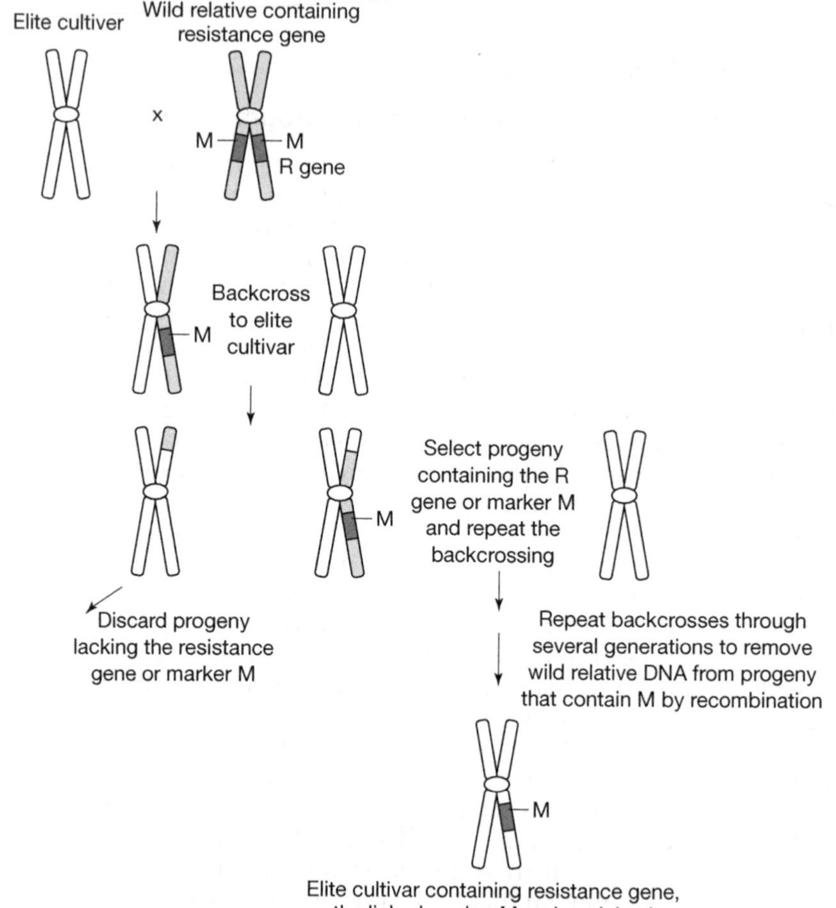

Figure 13.1

The rationale behind using molecular markers in breeding programmes. M is a molecular marker tightly linked to the resistance gene (see text for details). Following each cross, progeny containing the resistance gene are selected by pathogen screens or by detection of the tightly linked marker (M) if using marker-assisted breeding.

At each step, the progeny from the crosses will need to be tested with the pathogen to select for those individuals that have inherited the resistance gene, a time-consuming and often difficult process. Thus, an extensive 5–10 year breeding programme may be required before a cultivar with suitably improved characteristics is generated, which in turn may only last in the field for a few years before it is overcome by the pathogen and becomes obsolete. Fortunately, some resistances that have been bred into commercial crops have proved to be longer-lasting than others in the field. Resistance to *Fusarium oxysporum* f.sp. *conglutinans* in cabbage has been effective for over 90 years, the *Sr26* stem rust resistance genes in Australian wheats has been effective for over 30 years, and the *Lr34* brown rust resistance genes in wheat has also proved difficult for the pathogen to overcome in both North American, South American and Australian wheats. The pyramiding of additional resistance genes into cultivars (introducing them in particular combinations and deploying them based on the results of pathogen surveys) can improve the longevity of a cultivar. An additional approach can be to deploy resistance genes in mixtures and multilines. In a multiline variety, there is sufficient uniformity in the essential agronomic characteristics for the grower, but genetic heterogeneity for the resistance. However, producing such lines can involve numerous additional backcrosses in breeding programmes. Using mixtures, where cultivars with different resistance genotypes are grown together in the same field is simpler, but the agronomic differences between the cultivars can present problems. An alternative can be to use varietal diversification, where varieties of different genotype are grown in adjacent fields.

13.2 The use of tissue culture in plant breeding

The ability to grow many plant species in tissue culture and propagate whole plants from protoplasts and callus cultures has opened up a number of ways to alleviate some of the problems with conventional breeding programmes. Protoplast fusion can be used for plants that might be impossible to cross by classical breeding methods and can be a way to incorporate resistance genes from incompatible plants into commercial species. These plants can then be regenerated and taken through backcrossing programmes. Alternatively, tissue culturing has been shown in many cases to result in greater variability in the progeny plants, a poorly explained phenomenon known as somoclonal variation. Somoclonal variants may be produced that are more resistant to disease than their progenitors simply through the tissue culture procedure. For example, in experiments where plants were generated in tissue culture from a line of potatoes susceptible to *Phytophthora infestans* and *Alternaria solani*, 2.5% were found to be resistant to *P. infestans*, and 1% to *A. solani*.

13.3 Marker-assisted breeding

A significant problem with conventional breeding programmes is the time taken between generations and the need to test with pathogens to select for the resistant progeny after each cross. This is particularly problematic if the pathogen only infects mature plants, or is difficult to work with and maintain, perhaps because it is not an indigeneous organism and needs to be maintained under strict licence constraints. As a result there has been much research into using molecular markers to assist in breeding programmes (*Figure 13.1*). The aim has been to identify molecular markers that are tightly linked to the resistance genes, so that the presence of these markers in progeny can be used as an effective monitor for the presence of the gene without the need for pathogenicity tests. The molecular markers may be isozyme markers, or DNA-based markers that can easily be identified.

An example of the usefulness of this approach comes from studies in the USA into resistance in soybean to the soybean cyst nematode *Heterodera glycines*. Phenotypic assays based on infection with the nematode take 5 weeks to provide results and are labour intensive. A molecular marker, Satt309, located 1–2 cM from the *rhg1* resistance gene has proved to be 99% accurate for predicting the presence of the resistance gene in progeny from crosses and is being widely used in breeding programmes. Markers have also been developed for a number of resistance genes against pathogens, such as against *Ustilago hordei* smut in barley, and in wheat, a PCR-based marker has been developed for *Lr40* and *Lr21* leaf rust resistance based on a resistance-gene analogue. The ability to detect single-nucleotide polymorphisms (SNPs) (see Section 1.5.4) through use of probes that fluoresce when they hybridise to exact matches but not to single base-pair mismatches, opens up the possibility of being able to generate molecular markers for almost any sequence. For example, the resistance gene itself could be developed into the marker if a single nucleotide difference is identified between it and the recessive allele. One of the drawbacks with using molecular markers for marker-assisted breeding has been the inefficiency of screening large number of samples by gel electrophoresis. Breeders may generate tens or hundreds of thousands of new genotypes each year, and DNA preparation, PCR and gel electrophoresis on these numbers can become unmanageable. However, the automation of real-time PCR methods and ability to discriminate alleles without gel electrophoresis is starting to make such analyses more feasible.

In addition to the use in single-gene breeding programmes, marker-assisted breeding has the potential to monitor quantitative trait loci (QTLs), where these are identified as being components of disease resistance phenotypes. This approach is technically challenging. However, recombinant inbred lines have been developed for many plant species such as wheat, tomato and sunflowers that would facilitate these strategies. These involve crossing disease-resistant and -susceptible parents and then inbreeding the progeny through successive generations. A similar approach to identify molecular markers is to undertake a wide cross between the commercial plant and an unadapted relative (which would be highly polymorphic and therefore capable of producing molecular markers) and then carry out two generations of backcrossing for several hundred sibling lines. These lines would each retain a different segment of the wild germplasm, and could therefore be used to generate molecular markers for this segment. In effect, each of these lines becomes a near isogenic line that can be used to isolate the effects of a potential QTL of the elite parent. The integration of such a development of lines with marker-assisted breeding could facilitate the monitoring of QTLs in future breeding programmes. The rapid advances in genome-mapping projects for crop species such as rice, barley and wheat, coupled with the increased use of bioinformatics is also certain to help the development of molecular markers that can assist in breeding programmes.

13.4 The identification of novel resistance gene specificities

As well as aiding the breeding of resistant cultivars, molecular genetics can help with the identification of new resistance genes. This might be approached in one of two ways. Cloned resistance genes and primers to conserved regions of resistance genes can be used in PCR to identify similar genes in wild relatives that might be potentially useful. For example, homologues of the broad-spectrum *mlo* resistance gene have been identified in a number of plant species and these may prove to be functional and durable resistance genes that can be transferred into commercial crops.

A second approach, that has been used successfully to identify new resistance genes in wild relatives of tomato, has been to identify specific fungal proteins, such as the secreted ECP2 (extracellular) protein, that are required for full pathogenicity of most if not all races of the pathogen. The gene encoding this protein has been engineered into a transient expression vector (the PVX system discussed in Section 1.5.3) and screened on a collection of tomato genotypes, to identify those in which a hypersensitive response is induced (*Figure 13.2*). The hypersensitive response, due to recognition of the ECP2 protein in these cultivars, was found to be based on a single dominant gene, the *Cf-ECP2* gene. This gene can now be bred into commercial cultivars as a means of producing cultivars resistant against all the races that express the ECP2 protein. Furthermore, the ECP2-expressing PVX system

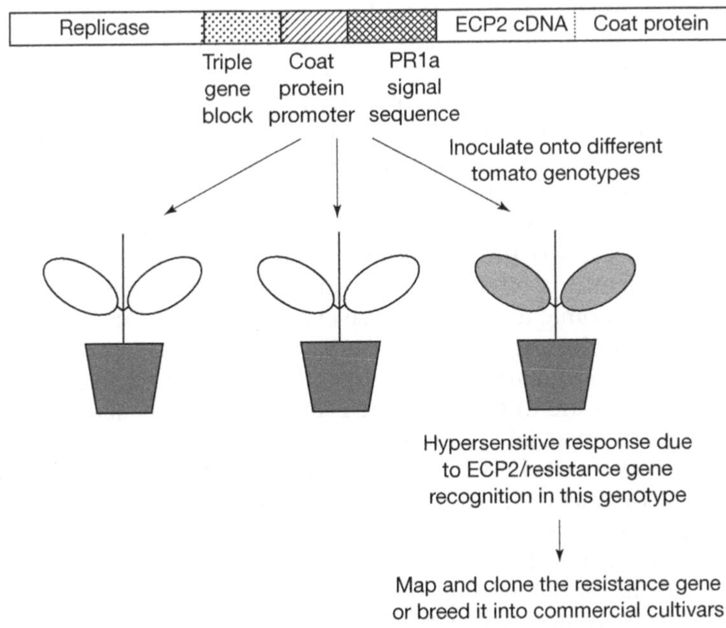

Figure 13.2

The use of the PVX transient expression system to isolate a resistance gene that recognises the ECP2 protein from *Cladosporium fulvum*. ECP2 cDNA has been fused to the PVX coat protein gene so that a fusion protein is expressed in the PVX system.

can be used at each step of the breeding and backcrossing programme to identify those individuals that have inherited the gene, removing the need for pathogenicity screens using the fungal pathogen. The development of similar virally based transient expression systems for other plant species is currently being undertaken, for example, using TMV, *Brome mosaic virus*, *Barley stripe mosaic virus* and others. This could potentially lead to the ability to screen a wide range of plant species for resistance mechanisms against virulence factors from a whole range of plant pathogens once these have been identified.

A third approach being used is to develop new specificities for resistance genes through DNA shuffling techniques (see Section 1.5.5). Essentially this involves using PCR-based techniques to create large numbers of recombinant resistance genes from cloned resistance genes and resistance-gene-like sequences. Screening methods, and particular transient expression systems, can be used to test the specificity of these recombinants. There is therefore enormous potential for molecular biology to be used in improvements to crop protection methods based on resistance mechanisms with or without the use of transgenic plants.

13.5 The use of chemicals for disease control

A wide array of chemicals, both inorganics and organics, are currently used in crop protection, and new agrochemicals are constantly being produced and evaluated. These chemicals include fungicides, bactericides, nematocides and virocides as well as insecticides that can be used to control vectors for viral and bacterial diseases and herbicides that can be used on parasitic plants and weeds. However, any new chemical must satisfy certain criteria before it can be used commercially. It must be effective and consistent at removing the pathogen, must not harm the plant, and it and its residues must not be harmful to other parts of the crop ecosystem nor to the consumer. At the same time, the chemical must be economic to produce, safe to handle, store and apply, and stable. Chemicals must therefore pass rigorous toxicological and efficacy tests to satisfy the legislative requirements of different countries. Because of the huge amount of time and money taken to develop new agrochemicals, often up to 10 years from initial screenings, it is generally only economic to develop these chemicals if they are effective for the control of a wide range of important pathogens of major crop species such as wheat, rice and maize. As a result, perhaps less than 1 in 20 000 of those chemicals that are initially screened for agrochemical activity eventually make it to commercial products.

As with the use of resistance breeding, persistent use of chemicals can influence the way the pathogen populations evolve and result in resistance developing or increasing in pathogen populations against the chemical, compromising the effectiveness of the control. Strategies for using mixtures of chemicals with differing modes of action, and combining the use of chemicals with methods such as disease-resistant cultivars into integrated control programmes, can increase the long-term effectiveness of chemicals and also reduce the quantities required. However, new, more efficient and more effective chemicals are constantly being sought. In particular, there is a drive towards developing chemicals that are able to eradicate diseases should they occur, as opposed to many of the current chemicals that tend to be preventive and are therefore used as a precautionary measure whether there is any disease pressure or not.

The modes of action of the major groups of fungicides have been detailed in *Table 4.2*. In the past, many of these chemicals have been identified by 'blue sky' synthesis, in which millions of chemicals are synthesised essentially at random and then taken through various stages of screening to identify those that satisfy the requirements outlined above. The mode of action of the chemical has generally been less important than the fact that it is effective. In many cases, these chemicals interfere with central metabolic processes such as the production of energy and ATP, inhibition of respiration or uncoupling of oxidative phosphorylation. Other chemicals may be more specific, such as the benzimidazoles, that target β-tubulin and inhibit the normal activity of the cytoskeleton including spindle formation and cell division. Others inhibit sterol biosynthesis (the DMIs), whilst more recent chemicals have been identified that block nucleic acid synthesis (the phenylamides) or mitochondrial electron transport (the strobilurins). A more recent approach for developing new chemicals has been to synthesise analogues of chemicals that have already been shown to be effective, sometimes those developed by competing companies ('patent-busting'). In addition, screening natural products produced by bacteria and fungi has become a source for new chemicals. Antibiotics have for many years been identified using this approach, and other chemicals such as the strobilurin fungicides are derived from natural chemicals that are the metabolites of certain mushroom species.

13.6 The use of molecular biology in agrochemical production

Rather than the somewhat hit-and-miss approaches that have been used in the past for agrochemical identification, molecular biology has the potential to be used in a more targeted approach. The aim would be to identify, potentially through insertion mutagenesis of pathogens (as discussed in Sections 1.5.3 and 1.5.4), metabolic pathways and biochemical processes that are essential for pathogenicity. For example, potentially important steps in fungal penetration of plants include the production of appressoria, melanisation and/or the metabolism that results in generation of turgor pressure. Alternatively, genes involved in toxin production, or biosynthesis of specific cell-wall components, for example chitin in fungi, may become targets. It is important that these processes are also non-essential in host plants and animals, to ensure that any agrochemical developed is specific to the pathogens and does not harm the plant or consumers.

Once a pathway and/or gene has been identified, three-dimensional modelling of the encoded protein(s) can be undertaken to help with identification of the active site and other domains that are significant. Computer based approaches and software programs for protein modelling are widely available and becoming increasingly sophisticated as more protein structures are identified and understood. Having identified the significant proteins and the important domains in these, chemical structures can be designed that might be effective at disrupting these domains, and such structures could be generated and tested. Alternatively, it may be possible to use recombinant DNA technology to identify structures that interact with and potentially inhibit the target protein. The phage display technique (see

Section 12.2.3) and yeast two-hybrid methods (see Section 1.5.6) are means of identifying proteins that interact with other proteins. By panning phage display libraries with target proteins, it would be possible to identify proteins that interact directly with the target. These target proteins could potentially be used for the production of plantibodies and transgenic approaches as discussed in Section 14.2, whilst from the agrochemical production point of view, they could be used in 3-dimensional modelling to identify domains that interact with the target proteins, which could then be reproduced in a chemical analogue.

13.7 Engineering chemicals that elicit defence responses in plants

Whilst the use of agrochemicals to target plant pathogens has been an effective control strategy and one that is likely to continue as new and improved chemicals that are more environmentally acceptable are produced, an alternative way to use chemicals is to elicit defence responses in plants. This approach of immunising plants chemically against disease, making use of natural defence mechanisms has been practised for many years. The discovery of phytoalexins resulted in strategies in which these were sprayed onto plants to control disease. Rishitin and phytuberin were used on potatoes to prevent potato blight, and isoflavonoid phytoalexins have been used to control a wide range of foliar pathogens such as rusts and mildews. However, whilst applications of these chemicals can prevent disease in some cases (though not as effectively as most fungicides), phytoalexins are ineffective as eradicants and can only be used in a preventative way. Some bacterial diseases have been controlled by similar approaches. The application of lysozyme to tomato leaves and proteinase inhibitors to cabbages has prevented some diseases. However, these approaches are not economically viable and also suffer from problems of reliability such that they are unlikely to be developed in the future.

More recently, attempts have been made to use elicitors to induce the battery of defence responses in plants, rather than using single components of defence pathways such as phytoalexins. Chitin, chitosan and other oligogalacturonides have been shown to elicit host defences akin to the hypersensitive response through rapid and transient membrane depolarisation, and this can effectively immunise the plants against a wide range of pathogens. Recently, harpin produced by *Erwinia amylovora* (see Section 6.7) has been developed into a commercial product in the USA.

As our understanding of the mechanisms of systemic defence mechanisms in plants has increased (see Sections 9.4 and 11.7) chemicals that are part of these mechanisms have been tried and tested, such as salicylic acid, jasmonic acid and ethylene. However, whilst these chemicals are effective in laboratory experiments, indeed many of the early experiments on systemic acquired resistance in tobacco were conducted by applying soluble aspirins to the soil, commercial applications have not been feasible. Along with the technical difficulties of applying these chemicals, the production of plants with permanently heightened defence systems, as occurs if systemic acquired resistance is permanently induced, results in plants that are generally lower yielding because of the diversion of biochemical pathways into defence. However, a benzathiodioazole (BTH) compound has been developed that mimics the effect of salicylic acid as an inducer of systemic acquired resistance, and this has been marketed under the tradename Bion as an effective chemical for the control of certain diseases in tobacco, tomato, *Arabidopsis*, wheat, rice and cucumber, and when appropriate concentrations are used there is no adverse effect on crop yield or appearance. Long-lasting protection has been found in rice against the rice blast fungus, and if treated at the seedling stage, resistance can occur throughout the growing season. Wheat can also be protected for an entire season against powdery mildew if treated early, and a 35% reduction in disease symptoms caused by *Puccinia triticina* and *Septoria* spp. along with an increase in yield of approximately 18% was also found. Unfortunately, the compound may not be as reliable as conventional fungicides and further work is required to produce additional and improved broad-spectrum agrochemicals that can be used to induce resistance in plants more effectively.

References and further reading

Asins, M.J. (2002) Present and future of quantitative trait locus analysis in plant breeding. *Plant Breeding* **121**: 281–291.

Brown, J.K.M. (2002) Yield penalties of disease resistance in crops. *Current Opinion in Plant Biology* **5**: 339–344.

Gebhardt, C. and Valkonen, J.P.T. (2001) Organisation of genes controlling disease resistance in the potato genome. *Annual Review of Phytopathology* **39**: 79–102.

Langridge, P., Lagudah, E.S., Holton, T.A., Appels, R., Sharp, P.J. and Chalmers, K.J. (2001) Trends in genetic and genome analyses in wheat: a review. *Australian Journal of Agricultural Research* **52**: 1043–1077.

Lassner, M. and Bedbrook, J. (2001) Directed molecular evolution in plant improvement. *Current Opinion in Plant Biology* **4**: 152–156.

Laugé, R., Joosten, M.H.A.J., Haanstra, J.P.W., Goodwin, P.H., Lindhout, P. and de Wit, P.J.G.M. (1998) Successful search for a resistance gene in tomato targeted against a virulence factor of a fungal pathogen. *Proceedings of the National Academy of Sciences, USA* **95**: 9014–9018.

Rafalski, A. (2002) Applications of single nucleotide polymorphisms in crop genetics. *Current Opinion in Plant Biology* **5**: 94–100.

Young, N.D. (1999) A cautiously optimistic vision for marker-assisted breeding. *Molecular Breeding* **5**: 505–510.

Transgenic approaches for crop protection

14

The enhanced understanding of plant diseases and defence responses at the molecular level has led to an explosion of interest in controlling diseases through the genetic modification of plants. That such approaches are technically feasible and can be effective in the laboratory has been clearly proven and it is the science that we will focus on in this chapter. The ongoing moral, ethical, commercial and legislative debate surrounding the use of genetically modified plants has been discussed at length elsewhere and will not be covered in detail here.

14.1 Pathogen-derived resistance

14.1.1 Coat-protein-mediated resistance

The concept of using mild strains of viruses to 'cross-protect' plants against more severe strains was first demonstrated in 1929 using TMV. For many years, attenuated strains of TMV were used to protect glasshouse grown tomatoes in the UK against more virulent strains of the virus. Mild strains of *Citrus tristeze virus* have similarly been used to protect citrus trees in Brazil, and *Apple mosaic virus* has been used in New Zealand in the same way. Attempts to understand the phenomenon led to the conclusion that expression of specific viral genes in plants might have the same effect. In 1986, this was demonstrated when the coat protein gene of TMV was expressed in transgenic tobacco plants under the control of the CaMV 35 S constitutive promoter. The result was delayed symptom development in all the transgenic plants, and 10–60% of the plants failed to develop any viral symptoms.

This form of resistance, now referred to as pathogen-derived resistance has subsequently been demonstrated to occur against a wide range of viruses and is recognised as an effective way of controlling specific viral diseases in plants. Initially, studies using coat protein genes were favoured, and it was believed that the viral coat protein produced *in planta* was the key component of the resistance. The expression of this protein was believed to block the normal uncoating and replication of invading viruses, and evidence to support this was provided by use of TMV pseudoviruses, in which the GUS reporter gene attached to the viral origin of assembly was packaged into intact viral particles. When these particles were applied to protoplasts derived from transgenic plants expressing the viral coat protein, the level of GUS activity was significantly reduced from that obtained when control plants were inoculated, indicating that the pseudovirion was unable to disassemble in the transgenic plants.

Although coat-protein-mediated resistance works effectively against closely related strains of viruses, in some cases requiring as little as 60% coat protein homology, it is ineffective against unrelated viruses and is therefore not the type of broad-spectrum resistance sought by commercial interests. Indeed, when commercial strains of potato possessing either PVX or PVY coat-protein-mediated resistance were produced in the 1980s that showed effective resistance against these viruses, the economic costs of registering the products made commercialisation unfeasible against the limited added-value to the grower. Furthermore, such coat-protein-mediated resistance mechanisms can be overcome when high concentrations of viral inoculum are used. There is also the risk that transencapsidation of infecting viral genomes by transgenically expressed coat proteins might occur, resulting in novel strains of viruses. In the case of *Zucchini yellow mosaic virus* (ZYMV), transencapsidation by the *Plum pox virus* (PPV) coat protein was shown to result in a non-aphid-transmissible strain gaining aphid transmissibility. How important this phenomenon might be in the field should transgenic plants be used is unclear, but there remains a real or perceived risk that transencapsidation would result in the formation of novel viruses at a more rapid rate than would occur through naturally occurring mixed infections of plants with viruses.

14.1.2 Replicase-mediated resistance

The risk of transencapsidation and lack of broad host range resistance led to the quest for alternative viral genes that might be effective in transgenic plants. Replicase-mediated resistance was first demonstrated against TMV, where the viral replicase was expressed in transgenic tobacco under the control of the 35 S promoter. Replicase resistance has subsequently been shown to be effective against a wide range of ssRNA viruses and some geminiviruses, and in many cases, deleted versions of replicase genes are sufficient for protection. The resistance is generally more effective than coat-protein mediated resistance against high inoculum doses, and in one case, broad spectrum resistance against a range of viruses was demonstrated when the TMV replicase gene containing a bacterial insertion element within it was used. Replicase-mediated resistance probably works through a post-transcriptional gene-silencing mechanism rather than through interference by the replicase protein. However, there have been a number of instances when resistance has been broken by a small number of point mutations within the infecting viral genome.

14.1.3 Movement protein-mediated resistance

Because of the importance of movement proteins (MPs) for cell-to-cell movement of viruses (see Section 8.6), it has been proposed that blocking plasmodesmata and viral movement may be a more effective means of preventing spread of both related and unrelated viruses, to result in a broader-spectrum resistance. Tobacco plants expressing the *Brome mosaic virus* (BMV) MP were found to have resistance against the unrelated TMV, and plants transgenic for a defective mutant of the TMV MP were resistant to a wide range of unrelated viruses. Attempts to produce transgenic plants with further MP mutations, based on the enhanced understanding of how these proteins work, are currently underway, in an attempt to produce more effective broad-spectrum resistance to viruses. Attempts have also been made to use other regions of viral genomes to make pathogen-derived resistant plants, such as proteases. These approaches have met with some success but because they are unlikely to provide the broad spectrum resistance that MPs might provide, they have been less extensively studied.

14.1.4 RNA-mediated resistance

The possibility that replicase-mediated resistance might be conferred by a post-transcriptional gene-silencing mechanism has led to the development of RNA-mediated means of protecting plants against viruses. This mechanism is supported by experiments in which transgenic plants with enhanced resistance have been generated by expressing satellite RNAs and defective interfering (DI) RNAs associated with certain viruses. However, this approach is limited to use against those few viruses that possess satellites and DIs, and the specificity of the sequence required makes it a very narrow host-range resistance. In addition, many viruses appear to possess a mechanism for suppressing PTGS, as discussed in Section 8.8.3. Ribozymes, which are self-cleaving RNAs based on certain satellite RNAs and viroids (see Section 7.8) have also been tested for their effectiveness in control of viral diseases. The aim has been to design ribozymes that would be expressed in transgenic plants and would bind to the viral RNA and cleave it, rendering it inactive. Similarly, antisense RNA approaches have been tested against some viruses. However, there are no results that suggest these mechanisms have significant advantages over the better-tested methods of coat-protein, replicase or movement protein-mediated resistance to date.

14.1.5 Pathogen-derived resistance against bacterial and fungal diseases

Attempts have been made to use pathogen derived resistance to control bacterial and fungal disease. Isolates of *P. syringae* pv. *phaseolicola* that produce phaseolotoxin encode a ROCTase (resistant to OCTase) enzyme to deal with the adverse effects of the toxin (see Section 5.6). The *argk* gene for ROCTase has been engineered into tobacco, targeted to the chloroplast (where OCTase is produced) by fusing it to a ribulose bisphosphate carboxylase small subunit transit peptide. Resultant plants were resistant to the toxin, and attempts are now being made to engineer such a system into beans, the natural host for this bacterium.

Some strains (roughly 1% of naturally occurring ones) of the maize smut fungus, *Ustilago maydis,* possess dsRNA mycoviruses that confer the killer phenotype, where the dsRNAs encode killer proteins and immunity to these toxins (see Section 4.11). These killer proteins are effective against isolates of *U. maydis* that lack the dsRNAs conferring immunity, including other species of *Ustilago*, such as *U. tritici*. This killer protein gene has recently been engineered into Swiss wheat cultivars susceptible to the stinking smut, *Tilletia tritici,* and the resultant plants had increased resistance to this fungus. However, whether such plants would be commercially useful is unclear, since one consequence of their use is likely to be a build up in the fungal populations of isolates that possess immunity to the toxin. Additionally, the killer proteins appear to be potent blockers of calcium channels and have some structural similarities to scorpion toxins, which may make their use in edible plants undesirable.

14.2 Plantibodies

Another approach to create plants resistant to pathogens might be to express antibodies that recognise the pathogen in plants, in such a way that they mop up and neutralise the infecting pathogen before it can cause disease. The production of antibodies has been discussed in Section 12.2. What is important for their deployment in plants is to be able to clone the genes encoding the antibodies and then ensure that they can be produced in a stable form in plants in the appropriate cellular compartments to recognise the invading pathogen. The use of phage display (*Figure 12.2*) has improved the ability to clone genes encoding specific antibodies, and panning phage display libraries with specific pathogen antigens, such as pathogenicity factors, can be used to identify antibodies that recognise the antigen and the relevant clone can then be isolated from the library. Targeting cloned antibodies to particular compartments in plants can also be accomplished by adding targeting sequences to the transgenes. For example, leader sequences can be added to target proteins for secretion to the apoplast, or KDEL sequences at the carboxy-terminal can be used for retention of proteins in the endoplasmic reticulum, which has often been shown to result in higher yields. Cytoplasmic expression of recombinant antibodies for viral inhibition in plants has, however, been more problematic. Antibodies (particularly those obtained by conventional means rather than by panning phage display libraries) are less stable in the cytosol than in other cellular compartments. For example, TMV-specific antibody levels were 300 times lower in the cytosol than when targeted to secretory pathways. However, methods to integrate antibodies into plasma membranes, for example by incorporating transmembrane domains such that the antibody faces the apoplast, have resulted in transgenic plants that are resistant to TMV.

Plantibody techniques have also been developed for use against bacteria, particularly phytoplasmas. These organisms are phloem restricted, so plantibodies have been targeted to the secretory system. Antibodies to a major membrane protein of the stolbur phytoplasma were expressed with a secretory signal in transgenic tobacco, and conferred resistance to stolbur. Use of plantibodies to target fungal diseases is less well developed.

14.3 Over-expressing defence genes

Many of the pathogenesis-related (PR) proteins that are induced by plant disease resistance responses have direct antimicrobial affects. For example, defensins, chitinases, glucanases, proteinase inhibitors and hydrogen peroxide-generating enzymes are all induced as part of the hypersensitive response, as well as enzymes involved in the biosynthesis of phytoalexins (see Chapter 9). Numerous attempts have been made to introduce genes encoding these into transgenic plants under the control of constitutive promoters to increase levels of the proteins and enhance resistance (*Figure 14.1*). In 1991, a bean chitinase gene introduced into tobacco under the control of the CaMV 35S promoter resulted in transgenic plants with enhanced resistance to *Rhizoctonia solani*, although there was no significant difference in resistance to other pathogens such as *Cercospora nicotinae* nor the non-chitin-containing oomycete *Pythium* spp. Addition of a glucanase transgene in combination with chitinase was subsequently shown to

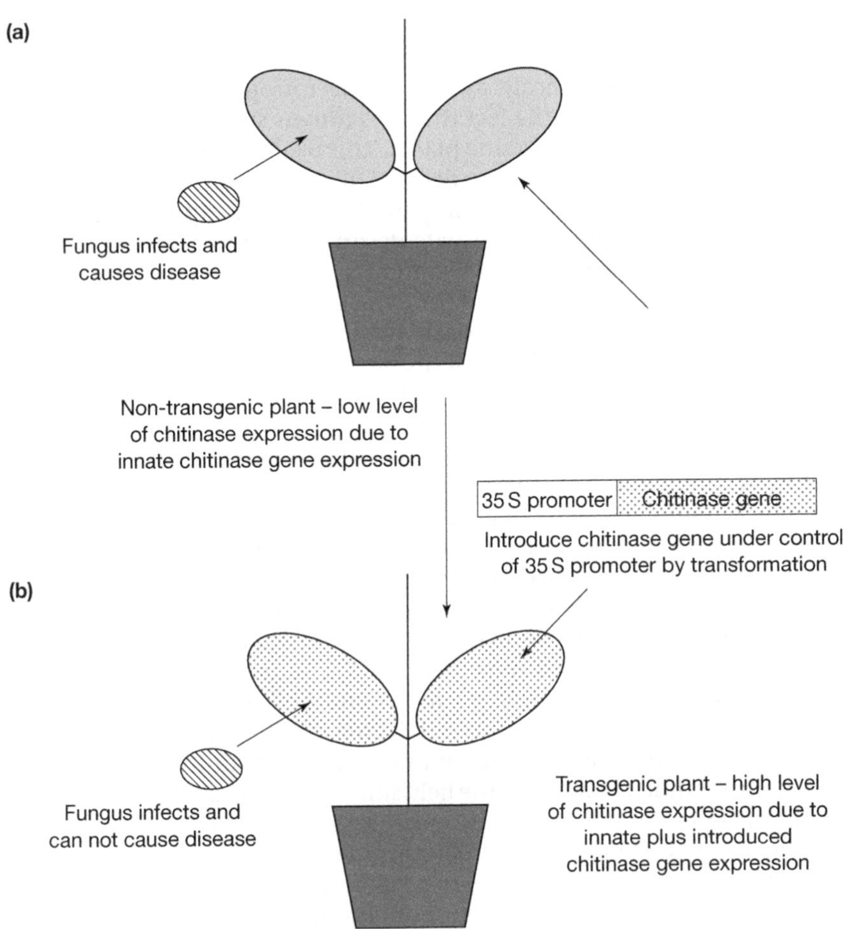
(a)
Fungus infects and causes disease
Non-transgenic plant – low level of chitinase expression due to innate chitinase gene expression
35 S promoter
Chitinase gene
Introduce chitinase gene under control of 35 S promoter by transformation
(b)
Fungus infects and can not cause disease
Transgenic plant – high level of chitinase expression due to innate plus introduced chitinase gene expression

Figure 14.1

General model for enhancing disease-resistance in transgenic plants. (a) The non-transgenic plant is susceptible to disease. (b) A constitutively expressed chitinase gene is introduced, and the high level of chitinase confers resistance to certain pathogens.

produce resistance to *C. nicotinae*. Similarly, expression of lysozyme in potatoes increased resistance to the bacterium *Erwinia carotovora,* whilst the inclusion of *PR-1a* in tobacco increased resistance to the oomycetes *Peronospora tabacina* (blue mould) and *Phytophthora parasitica* (black shank disease).

In an alternative approach to produce phytoalexins in plants, the grapevine stilbene synthase gene has been introduced into tobacco to confer resistance against *Botrytis cinerea*. This gene is involved in biosynthesis of the phytoalexin resveratrol in grapevine and some other plant species, but is absent in tobacco, although the chemical precursors, malonyl-coA and p-coumaroyl-coA are produced in tobacco. In transgenic plants, the stilbene synthase converts these precursors into stilbene, resulting in enhanced resistance. This strategy has also been used in alfalfa for resistance to *Phoma medicaginis,* and the approach of engineering plants to produce different phytoalexins may be applicable for other diseases as well, and is being tested as more biosynthetic pathways are elucidated.

An approach used specifically against the necrotrophic fungus, *Sclerotinia sclerotiorum,* makes use of the fact that this fungus synthesises large quantities of oxalic acid when infecting plants. This oxalic acid is an essential co-factor required by the pathogen to degrade plant cell walls and colonise the plant, and defective mutants are non-pathogenic. Over-expression of plant-derived oxalic acid oxidase or oxalate decarboxylase in transgenic plants such as soybean, oilseed rape and sunflower, has been shown to result in rapid breakdown of the oxalic acid from the invading pathogen and results in increased resistance to the pathogen. In the case of oxalic acid oxidase, an additional benefit appears to be that the breakdown of oxalate results in the production of hydrogen peroxide which may serve as a signal for systemic defence responses in the plant.

Many other defence-related genes have been expressed in transgenic plants and confer resistance to particular pathogens, as do some defence peptides (known as defensins) obtained from other organisms. Cercopins and Melittins are defensins in the giant silk-moth and bees respectively, and engineering a chimaeric gene between these into potato resulted in increased resistance to a range of bacterial, fungal and oomycetes pathogens. A Maganin-2 defence peptide from the skin of the African clawed frog (*Xenopus laevis*) has been engineered into the chloroplast genome of tobacco and confers resistance to *P. syringae* pv. *tabaci* and *Colletotrichum destructivum*. Chloroplast transformation is currently being developed as a general strategy for engineering plants because it offers the opportunity to contain the genetically modified material in the field since the maternally inherited DNA is not transferred in pollen and also high levels of gene expression may be possible due to the high copy of chloroplast genomes within cells. However, care must be taken in the selection of genes that are used if the products are intended for human and animal consumption, and whether any of these become commercial products will depend on whether they are as safe as non-transgenic equivalents, effective against a broad enough range of important pathogens, and offer durable long-term resistance, sufficient to make the registration costs economic.

14.4 Expressing defence genes under the control of inducible promoters

One of the fundamental problems with over-expressing defence genes in plants is that plant resources that might otherwise be used for increased yield production have to be diverted to maintain this high level of resistance. The constant production of defence gene products such as phytoalexins, whether the pathogen is present or not, might also mean that undesirable chemicals are present at much higher levels than normal, making the produce unsuitable for consumption. An alternative approach is to engineer defence genes, or genes that mimic the defence response, under the control of pathogen-induced promoters so that they are only expressed upon pathogen invasion. An example of this is with ribosome inactivating proteins such as the pokeweed (*Phytolacca americana*) antiviral protein. This particular protein has been used in human medicine as an antiviral agent (e.g. against AIDS) for a number of years. More recently, it has been engineered into tobacco, where it has been shown to protect plants against a diverse range of viruses, such as PVX, PVY and CMV. In a more elegant advance, a second ribosome-inactivating protein, dianthin, has been introduced into tobacco, but under the control of a geminivirus promoter. In this way, the gene and antiviral response is only induced in cells that are attacked by pathogens, such that there is no dianthin expressed in uninfected plants or cells away from the site of pathogen attack.

A similar strategy has been piloted in potato against *Phytophthora infestans*. In this case, the promoter from the potato *prp1-1* gene (which mediates gene expression in response to blight attack) has been linked to the *Bacillus* bacterial ribonuclease barnase gene. Expression of this gene in plants results in the formation of necrotic lesions analogous in many ways to the innate plant hypersensitive response. Because this gene was found to be expressed at a low level constitutively and to have a deleterious effect on the plants, a second *Bacillus* gene, the barstar inhibitor of barnase has been simultaneously introduced into the transgenic plants. The resultant transgenic plants have reduced levels of fungal sporulation when inoculated presumably because the infecting pathogen induces the production of barnase in infected cells to mimic the hypersensitive response and restrict the development of the pathogen.

A potential advance to make the resistance more broad based would be to identify a single promoter that is induced by a wide range of pathogens, and link this to genes such as barnase. Experiments have been undertaken to try to find such promoters, and a number of candidates have been identified, such as the promoter for the *fis1* gene of flax, which was identified as an up-regulated gene during compatible flax rust infections, and homologues of this have been identified in other plant species. The promoter from such a gene linked to barnase or other phytotoxic compounds would effectively mimic disease resistance genes by inducing a hypersensitive-like response when the plants are attacked by a wide range of pathogens to produce a broad-spectrum resistance. With the increasing use of genomics and microarrays, the identification of further suitable genes and promoters is likely to be enhanced.

14.5 The use of cloned resistance genes

The cloning of functional resistance genes has resulted in the capacity to transfer such genes directly into commercial crops without the need for extensive backcrossing programmes. For example, the *Xa21* gene from rice for resistance to *Xanthomonas oryzae,* has been transferred at the Rice Research Institute in the Philippines from the wild species *Oryzae longistaminata* into elite cultivars of rice for commercial use, a process that has been difficult to achieve by conventional breeding programmes. In addition, the *N* gene has been transferred from tobacco into tomato where it confers resistance to TMV. However, when the *Bs2* gene from pepper for resistance to *Xanthomonas* was cloned into species outside of the *Solanaceae,* it was ineffective, indicating that many *R* genes may have ‘restricted taxonomic functionality’ and not operate in unrelated plant species.

Such approaches are clearly feasible and may save a great deal of time in introducing genes into good agronomic cultivars, and these approaches could be combined with the identification of novel resistance specificities as discussed in Section 13.4. Alternatively, experiments are being undertaken in which domains of resistance gene and resistance gene analogues are being randomly recombined *in vitro* to attempt to generate resistance genes with novel specificities. Technologies are being developed to rapidly screen the products of these recombinations for their effectiveness as resistance genes, such that any that are potentially useful could then be engineered into transgenic plants.

14.6 Engineering broad-spectrum resistance

Introducing single cloned resistance genes into transgenic plants does not get round the problems identified in conventional breeding programmes of rapid pathogen evolution to overcome resistance genes. It is important that the practice of pyramiding resistance genes in particular formats that are likely to be effective is continued. Unless this is done in a rational, targeted way, merely deploying new resistance genes as and when they become available will continue to fuel the evolutionary arms race between crops and their pathogens. An additional priority must be to identify more broad spectrum and durable resistances that can be introduced into plants. The *Xa21* gene for example is capable of recognising avirulence proteins secreted by most if not all pathovars of *X. oryzae*, so that it effectively provides durable resistance. The ECP2 protein produced by *Cladosporium fulvum*, is also secreted by all known pathogenic races of this fungus, so targeting this protein by identifying and cloning the *Cf-ECP2* gene has increased the potential to develop durable resistance against this fungus in tomato. The barley *mlo* gene is also durable and attempts have been made to identify homologues of this in other plant species. In the future, it may be that genes that make a particular plant species into a non-host will be identified, and these could be cloned into current host plants to render them resistant. For example, the non-host resistance of tobacco to *P. infestans* has been correlated with response of tobacco to a particular peptide elicitor, INF-1 from *P. infestans*, and a resistance-gene-like protein has been identified in tobacco that appears to interact with this elicitor at least in yeast two-hybrid systems.

An additional transgenic approach to develop broad spectrum resistance based on resistance genes has been the genetically engineered acquired resistance (GEAR) system (*Figure 14.2*). This involves the incorporation into plants of a fungal avirulence gene in combination with the corresponding resistance genes inactivated by insertion mutagenesis. For example, tomato plants have been engineered that contain the tomato *Cf-9* resistance gene inactivated by a *Ds* transposable element in combination with the *C. fulvum* *Avr9* gene. Because the resistance gene is inactivated by the transposon, the plants grow normally. However, at a low frequency, the transposon excises in somatic cells, and in these cells, the function of the resistance gene is restored. The product of this gene interacts with the product of the avirulence gene to result in a localised hypersensitive response in and around those cells from which the transposon has excised. The result is small necrotic lesions and induction of a systemic resistance response throughout the plant, such that they become immune to infection by other races of *C. fulvum* and other pathogens including *P. infestans* (late blight) and *Oidium lycopersici* (powdery mildew). These resistance and avirulence genes were shown to work when transferred together into potato plants and conferred

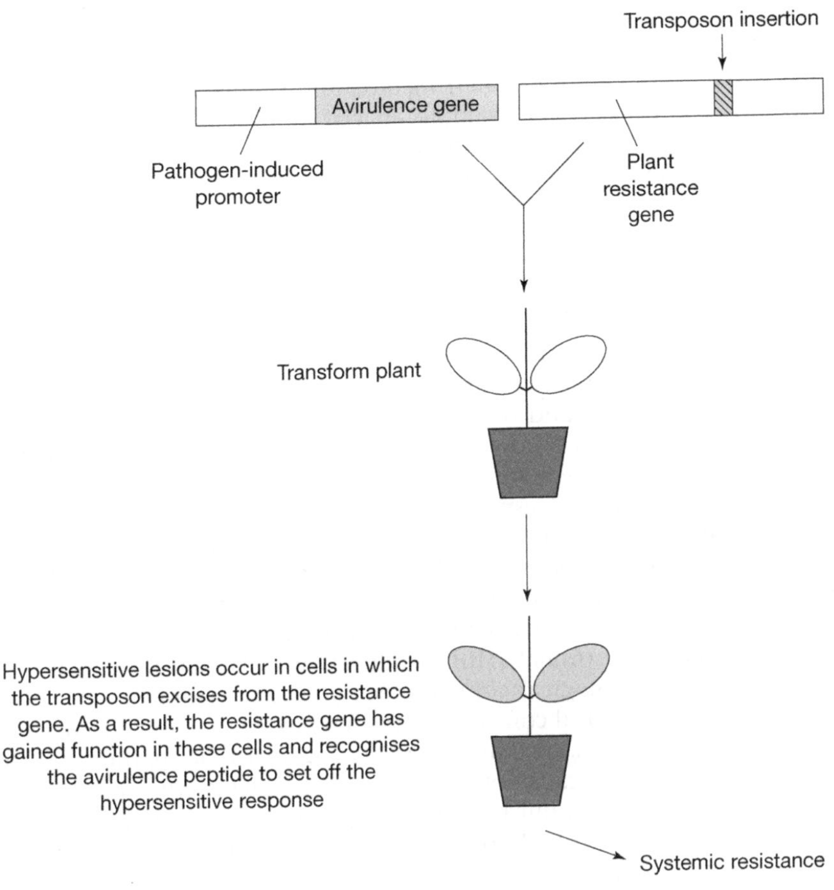

Figure 14.2

The principal behind genetically engineered acquired (broad host-range) resistance (see text for details).

enhanced resistance against potato blight. *Cf-9* and *Avr9* have also been engineered into *Brassica napus* (oilseed rape), where the non-lethal HR they induce confers enhanced resistance to *Leptosphaeria maculans*. Such systems could be engineered into other plant systems in which resistance and corresponding avirulence genes are identified. Furthermore, controlling the expression of the resistance gene or avirulence gene through the use of pathogen-induced promoters would ensure that the systemic resistance is only activated when plants are infected by pathogens.

A further approach to produce broad host range resistance has been to make use of genes that are part of the signal transduction mechanisms in local and systemic resistance pathways (see Chapter 11). *NPR1* is a key gene in regulating these pathways (see Section 11.10.2), and over-expression of NPR1 in *Arabidopsis* enhances resistance to a broad variety of pathogens such as *Pseudomonas* and *Peronospora* spp. with no obvious adverse phenotypic effects on the plant. In rice, expression of *NPR1* homologues resulted in increased resistance to *Xanthomonas* and *M. grisea*. However, expression of an *NPR1* homologue in wheat had the opposite affect, suppressing the expression of pathogenesis-related genes and enhancing disease susceptibility. Other genes in these signalling pathways are also being examined as potential candidates for conferring broad host range resistance, such as *RAR1*, *SGT1* and *Pad4*. For example, inactivation of *Pad4* in *Arabidopsis* results in extreme susceptibility to a range of fungal, oomycete and bacterial pathogens, so it may be possible that over-expression has the opposite effect and results in enhanced resistance.

14.7 Resistance based on antagonistic microbes

Antagonism between microbes is a common feature of natural ecosystems as they compete for nutrients and habitats. Many are capable of producing toxins, antibiotics and other antimicrobials capable of killing other microbes to enhance their own chances of occupying a particular niche. This has been exploited with the development of biological control microbes that can be used, especially in the rhizosphere, to target and combat plant pathogens. Isolates of the soil-borne fungus *Trichoderma*, or bacteria such as *Bacillus* species and *Pseudomonas fluorescens* have been developed into commercial biological control agents against a range of seedling diseases. These organisms are particularly useful because they can be produced relatively easily and formulated into products with relatively long shelf lives that give reliable biological control efficacy. The development of many other potential biological control agents, that have been shown to work in laboratory experiments, or that are targeted against foliar pathogens, has been less profitable because of the inherent problems with storage and appropriate means of application to give good reliable control. Without these properties growers will not be willing to use them to replace conventional control methods.

The mode of action of biocontrol microbes is variable. Some are able to occupy niches that would otherwise be occupied by the plant pathogen, e.g. non-pathogenic strains of *Fusarium oxysporum* can be used to prevent diseases caused by pathogenic strains of the fungus in this way. Others may produce antibiotics. *P. fluorescens* produces an antibiotic phenazine that is active against wheat take-all, *Gaeumannomyces graminis*. Some fluorescent pseudomonads also produce siderophores that scavenge iron from the environment and thus limit the growth of co-existing microbes including plant pathogens. Other organisms may be mycoparasites that colonise and kill the plant pathogenic organisms. The fungus *Aspelomyces quisqualis* is a mycoparasite that is often used in integrated control programmes against the grape powdery mildew fungus *Uncinula necator*. Spores of *A. quisqualis* are formulated in water-dispersible granules and germinate on the powdery mildew and the fungal germ-tube invades and kills the mildew fungus. A further group of microbes appear to work by activating defence responses in plants. The plant-growth-promoting rhizobacteria do this by inducing induced systemic resistance (see Section 9.4) and other organisms such as mycorrhizal fungi also appear to protect plants against pathogens by induction of defence responses.

Molecular genetics has the potential to improve biocontrol agents and might allow exploitation of the factors that they use. Natural products from these organisms could be developed into agrochemicals as discussed in Section 13.6. Alternatively, the genes that encode particular factors, for

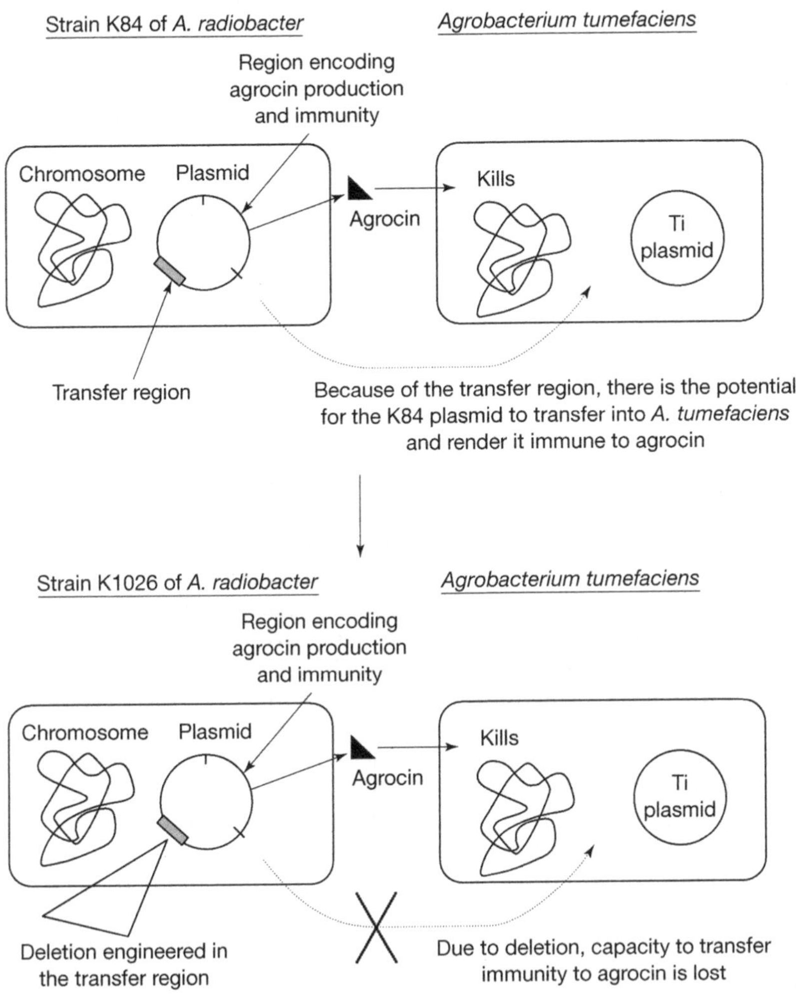

Figure 14.3

The development of strain K1026 of *Agrobacterium radiobacter* for biological control of *Agrobacterium tumefaciens*.

example the killer toxins of maize smut, could be utilised by cloning them into plants, as discussed in Section 14.1.5. Endochitinases from *Trichoderma harzianum* have been engineered into plants such as *Brassica* spp. where they confer good levels of resistance against some fungal pathogens. The cloning and sequencing of genes from these organisms is likely to result in additional methods and products in the future.

However, there are some examples of genetically modified microbes that are already being used as biocontrol agents, and one example is for the biocontrol of crown gall disease, *Agrobacterium tumefaciens* (*Figure 14.3*). A non-pathogenic soil-inhabiting related bacterium, *A. radiobacter* was identified that produces an antibiotic, agrocin, and for many years, a strain of this bacterium, K84 was used as a biocontrol agent against certain strains of *A. tumefaciens*. Seedlings of plants were dipped in suspensions of the bacterium prior to planting and the strain K84 would then colonise the soil around the roots and protect them against crown gall. The capacity to produce agrocin is encoded on a large plasmid, as is the immunity region that protects K84 from the effects of the agrocin. However, this plasmid also possesses a transfer region, and in laboratory experiments at least, it was shown that this transfer region enabled the agrocin-producing plasmid to be transferred into *A. tumefaciens*. Should this occur in the field, the result would be crown gall disease that is immune to agrocin. As a result, a new strain of *A. radiobacter* was engineered in the 1980s in which the transfer region was deleted. This new strain, K1026, has been sold commercially as a biocontrol agent in Australasia and more recently has been registered for use in parts of the USA, with no obvious detrimental effects on the rhizosphere.

An alternative approach being developed is based on *Bacillus subtilis*. This bacterium produces an enzyme AiiA, which is a metalohydrolase that inactivates the *Erwinia carotovora* acyl-HSL autoinducer involved in quorum sensing (see Section 5.1), and prevents the *Erwinia* from colonising plants. The *AiiA* gene has been engineered into tobacco and potato plants, where it confers resistance even to very high numbers of bacteria that are artificially inoculated into wounded tissues. Recombinant strains of *P. fluorescens* have also been developed as biocontrol agents. In particular, some have been developed by modification of regulatory genes, which express high levels of the antibiotic oomycin, which is particularly effective against *Pythium* root rot. A further use of genetic modification of microbes has been for the control of the chestnut blight fungus, *Cryphonectria parasitica*. This fungus infects chestnut trees via wounds, invades the bark and causes cankers and death of trees. Since the early 1900s, it has virtually eliminated chestnut as an economic tree in parts of the USA. It has also been a serious disease in Europe, but in the 1930s isolates of the fungus were found in parts of Italy that gave unusually mild symptoms on plants. Furthermore, these isolates could transfer this trait, referred to as hypovirulence, to aggressive strains of the fungus. As a result, they were developed into biocontrol agents. The factors required for hypovirulence were found to be cytoplasmic dsRNA molecules that could be transferred between isolates during hyphal contact. However, the naturally occurring hypovirulent isolates were incompatible with many of the aggressive isolates common in the USA. As a result, dsRNAs were cloned and transformed into isolates that were compatible with the aggressive isolates, and these strains are now used as biocontrol agents against chestnut blight.

14.8 Expression of vaccines in plants

Plants offer a cheap and reliable way of producing medicinal proteins that might prevent human diseases in a high-yielding and easily purifiable form. Approaches have been developed for using plants to produce vaccines. The strategy has been to engineer epitopes from animal viruses (e.g. Foot and Mouth virus) at specific positions into the coat protein genes of plant viruses such as *Cowpea mosaic virus*. The resultant hybrid virus has the foot and mouth epitope on the outer surface of its coat. This virus can then be grown in plants, purified and used as a vaccine. Similarly, an epitope from an outer membrane protein of *Pseudomonas aeruginosa* has been engineered into the TMV coat protein (*Figure 14.4*).

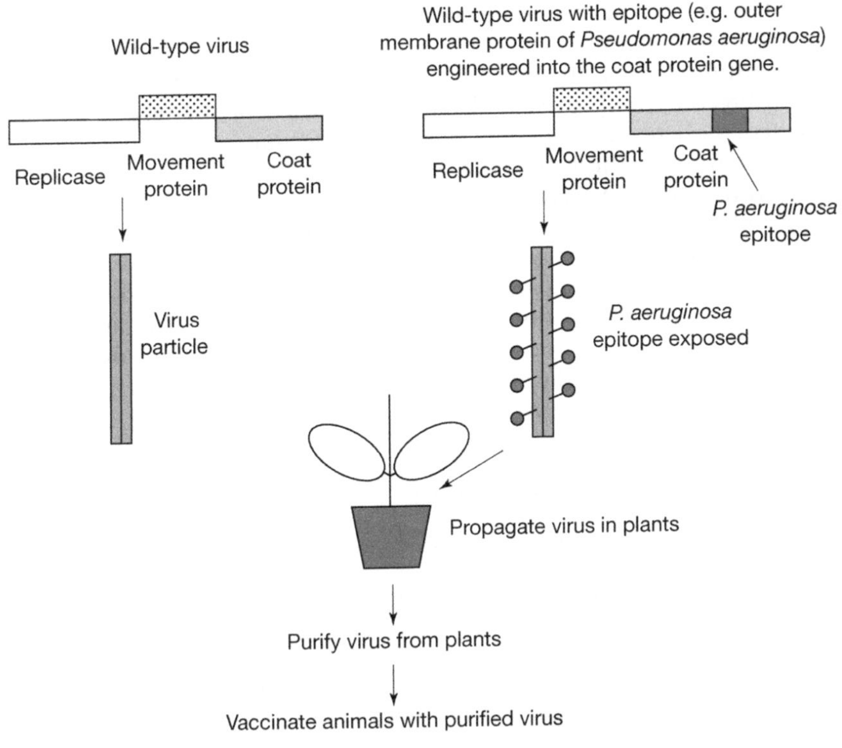

Figure 14.4

Engineering plant viruses for use as vaccines. cDNAs encoding epitopes are engineered into the plant virus coat protein gene so that they are exposed on the outer surface. This virus is then propagated in plants, purified and used to vaccinate animals.

14.9 Concluding remarks

In this chapter and throughout this book, we have detailed advances in the understanding of the molecular genetics of plants and their pathogens. We have examined how pathogens cause disease, how plants resist disease and how this might be exploited either by supporting conventional means of disease control or through biotechnological approaches. Whether the use of genetically modified plants will gain public acceptance worldwide is debatable. It is also not certain that these approaches will significantly alter the continued arms race between plants and pathogens. The additional selection pressure exerted on pathogens by more elaborate control methods may eradicate some of them, but it may also result in the development of superpathogens for which yet more elaborate control methods are required. Equally, non-intervention policies have their problems. Apart from reduced yields, many fungi produce mycotoxins, and if these are not adequately controlled, the produce has the potential to be more harmful to human health than if it contains fungicide residues.

There is no doubt that molecular biology has a place in the development of crop protection methods. It is only through advancing our understanding of the ways in which plants and microbes interact, and through dissecting the signalling pathways and changes in gene expression in greater detail, that we will be able to develop rational and integrated control programmes that maximise protection of plants from disease and consumers from harmful toxins and chemical residues. Whether through marker assisted breeding, improved molecular diagnostics or the other methods discussed, more research and an enhanced understanding will result in the development of methods that can be applied in the future. New problems will continue to arise. Perhaps new diseases will evolve or will spread to regions where they have previously been absent, and new techniques for studying these pathogens will be developed as technologies advance. The plant pathologist, who understands and keeps abreast of modern concepts about the nature of plant disease and the ways in which they can be controlled will be a vital resource for the agricultural, horticultural and forestry industries, just as the medic and the vet remain essential for the protection of human and animal health.

References and further reading

Gerhardson, B. (2002) Biological substitutes for pesticides. *Trends in Biotechnology* **20**: 338–343.

Pink, D. and Puddephat, I. (1999) Deployment of disease resistance genes by plant transformation – a mix and match approach. *Trends in Plant Science* **4**: 71–75.

Punja, Z.K. (2001) Genetic engineering of plants to enhance resistance to fungal pathogens – a review of progress and future prospects. *Canadian Journal of Plant Pathology* **23**: 216–235.

Rommens, C.M. and Kishore, G.M. (2000) Exploiting the full potential of disease resistance genes for agricultural use. *Current Opinion in Biotechnology* **11**: 120–125.

Schillberg, S., Zimmermann, S., Zhang, M-Y. and Fischer, R. (2001) Antibody-based resistance to plant pathogens. *Transgenic Research* **10**: 1–12.

Stuiver, M.H. and Custers, J.H.H.V. (2001) Engineering disease resistance in plants. *Nature* **411**: 865–868.

Tepfer, M. (2002) Risk assessment of virus-resistant transgenic plants. *Annual Review of Phytopathology* **40**: 467–491.

Index

T - #0648 - 101024 - C0 - 246/174/14 - PB - 9781859960448 - Gloss Lamination